73
75

ENVIRONMENTAL STRESS SCREENING

Books by Dimitri Kececioglu, Ph.D., P.E.

Reliability Engineering Handbook, Volume 1
Reliability Engineering Handbook, Volume 2
Reliability & Life Testing Handbook, Volume 1
Reliability & Life Testing Handbook, Volume 2
Maintainability, Availability and Operational
 Readiness Engineering Handbook, Volume 1
Environmental Stress Screening
 — Its Quantification, Optimization and Management

ENVIRONMENTAL STRESS SCREENING

ITS QUANTIFICATION, OPTIMIZATION AND MANAGEMENT

Dimitri Kececioglu, Ph.D., P.E.

Department of Aerospace and Mechanical Engineering
The University of Arizona

Feng-Bin Sun, Ph.D. Candidate

Department of Aerospace and Mechanical Engineering
The University of Arizona

For book and bookstore Information

http://www.prenhall.com

Prentice Hall PTR
Upper Saddle River, NJ 07458

Library of Congress Cataloging-in-Publication Data

Kececioglu, Dimitri.
   Environmental stress screening : its quantification, optimization
and management / Dimitri Kececioglu, Feng-Bin Sun.
      p.   cm.
   Includes bibliographical references and index.
   ISBN 0-13-324229-3
      1. Environmental testing. 2. Reliability (Engineering)
   3. Experimental design.   I. Sun, Feng-bin.   II. Title.
   TA171.K43 1995
   620'.0045—dc20   95-9917
                                                            CIP

Editorial/production supervision: *Kerry Reardon*
Cover design: *Lundgren Graphics*
Manufacturing buyer: *Alexis R. Heydt*
Acquisitions editor: *Bernard Goodwin*

 © 1995 by Prentice Hall PTR
Prentice-Hall, Inc.
A Simon & Schuster Company
Upper Saddle River, New Jersey 07458

The publisher offers discounts on this book when ordered
in bulk quantities. For more information, contact:

Corporate Sales Department
Prentice Hall PTR
One Lake Street
Upper Saddle River, NJ 07458

Phone: 800-382-3419
Fax: 201-236-7141
E-mail: corpsales@prenhall.com

Printed in the United States of America

10 9   8   7   6   5   4   3   2   1

ISBN   0-13-324229-3

Prentice-Hall International (UK) Limited, *London*
Prentice-Hall of Australia Pty. Limited, *Sydney*
Prentice-Hall Canada Inc., *Toronto*
Prentice-Hall Hispanoamericana, S.A., *Mexico*
Prentice-Hall of India Private Limited, *New Delhi*
Prentice-Hall of Japan, Inc., *Tokyo*
Simon & Schuster Asia Pte. Ltd., *Singapore*
Editora Prentice-Hall do Brasil, Ltda., *Rio de Janeiro*

To my wonderful wife Lorene,
daughter Zoe,
and son John.

— Dimitri B. Kececioglu

To my wife Le-Ling and daughter Spring.
To my dear parents.
To all my teachers.

— Feng-Bin Sun

TABLE OF CONTENTS

xviii   *CONTENTS*

PREFACE

As the first of our two sister texts, *Environmental Stress Screening - Its Quantification, Optimization and Management* and *Burn-in Testing - Its Quantification and Optimization* , this book results from a long-standing and growing demand from industry to quantify and optimize environmental stress screening (ESS). Burn-in and environmental stress screening (ESS) have become the primary approaches in modern electronic industry to precipitate and eliminate latent defects in products which are introduced mainly during the manufacturing, assembling and packaging processes. Many successful applications of burn-in and ESS have shown that scientifically planned and conducted burn-in and ESS will yield effective screening and thereby provide failure-free products with specified and desirable post-screen reliability, failure rate, mean life, and minimum life-cycle cost goals.

A distinction is made here between burn-in and ESS, as will be done also in our second text, *Burn-in Testing - Its Quantification and Optimization*. Burn-in is usually a lengthy process of powering a product at a specified operating or accelerated, *constant* temperature stress. ESS is an accelerated *process* of stressing a product in continuous cycles between predetermined environmental extremes, primarily temperature cycling, plus random vibration. Generally speaking, burn-in is conducted at the component level, while ESS is conducted at the assembly, module and system levels. Burn-in can be considered as a special case of ESS when only a constant temperature stress is applied. More explanation and clarification on the most common misconceptions will be found in Chapter 3 of this book.

This book attempts to summarize the knowledge derived from the studies of numerous workers in the field. It presents comprehensive coverage of the subject from the basic concept and historical evolution of ESS to the statistical and physical quantification of ESS, design-of-experiments (DOE) approach to multiple-stress screening evaluation, ESS planning, tailoring, monitoring, control and evaluation, and finally to ESS case histories. Emphasis is given to the quantification, or statistical and physical description and formulation, of the failure precipitation process during ESS, and the determination of the optimum stress regimen for a prespecified post-screen failure rate or post-screen remaining defects goal, and for minimum life-cycle cost goal. Mathematical derivations are provided wherever the authors believe necessary.

This book is intended to serve two kinds of readers: the practicing engineers, including reliability engineers, reliability and life testing engineers, and product assurance engineers, and the advanced undergraduate and graduate students. As an advanced topic in Reliability

Engineering and Life Testing, this book is an ideal sequel to the *Reliability Engineering Handbook*, by Dr. Dimitri Kececioglu, published by Prentice Hall, Inc., Englewood Cliffs, N.J. 07632, Vol. 1, 720 pp., and Vol. 2, 568 pp., in 1991, and now in its fourth printing, and to the *Reliability & Life Testing Handbook*, by Dr. Dimitri Kececioglu, also published by Prentice Hall, Vol. 1, 960 pp., in 1993, and Vol. 2, 900 pp., in 1994. It can be used as an one-semester college textbook for those students who already took at least one course in Reliability Engineering.

Chapter 1 is an introduction to *ESS*, its objectives, its general stress conditions, its economic benefits, major failure types it precipitates and a brief summary about the coverage of this book.

Chapter 2 describes the historical evolution of *ESS* which originated from environmental testing during World War II, then burn-in testing, environmental simulation testing, temperature cycling paired with low-frequency vibration, and finally the modern *ESS* concept based mainly on thermal cycling and random vibration. The development of *ESS* after the mid-1970's, in particular during the 1980's, is summarized.

Chapter 3 presents the basic definition and related concepts of *ESS*. Some common misconceptions about *ESS* in practice are discussed and clarified.

Chapter 4 summarizes the flaw-stimulus relations, typical *ESS* types, their parameters, their popularity and effectiveness. A distinction is made between *patent defects* and *latent defects*, and their detectability during *ESS*.

Chapter 5 presents an updated overview of major *ESS* military standards, directives, pamphlets and other guidance documents issued by the U.S. Department of Defence (DoD), Army, Air Force, Navy and the Institute of Environmental Sciences. The relationships and conflicts between these documents and the potential implications of using them as guides to *ESS* are also discussed.

Chapter 6 quantifies the *ESS* process statistically and presents various *ESS* cost models from different perspectives. First, a bimodal mixed-failure distribution function, describing the failure process of electronic components during screening, is presented. Its relevant characteristic functions, such as failure rate and renewal functions, and their relationships with screening duration, are discussed and quantified. The optimum screen time for a specified failure rate goal, or screen residual goal, is determined at both part and system levels and is illustrated by examples. The cost functions at part, system, and the combined part-system levels are developed based on the above renewal function. A procedure to determine the optimum screening time for minimum life-cycle cost is given and is illustrated by an example. To overcome the difficulty in determining the optimum screening time starting with the unknown parameters of the mixed distribution, a Bayesian approach is presented, which assumes the *beta* and *exponential* prior for the unknown parameters, respectively. The optimum screen time is then determined to minimize the expected total cost with respect to the joint prior distribution of the mixed life distribution parameters.

Chapter 7 offers a physical insight and quantification of thermal cycling screen. The standard Arrhenius model and the corresponding acceleration factor equations are presented first. The physical meaning of the activation energy is discussed. After illustrating the pitfalls of the conventional Arrhenius model, in particular the misconception that the activation energy is constant; i.e., temperature independent, two versions of the modified dynamic Arrhenius life-temperature model are presented. The least-squares estimates of the parameters of the modified Arrhenius model are developed. A general aging acceleration model for a typical temperature profile used in thermal cycling screen is derived based on each one of the two modified Arrhenius models. An example is given which illustrates the use of the developed aging acceleration models. Next, a math model is derived, which is based on the acceleration factor and the life distribution at the screening stress level, to determine the optimum thermal cycling time or the optimum number of thermal cycles for a given temperature profile and a specified field $MTBF$ goal. Finally, several useful physical models for thermal fatigue life prediction, for various failure mechanisms which may occur during thermal cycling, are summarized.

Chapter 8 offers a physical insight and quantification of random vibration screen. After a comprehensive review of the fundamental theory of a random process, the failure mechanisms and the corresponding structural reliability evaluation, under random vibration, are discussed. Great effort is spent and details are given to distribution determination of the cumulative damage and fatigue life, and fatigue reliability evaluation for both stationary narrow-band and wide-band random stressing processes. An algorithm is presented to determine the optimum screening length using the distributed stress-operating cycles $(S - N)$ curve approach.

Chapter 9 presents a combined-stress reliability analysis when the assembly is exposed to both thermal cycling and random vibration, simultaneously. The lognormal and the asymptotic truncated Weibull distributions are proposed for the fatigue lives under thermal cycling and random vibration screens, respectively. The combined failure probability under coupled thermal-vibration stressing is derived based on the failure surface concept and the associated variable transformations. The optimum number of thermal cycles and the optimum number of random vibration cycles are then determined so that the screening probability is maximized. This analysis is also illustrated by an example.

Chapter 10 presents a design-of-experiments (DOE) approach to ESS quantification when multiple stresses are applied. A hypothesis test is established with the null hypothesis of "all stresses accelerate the component independently" and the alternative hypothesis of "there are interactions among the applied stresses." The corresponding aging acceleration model under each hypothesis is developed and studied experimentally using a factorial experiment (FE) to see which model provides the best fit to the observed data, and then to decide if interactions are present or not. Finally, a general cost model is presented to determine the optimum multiple-stress regimen for an assembly-level stress screen; i.e., to determine the optimum stress level and duration of each prechosen stress such that the expected total cost during a screen and during the warranty period is minimized. A numerical example is given to illustrate the use of this model and its optimization.

Chapter 11 presents the definition of the screen strength and its relationship with the average defect failure rate during screen. The empirical equations, developed from industry, for both screen strength and the average defect failure rate, under different screening stresses, are summarized. The applicability of these empirical models are discussed. The procedures and equations for developing the most appropriate screening strength models based on your own data are given and illustrated with numerous examples.

Chapters 12, 13 and 14 cover the engineering procedures and techniques for ESS planning, tailoring, monitoring, control and evaluation according to DOD-HDBK-344(USAF) and RADC Guide to Environmental Stress Screening, etc. Chapter 12 provides methods, models and associated tables for the prediction of incoming and remaining defect densities, screening strength, test detection efficiency and test strength for various screening profiles. The procedure for quantitative ESS planning is also presented. Chapter 13 discusses the tailoring techniques in ESS which ensure that the selection of the screening parameters and methods of stress application are suited to the stress transmission characteristics and inherent strength or capability of hardware design to provide an effective screen without damaging good components. Chapter 14 presents a dynamic procedure during ESS, starting from data collection and failure classification, monitoring, and control using quality control charts, and finally process evaluation and appropriate corrective actions, to ensure that the planned remaining defect density goal is achieved.

Chapter 15 presents practical applications of the techniques covered in Chapters 12, 13 and 14 in developing a closed-loop dynamic ESS

program.

Chapter 16 presents numerous *ESS* case histories from U.S. industry, starting from an early application in the *Apollo* space program to the modern application of STRIFE in helping a company win the Baldrige National Quality Award. The various stress profiles used and the improvement in quality, reliability and goodwill achieved are highlighted with concrete numbers and pictures.

In developing this book, a number of computer programs in FORTRAN language were written. These are listed as appendices at the end of each relevant chapter. A conscientious effort has been made throughout to give credit to original sources. We regret any occasional oversights that may have developed during the "baking and shaking process" in generating this book.

ACKNOWLEDGEMENTS

The authors are deeply indebted to their families, particularly their wives and children, for their understanding, encouragement and support during the development of this book. We wish to thank Mr. Haofu Yin for his help in developing Chapters 12, 13 and 14. Thanks also go to many of our colleagues and graduate students at The University of Arizona, and industry for their inspiration and help in writing this book.

Dimitri Kececioglu
Feng-Bin Sun
Tucson, Arizona
April 23, 1995

ENVIRONMENTAL STRESS SCREENING

Chapter 1

INTRODUCTION

1.1 WHY *ESS*?

In our modern, growingly electronic world, the reliability of electronic systems is assuming paramount importance. $500 to $1,500 worth of electronics are used in each vehicle by automobile manufacturers [1]. About 60% of a military aircraft's cost now goes to its electronic systems, and many contracts require the manufacturer to provide service at a fixed price for product defects that occur during the warranty period [1].

Workmanship problems have been attracting more and more attention from the customers. "Never buy a car made on Monday or Friday," says the folk wisdom on U.S. automobiles. Over half the effort has been reportedly applied to rework in the United States, according to Willis J. Willoughby, director of reliability, maintainability, and quality assurance in the office of the Assistant Secretary of the Navy for Shipbuilding and Logistics [1]. He has estimated that the national average is about 15% to 20%. Though some defects due to design and workmanship problems are identified prior to delivery by quality control inspections, others often pass undetected, escape to the field and cause the early failures during the warranty period.

A common and widely-conducted practice of eliminating infant mortalities in electronic equipment manufacturing circles has been to "burn-in" a component population prior to product shipment. Scientifically planned and conducted burn-in will surface those items with part or workmanship defects and improve a product's reliability performance in the field. Usually burn-in involves aging (baking) items under normal operating conditions, which may take considerable time and resources [2; 3]. More effective screening can be achieved by introducing stressful environmental conditions over shorter durations,

1

TABLE 1.1– Cost comparison of semiconductor devices' defect removal at four stages in four different product markets [5, p. 305].

Product market	*Cost of defect removals at different stages, $			
	Incoming piece part	Board mount removal	System test	Field use
Commercial	3.00	7.50	7.50	75.00
Industrial	6.00	37.50	67.50	322.00
Military	10.50	75.00	180.00	1,500.00
Space	22.50	112.50	450.00	3×10^8

* Prices adjusted to 1978 costs in U.S. dollars.

which has been known as the so-called Environmental Stress Screening (*ESS*).

ESS is the tailored application of electrical and environmental stresses to electronic parts, modules, units and systems to identify and eliminate defective, abnormal or marginal parts and manufacturing defects [4]. Normally, *ESS* is conducted on 100% of the manufactured items. The objective of *ESS* is to accelerate early failures such that repair is accomplished at the most cost-effective stage. It is a generally accepted fact that the removal of defects from systems, starting at the lowest equipment level, is an economical approach for controlling the product's life-cycle cost [5]. The correction of defects at the manufacturer's facility is more economical than shipyard failure corrections, and shipyard failure corrections are more economical than post-delivery failure corrections during field operation. Table 1.1 [5, p. 305] offers a comparison of semiconductor devices' defect removal costs at four stages in four different product markets.

Table 1.2 [6, p. 296] is another table illustrating the economic value of screening at the lowest level of assembly. Dramatic field repair savings, reduced rework costs, minimization of scheduled delays, timely delivery of reliable products at reduced manufacturing and support costs, satisfied customers and increased profitability go hand-in-hand as production screening payoffs. The idea is to perform each screen at the lowest level of assembly consistent with the types of flaws involved.

TABLE 1.2– Repair cost in U.S. dollars per failure at
different assembly levels [6, p. 296].

Level of assembly	Historical nominal cost*	Major navy contractor**	Typical navy supplier**	Auto industry study**
Piece part	5	–	–	5
Module	5 – 30	382	110	30
Unit (Black box)	250 – 500	495	200	–
System	500 – 1,000	1,125	675	–
Field	5,000 – 15,000	15,345	–	300

* RADC Report TR-82-87, May 1982.
** Willougby, W. J., keynote address, IES National Conference &
Workshop, ESSEH, February 28 – March 2, 1979.

1.2 HOW *ESS* WORKS?

Figure 1.1 [7] presents a typical production flow with stress screening
and the corresponding repair cost per failure at each stage. The defects
at each stage are introduced by both the remaining defects from the
preceding lower level(s) of assembly and the defects induced at the
current level of assembly. The screening strength of the selected stress
screen at each stage governs the amount of defects eliminated and the
amount of defects escaping from this stage to the next higher level of
assembly.

However, to correctly and effectively plan, tailor, conduct and evaluate an *ESS* program requires the knowledge of why and how products
fail and the knowledge of how to stimulate, force, identify and finally
eliminate these failures before delivery. Of critical importance is being
able to address the types of product failure modes that may not be
discovered through inspection or testing [8], such as

1. Parameter drifts.

2. Printed circuit board shorts and opens.

3. Incorrect parts installation.

4. Wrong parts installation.

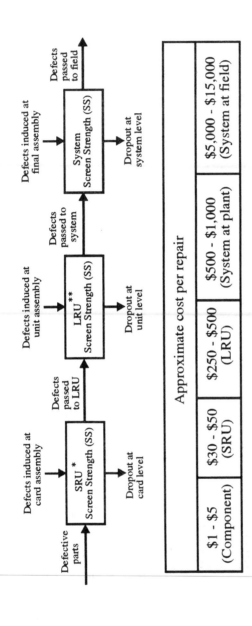

Fig. 1.1– A typical *ESS* production flow [7].

* Shop replaceable unit (SRU).

** Line replaceable unit (LRU).

4

5. Contaminated parts.

6. Hermetic seal failures.

7. Foreign material contamination.

8. Cold solder joints.

9. Defective parts.

Many distinct but complementary technical disciplines and a wide range of background experience are essential ingredients in the successful development and implementation of an *ESS* program. Although their titles and specific responsibilities may vary from one organization to another, the following engineering groups are found to possess the desired expertise [9]:

1. **Design Engineering,** which defines equipment design strength limits; establishes mechanical, thermal, and performance characteristics; and characterizes potential failure modes.

2. **Reliability Engineering,** which conducts failure analysis, accelerated life testing and evaluation, life-cycle-cost estimation, evaluation of correlation between in-house and field failure experience.

3. **Environmental Engineering,** which defines stress application methodology, conducts stress effectiveness assessment, specifies test facility performance requirements, and provides failure mode information.

4. **Test Engineering,** which defines equipment test tool requirements, production facility constraints, provides facility operator training, and documents equipment performance.

5. **Quality Engineering,** which reviews failure analyses, recommends workmanship and process corrective actions, and coordinates failure information with reliability engineering.

6. **Field Engineering,** which provides customer voice feedback, field performance insights, and early warning for potential problem areas.

Generally speaking, a complete *ESS* program consists of three major phases [9]; i.e.,

1. **Phase 1** – conceptual definition.

2. **Phase 2** – implementation.

3. **Phase 3** – effectiveness review.

These three phases and their corresponding tasks are summarized in Fig. 1.2 [9].

A successful *ESS* program involves a lot of effort. However, this effort will pay itself back. IBM estimates that it can save $20 in repairs for every dollar it spends on the *ESS* program for its Model 4234 printer [1]. *ESS* on switching logic units at Lockheed increased the *MTBF* from 100 hr to 500 hr [1]. *ESS* conducted at ZYTEC has improved the *MTBF* of its power supplies from less than 0.2 million hr to more than 1.5 million hr, reduced warranty costs by 50% from 1984 to 1990 and helped the company win the Baldrige Manufacturing Award [10]. The automated *ESS* optimization established at U.S. Army yielded over $1.1 billion in life-cycle-cost savings, 80 to 1 for return on investment (ROI), and 3 to 1 for field reliability improvements of production systems [11]. More and detailed *ESS* case histories will be presented in Chapter 16 of this book.

1.3   WHAT THIS BOOK OFFERS?

The objective of this book is to summarize the knowledge derived from the studies of numerous workers in the field. It presents comprehensive coverage of the subject from the basic concept and historical evolution of *ESS* to the statistical and physical quantification of *ESS*, design-of-experiments (*DOE*) approach to multiple-stress screening evaluation, *ESS* planning, tailoring, monitoring, control and evaluation, and finally to *ESS* case histories. Emphasis is given to the quantification, or statistical and physical description and formulation, of the failure precipitation process during *ESS*, and the determination of the optimum stress regimen for a prespecified post-screen failure rate or post-screen remaining defects goal, and for minimum life-cycle-cost goal. Mathematical derivations are provided wherever the authors believe necessary.

However, because *ESS* is a product-specific program, in which each application has its own unique set of variables, it is not possible to cover individual product applications in detail. Sources for this type of information are provided in the ADDITIONAL REFERENCES of the associated chapters. On the other hand, *ESS* theory has not reached its mature stage yet. Some general principles presented in this book only reflect the current state of the art. More explorations, verifications and improvements are needed in the future in various associated

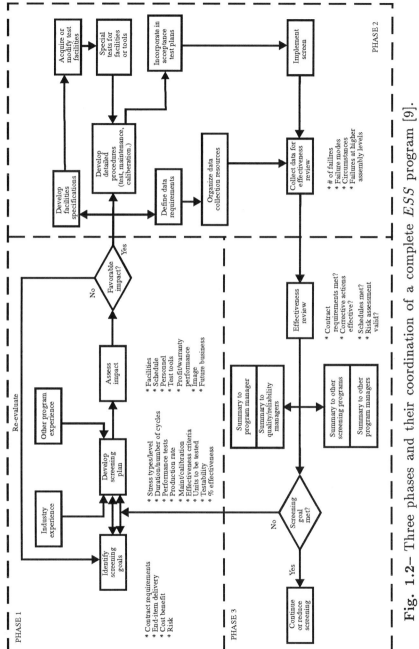

Fig. 1.2– Three phases and their coordination of a complete *ESS* program [9]. "Reprinted with permission from the Journal of the IES, Vol. 31, No. 2."

branches of *ESS*, such as more precise physical description and quantification of the combined-stress screening process, which may require cooperation among several scientific desciplines. These opportunities are open to every one who is willing to dedicate themselves to this more and more complicated world which demands more and more reliable products and services.

REFERENCES

1. Tustin, W., "Recipe for Reliability: Shake and Bake," *IEEE Spectrum*, pp. 37-42, December 1986.

2. Kececioglu, Dimitri and Sun, Feng-Bin, *Burn-in: Its Quantification and Optimization*, to be published by Prentice Hall, Englewood Cliffs, New Jersey, 1995.

3. Talbott, M., "R&M 2000 Environmental Stress Screening," *The Journal of Environmental Sciences*, pp. 17-20, January/February 1988.

4. Pennington, Duane R., "Environmental Stress Screening – Some Misconceptions," *The Journal of Environmental Sciences*, pp. 26-29, May/June 1986.

5. Arsenault, J. E. and Roberts, J. A., *Reliability & Maintainability of Electronic Systems*, Computer Science Press, 584 pp., 1980.

6. Mandel, C. E., "Environmental Stress Screening (ESS)," *Proceedings of the Institute of Environmental Sciences*, Orlando, Florida, pp. 294-302, 1984.

7. Saari, A. E., *Stress Screening of Electronic Hardware*, Hughes Aircraft Company, RADC-TR-82-87, May 1982.

8. Thermotron Industries, *The Environmental Stress Screening Handbook*, 291 Kollen Park Drive, Holland, MI 49423, 36 pp., 1988.

9. Caruso, H., "Environmental Stress Screening: An Integration of Disciplines," *The Journal of Environmental Sciences*, pp. 29-34, March/April, 1989.

10. Romanchik, D., "STRIFE Helps Zytec Win Baldrige," *Test & and Measurement World*, pp. 46-52, April 1992.

11. Huizinga, M. A., "Optimized Environmental Stress Screening of U.S. Army Electronic Hardware," *Proceedings of the Institute of Environmental Sciences*, San Jose, California, pp. 209-216, 1987.

ADDITIONAL REFERENCES

1. Alis, B., "The French Environmental Stress Screening Program," *Proceedings of the Annual Reliability and Maintainability Symposium, 31st Annual Technical Meeting*, Philadelphia, Pennsylvania, pp. 439-442, 1985.

2. Bateson, J. T., "Board Test Strategies – Production Testing in the Factory of the Future," *Test and Measurement World*, December 1984.

3. Bohan, E. M. and McGrath, J. D., "Shake and Bake – Shape Your Future," *Proceedings of the Institute of Environmental Sciences*, Los Angeles, California, pp. 234-238, 1983.

4. Capitano, J. L., "Innovative Stimulus Testing at the Lowest Level of Assembly to Reduce Costs and Induce Reliability," *Proceedings of the Institute of Environmental Sciences*, Orlando, Florida, pp. 303-305, 1984.

5. Capitano, J. L. and Feinstein, J. H., "Environmental Stress Screening (*ESS*) Demonstrates Its Value in the Field," *Proceedings of the Annual Reliability and Maintainability Symposium*, Las Vegas, Nevada, pp. 31-35, 1986.

6. Capitano, J. L., "R & M 2000 and Environmental Stress Screening," *IEEE Transactions on Reliability*, Vol. R-36, No. 3, pp. 346-350, 1987.

7. Caruso, H., "Significant Subtleties of Stress Screening," *Proceedings of the Annual Reliability and Maintainability Symposium*, Orlando, Florida, pp. 154-158, 1983.

8. Caruso, H., "An Overview of Environmental Stress Screening Standards and Documents for Electronic Assemblies," *Journal of Environmental Sciences*, Vol. 32, No. 4, pp. 15-25, July/August 1989.

9. Caruso, H., "Generic Environmental Stress Screening Requirements for Statements of Work," *Proceedings of the Institute of Environmental Sciences*, San Diego, California, pp. 410-414, 1991.

10. Caruso, H., "An Updated Overview of Environmental Stress Screening Standards and Documents for Electronic Assemblies," *Journal of Environmental Sciences*, Vol. 34, No. 4, pp. 49-61, March/April 1992.

11. Caruso, H., "Environmental Stress Screening of Spares and Repairs," *Proceedings of the Institute of Environmental Sciences*, Nashville, Tennessee, pp. 476-481, 1992.

12. Cerasuolo, D., "Development of Military Spares Screening Program Including Investigation of Subassembly Energization," *Proceedings of the Institute of Environmental Sciences*, Anaheim, California, pp. 207-212, 1989.

13. Chan, H. A., "A Formulation of Environmental Stress Testing and Screening," *Proceedings of the Annual Reliability and Maintainability Symposium*, pp. 99-104, 1994.

14. Chenoweth, H. B. and Bell, J. M., "Semiconductor Industry Screening Performance," *Proceedings of the Institute of Environmental Sciences*, Los Angeles, California, pp. 253-257, 1983.

15. Curtis, A. J., *Reliability Testing and Environmental Stress Screening*, Hughes Aircraft Company, 34 pp., 1989.

16. Dane, A. J., "Profitability of Planning for Stress Screening," *Proceedings of the Institute of Environmental Sciences*, Los Angeles, California, pp. 230-233, 1983.

17. DeCristoforo, R. J., "Environmental Stress Screening – Lessons Learned," *Proceedings of the Annual Reliability and Maintainability Symposium*, San Francisco, California, pp. 129-133, 1984.

18. Diekema, J., "Beyond ESSEH," *Evaluation Engineering*, pp. 84-89, March 1991.

19. DOD–HDBK–344(USAF), *Military Handbook: Environmental Stress Screening (ESS) of Electronic Equipment*, 126 pp., 20 October 1986.

20. Douglas, P., Johnson, A. and Mixer, M., "Continuous-Flow *ESS* Minimizes Defects," *Electronics Test*, pp. 42-44, August 1990.

21. Drees, D. and Winski, G., "Low Level Environmental Stress Screening," *Proceedings of the Institute of Environmental Sciences*, Los Angeles, California, pp. 242-246, 1983.

22. Emerson, D. A. and Buck, R. A., "Non-operating Temperature Cycling - an Effective Screen," *Proceedings of the Institute of Environmental Sciences*, Los Angeles, California, pp. 239-241, 1983.

23. "Environmental Stress Screening Guidelines for Assemblies," Panel Discussion, *Journal of the Institute of Environmental Sciences*, pp. 33-47, September/October, 1990.

24. Eustace, R. H., "Environmental Stress Screening (*ESS*) – Enthusiasm with Limited Understanding," *Proceedings of the Institute of Environmental Sciences*, San Diego, California, pp. 384-392, 1991.

25. Fedraw, K. and Becker, J., "Impact of Thermal Cycling on Computer Reliability," *Proceedings of the Annual Reliability and Maintainability Symposium*, Orlando, Florida, pp. 149-153, 1983.

26. Fiorentino, E., *RADC Guide to Environmental Stress Screening*, *RADC TR-86-138*, August 1986.

27. Gabriel, C. A., "Reliability and Maintainability Action Plan, R&M 2000," USAF, February 1985.

28. Garry, W. J., "Developing an *ESS* Automation Tool," *Proceedings of the Annual Reliability and Maintainability Symposium*, Atlanta, Georgia, pp. 495-501, 1989.

29. Geniaux, B., "*ESS* at Final Assembly Level – Actual Results – Optimization Method," *Proceedings of the Institute of Environmental Sciences*, San Jose, California, pp. 195-201, 1987.

30. "Global Competitive Streamlining of *ESS* for the 1990's," Executive Panel Discussion on *ESS*, *Journal of the Institute of Environmental Sciences*, pp. 35-44, May/June, 1991.

31. Golshan, S. and Oxford, D. B., "A Quality Correlation Program," *Proceedings of the Institute of Environmental Sciences*, King of Prussia, Pennsylvania, pp. 58-61, 1988.

32. Gould, D., "Tough Environments Call for Tough Tests," *Test & Measurement World*, pp. 49-56, February 1990.

33. Hnatek, E. R., *Effectiveness of Various Environmental Stress Screens*, Viking Labs/Honeywell, Mountain View, California, 56 pp., 1990.

34. Hobbs, G. K., "Development of Stress Screens," *Proceedings of the Annual Reliability and Maintainability Symposium*, Philadelphia, Pennsylvania, pp. 115-118, 1987.

35. Hobbs, G. K., "Evaluation of Stress Screens," *Proceedings of the Institute of Environmental Sciences*, King of Prussia, Pennsylvania, pp. 47-49, 1988.

36. Hobbs, G. K., "Modern Methods in Stress Screening," *Test & Measurement World*, pp. 47-48, June 1991.

37. Hobbs, G. K., "Highly Accelerated Stress Screens – HASS," *Proceedings of the Institute of Environmental Sciences*, Nashville, Tennessee, pp. 451-457, 1992.

38. Institute of Environmental Sciences, *Environmental Stress Screening Guidelines*, 940 East Northwest Highway, Mount Prospect, Illinois, 60056, 1981.

39. Institute of Environmental Sciences, *Environmental Stress Screening Guidelines*, 940 East Northwest Highway, Mount Prospect, Illinois, 60056, 1984.

40. Institute of Environmental Sciences, *Environmental Stress Screening Guidelines for Parts*, 940 East Northwest Highway, Mount Prospect, Illinois, 60056, 1985.

41. Jacob, G., "Benefiting from *ESS* Experience," *Evaluation Engineering*, pp. 45-48, April 1992.

42. Jacob, G., "The Many Faces of *ESS*," *Evaluation Engineering*, pp. 58-59, April 1992.

43. Jawaid, S. and Crook, K., "Linear Ramp Chambers and Thermal *ESS*," *Evaluation Engineering*, pp. 70-79, June 1992.

44. Kallis, J. M. et al, "Techniques for Avionics Thermal/Power Cycling Reliability Testing," *Proceedings of the Institute of Environmental Sciences*, Nashville, Tennessee, pp. 427-436, 1992.

45. Killion, R. E., "An Overview and Critique of Environmental Stress Screening Theory and Practice," *Proceedings of the Institute of Environmental Sciences*, Orlando, Florida, pp. 289-293, 1984.

46. Lascaro, C. P. and DiGiovanni, F., "Environmental Stress Screening (*ESS*) for VRC-12 Overhaul Process, A Case Study," *Proceedings of the Institute of Environmental Sciences*, Anaheim, California, pp. 217-224, 1989.

47. Littlefield, J. W., "R & M 2000 Environmental Stress Screening," *IEEE Transactions on Reliability* , Vol. R-36, No. 3, pp. 335-341, 1987.

48. McLean, H., "Highly Accelerated Stressing of Products with Very Low Failure Rates," *Proceedings of the Institute of Environmental Sciences*, Nashville, Tennessee, pp. 443-450, 1992.

49. Meyers, R. and Randazzo, E., "Analytical Spares Screening Evaluation Technique (ASSET)," *Proceedings of the Annual Reliability and Maintainability Symposium*, Philadelphia, Pennsylvania, pp. 120-124, 1987.

50. MIL-STD-810D, *Environmental Test Methods and Engineering Guidelines*, 19 July 1983.

51. Moen, G., "Trials and Tribulations of Implementing *ESS*," *Proceedings of the Institute of Environmental Sciences*, Nashville, Tennessee, pp. 437-442, 1992.

52. Nagle, A. L., "Cost-Effectiveness of Environmental Stress Screening (ESS)," *Journal of the IES*, pp. 35-38, November/December, 1991.

53. *NAVMAT* P-9492, *Navy Manufacturing Screening Program*, Department of the Navy, May 1979.

54. Neumann, B., "Automation and Integration of Environmental Stress Screening in Continuous Flow Production," *Proceedings of the Institute of Environmental Sciences*, San Diego, California, pp. 403-409, 1991.

55. Pellicione, V. H. and Popolo, J., *Improved Operational Readiness Through Environmental Stress Screening*, Final Technical Report RADC-TR-87-225, C-51 pp., November 1987.

56. Perlstein, H. J. and Bazovsky, I., "Identification of Early Failures in *ESS* Data," *Proceedings of the Institute of Environmental Sciences*, Anaheim, California, pp. 194-197, 1989.

57. Phaller, L. J., "How Much Environmental Stress Screening is Really Required?" *Proceedings of the Institute of Environmental Sciences*, Orlando, Florida, pp. 306-311, 1984.

58. Punches, K., "Stress Screening Failure Sources," *Proceedings of the Institute of Environmental Sciences*, San Jose, California, pp. 192-194, 1987.

59. Reeve, P., "Whole Lotta' Shakin' Goin' On," *Test & Measurement World*, pp. 45-50, February 1991.

60. Ross, J. D., "Product Assurance, AMC Environmental Stress Screening," *US Army, AMC-R 702-25*, March 1986.

61. Ryerson, C. M., "Principles of Screening and Cost Effective Product Assurance," *Microelectronics and Reliability*, Vol. 20, No. 5, pp. 693-715, 1980.

62. Saari, A. E. et al, *Environmental Stress Screening, RADC TR-86-149*, September 1986.

63. Saari, A. E., "On the Implications of R & M 2000 Environmental Stress Screening," *IEEE Transactions on Reliability*, Vol. R-36, No. 3, pp. 342-345, 1987.

64. Schmidt, R. E. et al, "Making *ESS* a Dynamic Process Using the Procedures of DOD-HDBK-344," *Proceedings of the Institute of Environmental Sciences*, Nashville, Tennessee, pp. 458-465, 1992.

65. Screening Systems, Inc., *Environmental Stress Screening, A Tutorial*, 7 Argonaut, Laguna Hills, California 92656, 69 pp., revised May 1987.

66. Screening Systems, Inc., *Environmental Stress Screening, A Tutorial*, 7 Argonaut, Laguna Hills, California 92656, 77 pp., revised April 1990.

67. Screening Systems, Inc., *Environmental Stress Screening, A Tutorial*, 7 Argonaut, Laguna Hills, California 92656, 82 pp., revised April 1991.

68. Shalvoy, C. E., "Finding the Most Effective Stress-Screening Strategy," *Electronics Test*, pp. 42-46, March 1989.

69. Sly, L. D., "Stress Screening Improves AN/AYK-14(V) Computer Productivity and Reliability," *Proceedings of the Institute of Environmental Sciences*, Los Angeles, California, pp. 247-252, 1983.

70. Smithson, S. A., "Shock Response Spectrum Analysis for *ESS* and STRIFE/HALT Measurement," *Test Engineering & Management*, pp. 10-14, December/January 1991-1992.

71. Tustin, W., "Stress Screening: Its Role in Electronics Reliability," *Quality Progress*, pp. 18-22, June 1982.

72. Tustin, W., "Shake and Bake the Bugs Out," *Quality Progress*, pp. 61-64, September 1990.

73. Tustin, W., "Starting Up *ESS*," *Quality*, pp. 57-60, April 1992.

74. Williams, C. L., "Product Verification While Stress Screening," *Proceedings of the Institute of Environmental Sciences*, Anaheim, California, pp. 213-217, 1989.

75. Willoughby, W. J., "Navy Manufacturing Screening Program," *Department of the Navy, NAVMAT P-9492*, May 1979.

76. Willoughby, W. J., "Best Practices," *Department of the Navy, NAVSO P-6071*, March 1986.

77. Wong, K. L., "Unified Field (Failure) Theory – Demise of the Bathtub Curve," *Proceedings of the Annual Reliability and Maintainability Symposium*, Philadelphia, Pennsylvania, pp. 402-407, 1981.

78. Wong, K. L., "Off the Bathtub onto the Roller-Coaster Curve," *Proceedings of the Annual Reliability and Maintainability Symposium*, Los Angeles, California, pp. 356-363, 1988.

79. Wong, K. L., "A New Environmental Stress Screening Theory for Electronics," *Proceedings of the Institute of Environmental Sciences*, Anaheim, California, pp. 218-224, 1989.

80. Wong, K. L., "Demonstrating Reliability and Reliability Growth with Environmental Stress Screening Data," *Proceedings of the Annual Reliability and Maintainability Symposium*, Los Angeles, California, pp. 47-52, 1990.

81. Zimmerman, W., "Screening Tests to Monitor Early Life Failures," *Proceedings of the Annual Reliability and Maintainability Symposium*, Orlando, Florida, pp. 443-447, 1983.

Chapter 2

HISTORICAL PERSPECTIVE OF
ESS

Emerged from a strong consumer movement within the military services and the Department of Defense [1], *ESS* has its origins in environmental testing; basically, a means of exposing samples of product assemblies to one or more simulated field conditions [2]. Driven by the military, environmental testing became prominent during World War II, when the proliferation of sophisticated weaponry, aircraft, and communications systems demanded a less costly and time-intensive method of demonstrating the reliability of equipment than by use-stress level testing. Laboratory experimentation with small-scale hardware was conducted. An insulated chamber, equipped with the technology to simulate environments such as temperature, humidity, altitude and others was used. These tests were performed during developmental stages to verify design, and on a manufacturing audit basis to measure design compliance.

After the war, new electronic technologies became available to the consumer market, a situation that created a different need. As products were downsized, their complexity increased, involving unique processing and assembly procedures. The accumulation of field failure data showed that design compliance was no longer sufficient evidence of reliability. It also led to the discovery that a number of unpredictable factors involved in parts design, product workmanship and manufacturing processes were contributing to higher than desired failure rates. As many of these failures were occurring during the product's infancy, it was determined that a testing methodology that could mimic the product's infancy stage would provide the cure.

In an attempt to pass products through infancy, a process of powering products for an extended length of time was introduced. Referred to as burn-in, the process also generated a high degree of heat, a stress which many believed would have an added impact in precipitating early product failures. In the 1960's, manufacturers began to "burn in" electronic components, that

15

is, bake them for a specified period of time [3]. Resistor and capacitor producers, for example, would either conduct the tests themselves or contract outside test laboratories for the work.

Burn-in testing did succeed in forcing a small number of infancy failures, but not enough to significantly alter field failure rates. In addition, the process was time consuming, a factor that slowed production, increased unit cost and delayed delivery.

An answer came, again from the military, with the introduction of its newly devised approach of environmental simulation testing in the form of military standards. These standards required that products be operated at specific environmental extremes. The logic was that field operation would demand that products be used in a variety of environments, as well as be exposed to cyclical changes within these environments. For example, a jet climbing from a desert runway to altitude would be exposed to a significant temperature change in a relatively short period. A static testing process, such as burn-in, could not provide the proper environmental simulation. To address this problem, environmental simulation through the application of a combination of stresses was introduced. This early application involved temperature cycling between hot and cold extremes paired with low-frequency vibration through a mechanical shaker.

These basic concepts led to the development of mission profile testing, in which products were exposed to environments that simulated actual use conditions. Mission profiling produced some important results. First, it was discovered that a number of products could operate successfully at the temperature extremes, but would fail while being taken through a multiple of temperature cycles. This process, known as thermal cycling, was the basis for *ESS* theory. Second, a significantly higher number of failures were forced than with burn-in, particularly the types of failures that were occurring in the field. Finally, thermal cycling, precipitated failures in a much shorter time. It also became evident that testing of this type produced stresses, through changing temperatures and the resultant expansion and contraction this caused, that would force previously undetectable latent defects into product failures.

From these discoveries, the concept of *ESS* was born and the distinctions between this discipline and conventional test methodology became more defined. Where environmental testing was used primarily for design validation, environmental screening could be used to qualify product materials and workmanship. But until the mid 1970's, little public information on stress screening was available in the U.S.A.

One early application of *ESS* was by the U.S. National Aeronautics and Space Administration (NASA) during the *Apollo* Space Program for the acceptance testing of electronic equipment [3]. Since the major sources of vibration for products used in space are random, NASA was receptive to introduce *ESS* for acceptance testing. The Grumman Corporation in Bethpage, New

York, started the pioneering work in 1970 with tests on the design of the lunar module for the *Apollo* man-on-the-moon program. The company was able to precipitate and eliminate 85% to 90% of all workmanship defects from the equipment subject to random vibration and thermal cycling. Later the *ESS* research was extended in the Grumman Corporation to Airborne Electronic Systems for military aircraft.

The *ESS* work conducted early at the Grumman Corporation was one of the main bases for the U.S. Navy's document P-9492, "Navy Manufacturing Screening Program," which was one of the first military documents to suggest effective temperature profiles and random vibration spectra for electronic equipment screening.

As interest developed in *ESS*, a number of companies became involved in experimentation with different forms of stresses, and made attempts to establish specific guidelines for each type of a product. As a result, some of the early distinctions drawn between *ESS* and conventional testing were blurred, creating confusion in industries where *ESS* had gained only preliminary interest. In an attempt to clarify, the Institute of Environmental Sciences (IES) undertook a comprehensive study of *ESS* and held conferences in 1979, 1981, 1984 and 1985 on the subject of *ESS*. Its books, *Electronic Stress Screening of Electronic Hardware (ESSEH)*, which was published in 1981 [4] and 1984 [5], and *Environmental Stress Screening Guidelines for Parts*, which was published in 1985 [6], set out to demystify the process and provide some general guidelines for *ESS* applications.

During the 1980's the following developments occurred:

1. Stimulation concept replaced simulation.

2. Response was recognized as the key to success, not input.

3. *ESS* was recognized as hardware unique.

4. Tailoring replaced use of arbitrary parameters.

5. Random vibration proved superior to sine-swept vibration.

6. High ramp-rate temperature cycling proved superior to burn-in.

7. The need to isolate defect cause became apparent.

8. Stress screening was recognized as

   (a) an essential process-control tool of manufacturing and repair,

   (b) a diagnostic tool for intermittent failures, and

   (c) a useful tool for proof of design and reliability growth.

The United States Air Force, Army, and Navy have committed major resources to obtain reliable electronic equipment and established the need for Environmental Stress Screening. The Air Force issued its policy statement in R&M 2000 in 1985 [7; 8; 9; 10]. The Navy issued NAVMAT P-9492 in 1979 [11] and Best Practices NAVSO P-6071 in 1986 [12]. The Army issued its statement AMC-R 702-25 in 1986 [13]. The intent of each of these documents is to convey to industry the message that reliability of fielded products is vital to weapon system readiness.

All these efforts, and followup studies conducted by the IES and other sources, provided solid evidence that *ESS* is a highly successful method of enhancing product reliability.

REFERENCES

1. Caruso, H., "A Consumer Guide to *ESS*: Making Sense Out of Environmental Stress Screening Standards and Requirements," *The Journal of Environmental Sciences*, pp. 13-17, January/February 1987.

2. Thermotron Industries, *The Environmental Stress Screening Handbook*, 291 Kollen Park Drive, Holland, MI 49423, 36 pp., 1988.

3. Tustin, W., "Recipe for Reliability: Shake and Bake," *IEEE Spectrum*, pp. 37-42, December 1986.

4. Institute of Environmental Sciences, *Environmental Stress Screening Guidelines*, 940 East Northwest Highway, Mount Prospect, IL 60056, 1981.

5. Institute of Environmental Sciences, *Environmental Stress Screening Guidelines*, 940 East Northwest Highway, Mount Prospect, IL 60056, 1984.

6. Institute of Environmental Sciences, *Environmental Stress Screening Guidelines for Parts*, 940 East Northwest Highway, Mount Prospect, IL 60056, 1985.

7. Gabriel, C. A., "Reliability and Maintainability Action Plan, R& M 2000," USAF, February 1985.

8. Littlefield, J. W., "R & M 2000 Environmental Stress Screening," *IEEE Transactions on Reliability* , Vol. R-36, No. 3, pp. 335-341, 1987.

9. Saari, A. E., "On the Implications of R & M 2000 Environmental Stress Screening," *IEEE Transactions on Reliability*, Vol. R-36, No. 3, pp. 342-345, 1987.

10. Capitano, J. L., "R & M 2000 and Environmental Stress Screening," *IEEE Transactions on Reliability*, Vol. R-36, No. 3, pp. 346-350, 1987.

11. Willoughby, W. J., "Navy Manufacturing Screening Program," *Department of the Navy, NAVMAT P-9492*, May 1979.

12. Willoughby, W. J., "Best Practices," *Department of the Navy, NAVSO P-6071*, March 1986.

13. Ross, J. D., "Product Assurance, AMC Environmental Stress Screening," *U.S. Army, AMC-R 702-25*, March 1986.

Chapter 3

ESS DEFINITIONS, BASIC CONCEPTS AND COMMON MISCONCEPTIONS

As is true throughout the electronics industry, the terms used to define or describe a discipline can vary widely. In understanding *ESS*, there are a number of critical areas where these variations may cause confusion. The terms "screening", "aging", and "burn-in", for example, are sometimes used interchangeably in industry. In the interest of clarity, this chapter makes some needed distinctions.

3.1 WHAT IS *ESS*?

The *ESS* definition given by DOD-HDBK-344 (USAF) [1, p. 1] is as follows:
"ESS is a process or series of processes in which environmental stimuli, such as rapid thermal cycling and random vibration, are applied to electronic items in order to precipitate latent defects to early failures."
The *ESS* definition given by Thermotron Industries [2, p. 5] is the following:
"ESS is a means of screening products, ideally at the most cost-effective point of assembly, to expose defects that can't be detected by visual inspection or electrical testing. These defects typically are related to defective parts, workmanship or process, and are major contributors to early product field failure."
The *ESS* definition given by Mandel [3, p. 294] is as follows:
"ESS is a process which involves the application of a specific type of environmental stress, on an accelerated basis, but within design capability, in an attempt to surface latent or incipient hardware flaws which, if undetected,

21

would in all likelihood, manifest themselves in the operational or field environment."

Two key elements of this definition are "accelerated stress" and "design capability." The bottom line is that the accelerated stresses should not exceed the hardware design capability. Otherwise the hardware will be overstressed or even damaged and its useful life shortened.

The *ESS* definition given by RADC [4, p. A-4] is the following:

"ESS is the process of applying mechanical, electrical and/or thermal stresses to an equipment item for the purpose of precipitating latent part and workmanship defects."

The *ESS* definition given by Pennington [5, p.26] is the following:

"ESS is the tailored application of electrical and environmental stresses to electronic parts, modules, units and systems to identify and eliminate defective, abnormal or marginal parts and manufacturing defects."

Through an *ESS* program, 100% of a group of products is subjected to an environmental stimulus, or a set of stimuli, for a predetermined time, for the purpose of forcing failures which are likely to occur in the hands of the user, to occur in house.

Critical to an understanding of *ESS* is that, within the program, failures are expected, normal and unavoidable. In this sense, *ESS* is radically different than conventional certification testing, which requires failure-free operation as proof of reliability. It is equally important to realize that *ESS* is not simply an amplified certification test forcing greater, and possibly damaging, stress on the product. In fact, *it is not a test at all*, but a program used to upgrade product reliability in ways conventional testing cannot.

Toward this end, two important factors are key to proper implementation of *ESS*:

1. An optimum level of stress must be applied to the product to force latent defects into failure.

2. The stress environment must not exceed the electrical or mechanical limits of the product, forcing needless failures or reducing the product's useful life.

3.2 DEFINITIONS AND ACRONYMS OF FREQUENTLY USED TERMS IN *ESS*

The following is a collection of definitions, in alphabetic order, based on [1; 2; 4; 6; 7; 8; 9], for some frequently used terms in *ESS* literature and documents:

Accelerated aging – Increasing certain stress parameters, such as temperature or voltage, above their normal operating values to observe deterioration in a relatively short period and infer to expected service life under normal conditions in a longer calendar age.

Accelerated test – A test which uses stress levels higher than normal to reduce test time.

Acceleration factor – The multiplier by which the calendar time, or other usage, of an accelerated test will be factored to calculate the equivalent test time, or usage, under less stress or normal stress.

Acceleration spectral density – Also called power spectral density, a measure of the variance or mean-squared acceleration per unit of spectrum bandwidth representing the spectral distribution of the average energy.

Acceleration test – A test where the test article is subjected to an acceleration force in one plane.

Acceptance test – A test used to demonstrate compliance of a product to specified criteria as a condition of acceptability for "next" usage (next assembly, customer acceptance, etc.).

Acoustic vibration – Excitation of an article by aerodynamic pressure fluctuations or acoustically propagated noise.

Activation energy – An energy input, in electron volt units, which is required to cause a molecule of a constituent to participate in the reaction and which determines the relationship between a component's temperature and its lifetime, for certain temperature dependent kinds of failure modes of chemical nature. This energy may be determined experimentally by observing the times to failure of different batches of components at different temperatures.

Aging/Conditioning/Life-aging/Product maturing – Accumulation of stressful usage experienced by a device whose failure distribution is a function of that usage.

Airborne – A classification for the most general aircraft equipment environment.

Airborne inhabited (AI) – A classification for the aircraft equipment environment found in inhabited areas that does not experience environmental extremes.

Airborne uninhabited (AU) – A classification for the aircraft equipment environment found in cargo storage areas, and in wing and tail installations in which the following typical conditions are often present:

1. Contamination from engine exhaust hydraulic fluid oil.

2. Extreme pressure.

   3. Temperature cycling.

   4. Vibration cycling.

Ambient environment – Room conditions with no special controls imposed except for worker comfort.

Ambient test – A test, in contrast to an **Environmental test**, that is conducted under existing static conditions of the laboratory, factory floor, or office environment in which no environmental condition is applied.

Arrival quality – The percent of products found to be defect free upon installation, or first use, at the customer's usage site.

Assembly/Module – A number of parts joined together to perform a specific function and capable of disassembly, such as a printed circuit board. An assembly of parts designed to function in conjunction with similar or different modules when assembled into a unit; e.g., printed circuit assembly, power supply module, core memory module.

Assembly level – The level at which a screen is applied, such as module, unit, system, etc.

Attribute – Qualitative property that a product has or does not have.

Attributes testing – A test procedure to classify items under test according to qualitative rather than quantitative characteristics, such as **Go/No-Go testing**.

Autopsy – Failure analysis to determine the root cause of a removed item's last wearout failure when this is the primary reason for the item being decommissioned or scrapped. Secondary failure modes and degradation may be noted.

Average outgoing quality level – Average percent defective of product that could leave an inspection, including both accepted lots and rejected but screened lots.

Axis – The direction along which a mechanical stress is applied.

Benign environment – Conditions external to a unit which induce either minimum or no stress on it.

Black-box testing – Testing, in contrast to **White-box testing** and **Gray-box testing**, which is conducted for acceptability of the external output given specified external inputs and without concern for intermediate internal functions.

Board level – A level at which circuit boards are screened prior to sub- or final assembly. Components are in place and the product can be operated with proper interconnections. The only added complexity is usually in packaging.

Bogey test – Also called **Prestress testing**, a single-step, go/no-go, mechanical part test to a predetermined number of cycles run and stress, that is calculated to equal some given percentile of worst case customer usage for one lifetime. This kind of test is intended to detect or screen manufacturing defects and used as the first stress step of some types of *step-stress tests*.

Broad-band vibration – Vibration which contains significant levels within a broad frequency range, as opposed to narrow-band vibration, where the significant vibration levels are concentrated around a single frequency or a very narrow band of a frequency range.

Bulk defect – A time- or temperature-dependent, intrinsic or processing imperfection, impurity, or crystalline defect in semiconductor material.

Burn-in/Steady-state burn-in – The process of continuously powering a product, usually at a constant temperature, in order to accelerate the aging process.

Burn-in effectiveness – The percentage increase of temperature induced failures that will be experienced in a given time for a specified product at a given higher temperature than at a lower reference temperature.

Chamber/Temperature chamber – A cabinet in which products are placed to subject them to a stress or stresses. This usually consists of an insulated unit equipped with air circulation and temperature conditioning equipment with which ambient temperature can be varied rapidly and controlled in a prescribed manner.

Circuit board – An assembly containing a group of interconnected parts which are mounted on a single board.

Cold cracking – Crack defects caused by exposure to low temperature.

Cold soak – The part of the cycle of *environmental testing* at the lowest temperature. Contrast with **Temperature soak.**

Cold solder connection/Cold solder joint/ High-resistance joint – A failure mode in which a soldered electrical connection has solidified unevenly, forming poorly conductive microscopic cracks, which may lead to open or intermittent open failures in the field under slight mechanical stress.

Cold starts – Application of power following low-temperature stabilization.

Combined environmental test – A test performed during which more than one environmental stress is imposed.

Commercial – Uses for which military specifications are not imposed.

Common cause failures – Several items malfunctioning due to a single cause.

Component – A nonrepairable throwaway item (e.g., integrated circuit, resistor, capacitor, diode, transistor, transformer, hybrid, etc.).

Component defect – A defect caused by the failure of a component to meet its design specifications.

Component level – A level at which components are screened before being placed on circuit boards. Components may be screened by the manufacturer or by the user upon receipt.

Conditioning – Screening for a specified period of time.

Control chart – A graphical method for evaluating whether a process is in a state of statistical control; i.e., the plotted statistical measures are within the prespecified control limits.

Control limit – Limit on a control chart for judging whether a statistical measure, obtained from the sample, falls within acceptable bounds.

Corrective action – A process of correcting the root cause of a defect or failure.

Corrective action effectiveness – A percentage estimate of the fraction of the failures of a particular type that a corrective action was to eliminate and that will no longer be present in a new design.

Cost effectiveness – Monetary difference between the cost and the benefits derived from it.

Cost of performance – Cost of conformance plus cost of failure.

Crazing – Minute crack, or cracks, on or near the surface of materials such as ceramics and plastics caused by different rates of expansion or contraction of different layers.

Creep – A gradual change in the dimensions of a material under a mechanical load, or high stress, usually at high temperature.

Damage – Reduction in strength due to stress degradation often expressed as the ratio of the "number of usage cycles applied" to the "number of usage cycles to failure."

Damaging overstress – An induced or natural overstress which exceeds the design capability of the article being screened, or under test, and which causes partial or catastrophic failure of the item.

Defect/Latent defect – A flaw in an item that would eventually prevent it from meeting its functional requirements when operating within its specified environment and within its expected lifetime. Stated equivalently, any nonconformance of a unit of product with specified requirements.

Defective/Defective unit – A unit of a product which contains one or more defects.

Defect density – Average number of latent defects per item. Symbols, such as D_{IN}, D_{OUT}, D_R and D_O, stand for incoming, outgoing, remaining and observed defect densities, respectively.

Defect-free – That portion of a test or screening sequence which must be completed without a defect.

Degradation – Gradual deterioration in performance as a function of time and/or stress.

Degradation factor – The factor by which the *inherent reliability* is reduced due to environmental stresses the equipment experiences when being manufactured, transported, maintained, and used throughout its life cycle.

Degradation failure – A deterioration of one or more parameters beyond specification limits in two or more partial steps, such as electronic part drift, lubricant aging, inelastic or plastic strain, metal corrosion, and wearing away by abrasion.

Dependent failure – Failure caused by the failure of an associated item or items.

Derating – Upgrading component reliability by intentionally reducing the stress level in the application of an item.

Derating factor – Factor by which an equipment is derated to achieve a reliability *safety margin*.

Design defect – A defect caused by a faulty product or process design.

Design deficiency – One or more of the following unreliability causes:

1. Actual mistakes in the design.
2. Incapable state-of-the-art components.
3. Oversights.
4. Unavoidable complexities and undesirable conditions.
5. Unforeseen material incompatibilities and unforeseen conditions.
6. Unknown environments.

Design fault – Faults generated during design and not corrected.

Design margin – The self-imposed restriction on a design more severe than either specified or operational use requirements.

Design of experiments (DOE) – A branch of applied statistics dealing with planning, conducting, analyzing, and interpreting controlled tests to evaluate the effects and interactions of various factors which control the value of one or more parameters.

Designed test – A controlled test to evaluate a parameter or group of parameters.

Design weakness – The inability of a product to survive such rigors as handling, transport and service.

Detectable failure – A failure that can be detected with 100% test detection efficiency.

Detection efficiency/Test detection efficiency – A characteristic of a test measured by the ratio of the number of failure modes detectable to the total number of potential failure modes.

Electromigration – A failure mode for a very large-scale integrated circuit due to high current densities and operating temperatures which cause or enlarge vacancies or *puddles* in the aluminium crystal lattice grain boundaries of conductor leads. This failure mode has symptoms of open contacts and conductor lines, and shorts or leaks between and within layers and junctions.

Endurance strength/Fatigue strength – Maximum repetitive or cyclic damage that a material can withstand before fracturing. For mechanical components it corresponds to the strength distribution at and beyond the knee of the $S - N$ diagram.

Environmental cycle – A single complete submittal of the equipment to all of the specified environments.

Environmental sensitivity – The change in a specified parameter of a part, assembly, unit, or system that results from exposure to the environment.

Environmental stress screening/Stress screening – A process in which 100% of products are subjected to one or more stresses (jointly or separately), such as thermal cycling, random vibration, high-temperature burn-in, electrical stress, thermal shock, sine-fixed-frequency vibration, low temperature, sine-wave-swept-frequency vibration, altitude or humidity, etc., with the intent of forcing latent defects to surface as early failures.

Environmental testing – A process in which a sample of a product is subjected to environmental simulation. Testing can be used in a variety of ways:

1. Testing on a prototype at extreme temperatures to confirm the range in which it was designed to function.

2. Testing on a randomly selected product at extreme temperature ranges to confirm continuing design and production process compliance.

3. Life-testing a product to determine its mean time between failures ($MTBF$).

4. Testing the product under the simulated environments which the product will encounter in transportation and operation.

Equipment power on-off cycle – The state during which an electronic item goes from zero electrical activation level to its normal design system activation level and back again to the zero activation level.

Escapes – A proportion of incoming defect density, referred to as D_{OUT}, which is not detected by a screen test and which is passed on to the next level.

ESSEH – Environmental stress screening of electronic hardware.

ESS **profile** – The sequence and duration of stress environments to which items are to be subjected.

Experiment – Same as designed test.

Factorial experiment – An experiment in which all possible treatment combinations formed from two or more factors, each being studied at two or more levels, are examined to estimate the main effects and the interactions.

Factory failure cost savings – Costs that are avoided by the use of screening techniques to reduce the number of failures occurring at the manufacturer's facility.

Failure – Lack of proper operation. Failures can be classified into the following most common types:

1. Critical failure – A case where the product is unable to operate under the conditions it is expected to.

2. Noncritical failure – Failure that occurs only outside the normal operating range of the product.

3. Hard failure – A critical failure where the product stops and does not resume functioning.

4. Soft failure – A critical or noncritical failure where the product stops functioning under certain conditions but then resumes under others.

5. Infancy failure – Failure that occurs in the early stages of a product's life.

6. Electrical failure – An electrical malfunction caused by electrical components, connections, switches or related devices.

7. Mechanical failure – A mechanical malfunction caused by cracking, displacement or misalignment of assemblies.

8. Process failure – Failure caused by the manufacturing process.

Failure-free period – A contiguous period of time during which an item is to operate without failure while under environmental stresses.

Failure-free test – A test to determine if an equipment can operate without failure for a predetermined time period under specific stress conditions.

Fallout – Failures observed during, or immediately after, and attributed to stress screens.

Fatigue ductility coefficient – The minimum true strain required to cause failure in one reversal cycle of a test which applies cyclic tension.

Fatigue ductility exponent – A material fatigue property, nearly a constant ranging from about -0.50 to -0.70 for metals, which is equal to the log of the fraction of a given plastic strain range which will cause a tenfold increase in the median fatigue life.

Fatigue failure – The breakage failure of a component subjected to cyclic stresses, due to the spreading of a fracture which starts from a weak point.

Fatigue resistance – Resistance to metal crystallization and embrittlement from flexing, which would lead to metal fracture.

Fatigue strength – See **Endurance strength**.

Fatigue strength coefficient – A material fatigue property, ranging between about 100 ksi and 500 ksi for heat-treated steels, which is equal to the true stress required to cause a fracture in one reversal of a destructive cyclic tension test.

Fatigue strength deviation – A measure of inherent scatter in fatigue strength, which is equal to the standard deviation of the fatigue strength from the mean fatigue strength over a range of fatigue lives.

Fatigue strength exponent/Basquin's exponent – A material fatigue property, ranging from about -0.05 to -0.12, which is equal to the log of the fraction of a given true stress range which will cause a tenfold increase in the median fatigue life.

Fatigue strength limit – The maximum alternating stress, ranging between 20 ksi and 200 ksi, below which there will be no fatigue failure regardless of the number of operating cycles.

Fault detection – One or more tests conducted to determine whether or not any malfunctions or faults are present in a unit.

Fault detection coverage/Fault detection efficiency/Diagnostic coverage – Percentage of failures which can be detected and diagnosed correctly.

Fault isolation – Tests conducted to isolate faults within the unit under test.

Field – The place where a product is ultimately used.

Field failure cost savings – Costs that are avoided by the use of screening techniques to reduce the number of failures occurring during the operational life of the equipment.

Field replaceable unit (FRU) – See **Line replaceable unit (LRU)**.

Field warranty rate – The rate of failure during the time a warranty is in effect.

Final assembly level – A screening level at which *ESS* is applied to finished products.

Fixed sinusoidal vibration – Vibration excitation with a constant level and frequency, and with a waveform of a sinusoid.

Fractional defective – Proportion of defective units expressed as a decimal rather as a percentage.

Fractional factorial design/Fractional replicates – A factorial experiment in which only an adequately chosen fraction of the treatment combinations required for the complete factorial experiment is selected to be run.

Freak failures – Anomalous component, subsystem, or system failures usually distributed in time, centered near a specified percentage of useful life such as 10%, on less than or equal to another specified percentage of a population, such as 20%.

Functional test – A test which measures a limited number of critical parameters to assure that the test article is operating properly.

Fixture/Fixturing – An intermediate structure to attach and secure items on a shaker or within a chamber in preparation for running screens.

Ghost – An *intermittent* class of failure, the cause of which can neither be diagnosed nor effectively repaired.

Go/No-Go – The binomial attribute condition or state of operability of a unit which can only have the following two parameters:

1. Go or functioning properly.
2. No-go or not functioning properly.

Go/No-Go testing – Test method using a fixed measure for *inspection by attributes* to determine whether or not a measurement conforms with specifications.

Gray-box testing – Testing, in contrast to **Black-box testing** and **White-box testing**, which is conducted for acceptability of the internal output given specified internal inputs.

Ground benign environment (GRB) – A classification for the least stressful component or system environment.

Ground fixed environment (GRF) – A classification for a maintained installation condition somewhat less ideal than **Ground benign** but more ideal than **Ground mobile**, such as the permanent installation in air traffic control, in communication facilities, in ground support equipment, in racks with adequate cooling air, in radar, in unheated buildings, etc.

Ground mobile environment (GRM) – A classification for an equipment environment which is more severe than **Ground fixed** environment mainly due to the following reasons:

1. Less uniform maintenance.
2. More limited cooling supply.
3. Natural outdoor environmental stresses.
4. Shock.
5. Vibration.

Hard failure – A failure, in contrast with **Intermittent** and **Ghost failure**, which is always repeatable or continuously observable.

Heat endurance – The total time at a specific heat condition which a material can withstand before failing a specific physical test.

Heat shock – A form of **Thermal shock** due to sudden exposure to a high temperature for a short period.

High reliability program – A series of tasks performed to assist in developing an equipment which has an extremely low probability of failure.

High-temperature reverse bias (HTRB) – Fault avoidance methodology in which burn-in of diodes and transistors is conducted with the junctions reverse biased, in order to force any failures to occur which are likely to be caused by ion migration in bonds of dissimilar metals.

Hot start – Application of power following high-temperature stabilization.

Imperfection – A departure of a quality characteristic from its intended level or state without any association with conformance to specification requirements or to the usability of a product or service.

Incipient failure – A failure which has not occurred at a time when the damage, which must lead to failure, is first detected, but which will undoubtedly occur during the useful life of the specific item in an operating environment.

Infant failures – Failures which occur early in the operating life of a component, module, or unit.

Inherent defect – A failure or defect that is a function of the intended design application of the item, when operated in its intended operational and logistic support environment.

Intermittent failure – A nonpermanent item failure which lasts for a limited time period, followed by recovery of its ability to perform within specified limits.

Irrelevant failure – A failure that is not charged against the equipment.

Latency – A state in which a fault remains undetected; i.e., being **Incipient**.

Latency time – Time period between the occurrence of a failure to its detection.

Latent defect/Random defect – An inherent or induced weakness, not detectable b ʳ ordinary means, which will either be precipitated as an early failure under environmental stress screening conditions or will eventually fail in the intended use environment.

Level of assembly – The hierarchy of items within a larger item which are functionally or physically grouped and considered together for various purposes, such as system, subsystem, module, board, and component.

Life-cycle testing – A process through which a small percentage of products are subjected to stresses similar to those they will experience during their lives. This process is usually used to determine a product's anticipated life expectancy and its *MTBF*.

Life test – A test performed on a group of items (parts, assemblies, units, or systems) continued until a specified percentage of this group fails or some predetermined minimum operating time has elapsed.

Limiting quality level (LQL) – See **Lot tolerance percent defective**.

Line replaceable unit (LRU) – A unit normally removed and replaced as a single item which consists of assemblies (SRU's), accessories, and components that collectively perform a specific functional operation.

Lot – A group of products manufactured or processed under substantially the same conditions.

Lot tolerance percent defective (LTPD) – Value of lot percent defective on an *operating characteristic curve* corresponding to the value of the customer's risk.

Main effect – A term describing the contribution of a factor at each level to process responses averaged over all levels of other factors in the experiment.

Malfunction – Inability to perform a required operation, in particular, an **Intermittent failure.**

Manufacturing defect – A flaw caused by in-process errors or uncontrolled conditions during assembly, test, inspection, or handling during manufacturing.

Margin of safety – See **Safety margin.**

Marginal failure – A failure which may cause minor injury in a military system, minor property damage, or minor system damage which will result only in delay, loss of availability, or mission degradation.

Module – An assembly of parts usually packaged in a plug-in form for ease of maintenance of the next higher level of assembly, such as printed circuit assembly, power supply module, core memory module, etc. It is designed to function in conjunction with similar or different modules when assembled into a unit.

Monitoring/Condition monitoring/ Performance monitoring – Observing or keeping track of output measurements, or obtaining data by any appropriate means, to signal changes of operating conditions or of outputs and the necessity of corrective actions to eliminate any potential failures.

Multiple failures – The simultaneous occurrance of two or more *independent* failures.

Multiple fault condition – More than one fault occurring during the same fault detection process.

Next higher-level effect – The consequences a failure mode has on the operation, functions, or status of the items in the next higher level of assembly.

Nonchargeable failure – A failure which is either irrelevant, or relevant but is caused by a condition prespecified as not the responsibility of any given organization, and therefore not charged.

Nonconformance/Nonconformity – A departure of quality characteristic from its intended level, or state, that occurs with a severity which is sufficient enough to cause an associated product, or service, not to meet a specification requirement.

Nonconforming unit – A unit of product containing at least one nonconformity.

Nonenvironmental stress screening failures – Failures which occur during *environmental stress screening* but are not considered as screened *incipient failures*, such as misinterpreted failures, repair-induced failures, operator-induced failures, dependent failures, and failures due to improper facility installation, etc.

Nonrelevant failures – Equipment failures which are not expected to occur during field service.

Notching – A technique of reducing the input power spectral density (PSD) level over a small frequency bandwidth, in particular around an equipment resonant frequency, so that the equipment is not damaged due to overstress at resonant frequency.

On-off cycling – Switching of equipment primary power on and off at specified intervals.

Orthogonal axis vibration – Excitation applied at right angles to an axis.

Parameter drift – A form of wearout in which electrical and mechanical components, which are designed to certain tolerance limits, drift out of or exceed their tolerance limits, as a function of operating time and stress level; however, the product recovers its normal function after a rest period.

Part – An element of a subassembly, or an assembly, so constructed as to be impractical to further disassemble for maintenance.

Part fraction defectives – The number of defective parts contained in a part population divided by the total number of parts in the population expressed in PPM.

Patent defect – An inherent, or induced, weakness which can be detected by inspection, functional test, or other defined means without the need for stress screens.

Percent defective – The number of defective units, divided by the number of units of the product times 100.

Periodic conformance test – A test performed at regular intervals to verify continued compliance with specified requirements.

Power cycling/Operational cycling – The process of continuously turning the product on and off at predetermined intervals. Power cycling adds internal heat and ages the product much faster than continuous power on and off. Power cycling also allows monitoring of products. This provides diagnostic information used to determine reasons for soft failures.

Precipitation of defects – The process of transforming a latent defect into a patent defect through the application of stress screens.

Predefect-free – That portion of a screening sequence prior to the disclosure of a defect. Contrast with **Defect-free**.

Prestress screening – Eliminating *incipient failures* which would probably occur during early life period by applying an environmental and/or use stress at a peak abuse level or worst case or user stress.

Prestress test – See **Bogey test**.

Printed circuit board (PCB) – Same as circuit board.

Printed wiring board (PWB) – Same as circuit board.

Producibility – Ease of fabrication and assembly.

Product assurance – A discipline that assures that products are designed and manufactured to be highly reliable, easy to maintain and safe to operate.

Production flaw – A latent defect in a product, such as a poorly soldered connection, a loose screw, or a contaminant like scrap wire, foreign particles, dust, moisture, etc.

Production lot – A group of items manufactured under essentially the same conditions and processes.

Production reliability test – A test performed to measure the reliability of production items.

Production sampling – Sample testing to determine compliance to specifications.

Productivity – Value achieved for the resources expended.

Proof-of-design test – A test performed during the design cycle to demonstrate compliance with specified design criteria.

Qualification test – A test to verify that a product's design meets contractual requirements.

Quality assurance – The total effort of *quality control* and *quality engineering*.

Quality defect – Any failures, errors, flaws, interruptions, and faults occurring in a product, its subassemblies, and *field replaceable units* during manufacturing and early field operation but before the end of the specified early life or promised warranty period. Contrast with **Reliability defect**.

Quality engineering (QE) – Analysis of the manufacturing system at all stages of development to maximize ultimate process and product quality.

Quality function deployment (QFD) – A structured method to identify customer needs, translate them into a realizable product or service parameters, and guide the implementation process in such a way as to bring about a competitive advantage.

Quality management – The totality of functions involved in the determination and achievement of quality.

Quasi-random vibration (QRV) – QRV may be described as an equally-spaced line spectrum whose fundamental frequency varies randomly with time within a restricted frequency range and, in turn, causes a random fluctuation in the amplitudes of the spectral lines. The fundamental frequency fluctuates sufficiently to produce an essentially continuous spectrum when averaged over a suitable interval.

Random vibration – Vibration excitation where magnitude and frequency are specified by a probability distribution function.

Relevant failure – Equipment failure which can be expected to occur in field service. Contrast with **Irrelevant failure**.

Reliability defects – Any failures, errors, flaws, interruptions, and faults occurring in a product, its subassemblies, and *field replaceable units* after the end of the specified early life or promised warranty period. Contrast with **Quality defect**.

Reliability demonstration – A test to determine a product's reliability, failure rate, *MTBF*, or percentile life.

Repair time – The period required to locate the cause of failure, perform the repairs required, and verify that the repair action was effective.

Resonant dwell – Vibration excitation of a test article at one of its resonant frequencies for a sustained period of time.

Run-in – A process, often confused with *burn-in* and *ESS*, of simulating or duplicating but not exceeding the stresses normally expected in the customer environment to *surface* all nonworking functions.

Safety allowance – Same as design margin.

Safety factor – Same as design margin.

Safety margin – Difference between the designed-in strength (stress at failure) and the operating stress level expressed in units of the rating, or the safety factor minus one (1) in mechanical component design.

Sample test – A test performed on a limited number of items during the production process.

Screen effectiveness – Generally, a measure of the capability of a screen to precipitate latent defects to failure. Sometimes used specifically to mean screening strength.

Screen level – The assembly level at which an *ESS* program is implemented; such as component, board, subassembly or final assembly.

Screen parameters – Parameters in screening strength equations which relate to screening strength, such as vibration g-levels, temperature rate of change, duration, etc.

Screenable latent defect – A latent defect which has an inherent failure rate of greater than 10^{-3} failures per hour under field stress conditions.

Screening – A process or combination of processes applied to 100% of a lot or group of like items to identify and eliminate defects.

Screening attrition costs – Costs associated with replacing and rescreening parts that fail during part level environmental screening, or repairing items at a higher level of assembly that fail during higher-level environmental screening.

Screening complexity – The degree to which a product is screened, such as the following:

1. Static – The application of a single stress to an unpowered product.

2. Dynamic – Power, and on occasion an additional screen, applied to a product during the initial screen.

3. Exercised – Power and monitoring, or loading inputs and outputs, are applied during the screen. The application of signals or loads is sometimes necessary to avoid undue product damage.

4. Full functional – The product is operated during screening as though in actual use, with all circuitry exercised.

5. Monitored – Providing monitoring devices that either display or record screen performance which is also helpful in providing field failure analysis data.

The screen complexity is dependent on the need to identify or force specific types of failures, the product's anticipated operating environment, the level of assembly at which the product will be stressed and, in general, consideration of the degree of stress that is both necessary and cost justifiable to gain a measurable result.

Screening cost effectiveness – The dollar difference between the cost of screening and the benefits derived.

Screening-induced degradation – See **Stress screen degradation.**

Screening regimen – A combination of stress screens applied to any level of complexity, such as module, unit or system, identified in the order of application.

Screening sequence – Chronological order of activities to conduct *ESS.*

Screening strength – The probability that a specific screen will precipitate a latent defect to failure, given that a latent defect susceptible to the screen is present. Symboled *SS*, screening strength is usually measured by the percentage of total defects identified by a specific screen.

Seasoning – A process, often confused with *burn-in*, of energizing a consumable item at its rated usage stress for a period of time equal to a specified percentage of its average rated laboratory life, often chosen to be 1%, in order to reduce fluctuation.

Secondary damage – Damage to an item due to failure of another item within the same configuration.

Secondary damage effects – Consequences indirectly caused by the interaction of a damage mode with a system, subsystem, or one of their components.

Secondary failure – A dependent item failure either indirectly or directly due to the failure of another item.

Selection and placement – The process of systematically selecting the most effective stress screens and placing them at the appropriate levels of assembly.

Sensor – A device which observes physical condition of interest and outputs an appropriate electrical signal related to the condition.

Shakedown – A form of *run-in* using actual operation under rigorous but typical usage merely to show that a unit is capable of operation.

Shaker – A machine usually powered by electricity, hydraulic pressure, or compressed air, providing a controlled source of vibration to a product.

Shock test – Subjection of a test article to one or more singly applied acceleration pulses at a frequency of less than 1 Hz.

Shop-replaceable unit (SRU)/Shop replaceable assembly – An assembly or any combination of parts, subassemblies, and assemblies mounted together, normally capable of independent operation in a variety of situations and repairable at maintenance levels above the field, such as on the bench, in a depot repair facility, or in a logistic center.

Sine vibration/Sine fixed frequency vibration – The vibratory excitation is sinusoidal with a prespecified fixed frequency.

Sine wave, swept frequency vibration – The vibratory excitation is applied with a gradually increasing or decreasing strength and frequency.

Single-frequency vibration – Sine vibration at only one frequency.

S-N curve – A graph of applied stress, S, versus the corresponding cyclic fatigue life, N.

Sneak circuit analysis – A procedure of identifying *latent* circuit paths which may cause occurrence of *unwanted* functions or *inhibit* desired functions assuming all components are functioning properly.

Sneak condition – Latent paths, timing indications, or labels in electrical hardware or computer logic.

Sneak indicators – False or ambiguous system status that can result from improper connection or control of display devices and their sensors.

Sneak labels – Potential causes of erroneous operator actions which can arise from a lack of precise nomenclature on controls or operating consoles.

Sneak paths – Latent current paths that cause an unwanted function to occur or inhibit a desired function independent of component failures.

Sneak software analysis – Analysing software sneak conditions using electrical/software analytical analogies.

Sneak timing – Inappropriate system response which may be due to incompatible hardware or logic sequences.

Soft failure – See **Intermittent failure**. Contrast with **Hard failure**.

Solid failure – A form of *hard failure* reported by a maintenance specialist who observes a user perceived continuous, permanent, or easy-to-duplicate failure condition during corrective maintenance in correcting that failure.

Sonic fatigue – Mechanical wearout due to vibration transmitted through a gaseous medium.

Sonic vibration – Potentially stressful mechanical vibration waveforms induced sonically in the frequency band between 1 Hz and 20 kHz.

Spectrum analyzer bandwidth – The interval separating the upper and lower frequencies observed at any given time during verification of a random vibration spectrum.

Standard environmental profile – A *cookbook* or standard testing environment described in a general reference which is not tailored to a specific application. Contrast with **Tailored environmental profile**.

Step stress – An experimental process of increasing stress levels incrementally so that the effects of each step may be evaluated.

Stimulus – Any energy applied to a device to produce a measurable response.

Strain – Solid body deformation measured by changed distances between points on the body.

Strain gage – Any device which can sense the deformation of a solid body and transmit the signal into a measurable output.

Strength – The inherent ability of resisting failures. The stress magnitude at failure.

Stress – Electrical, thermal, or mechanical forces caused by a stimulus being applied to a product.

Stress screen degradation – Reduction of component strength in a good component due to the application of a stress screen.

Stress screening – The process of applying mechanical, electrical and/or thermal stresses to an equipment item for the purpose of precipitating latent part and workmanship defects to early failure.

Subassembly level – The stage of production prior to final assembly, but more advanced than the circuit board level. The goal of screening at this level is to find latent defects in packaging and interconnections.

Substandard item – An item which may perform all its functions but not all to the minimum specified requirements.

Swept-sinusoidal vibration – Vibration excitation with amplitude progressively varying over a specified frequency range.

System/Equipment – A group of units interconnected or assembled to perform some overall electronic function (e.g., electronic flight control system, communications system).

Tailoring – The process of selecting an initial stress screening program and the subsequent changes to the methods and severities of the screens to achieve maximum effectiveness.

Tailored environmental profile – Duplicating a known or anticipated field environmental profile as closely as possible. Contrast with **Standard environmental profile**.

Target flaw – The latent defect or incipient failure type that the corresponding specific stress screen is chosen to precipitate.

Temperature cycling – A process through which a product is subjected to a predetermined temperature change rate between established temperature extremes for a specified period of time. This process can verify that the product will operate at various temperatures, as well as its ability to withstand the rates of change between these extremes. In ESS thermal cycling, the ability to develop fast rates of changes will determine how many cycles are required to force the greatest number of latent defects into failure.

Temperature shock – Repeatedly subjecting an item (part, module, unit, system) to high and low temperature limits with a high rate of change between them, such as equal to or greater than 25°C per minute.

Temperature soak – Portion of the thermal cycle in an environmental chamber at which the highest stable temperature is induced. Contrast with **Cold soak**.

Temperature stabilization period – This is the amount of time required to reach thermal equilibrium, at which item and chamber temperatures are identical, starting from the end of the prior stabilization period.

Test – A process of subjecting an item to conditions designed to determine whether or not the item meets specified performance requirements, such as qualification test, reliability demonstration test, acceptance test, etc. It should be pointed out that the conditions of a test may duplicate those of a screen. However, the intent of a screen is to precipitate failures while the intent of a test is to demonstrate how well the item meets its specified requirements. The goal of a test is usually zero failures. The goal of a screen is to force defects to surface during the screen rather than during subsequent testing or operation.

Test detection efficiency – A measure of test thoroughness or coverage which is expressed as the fraction of patent defects detectable, by a defined test procedure, to the total possible number of patent defects which can be present. Symbol (DE) used synonymously as the probability of detection.

Test specification – A document defining tests to be performed and the specified parameter limits.

Test strength – The product of screening strength and test detection efficiency. The probability that a defect will be precipitated by a screen and detected in a test. Symbol (TS).

Test to failure – Testing continued until the item ceases to function within specified parameter limits.

Thermal aging – A form of *accelerated aging* by exposing items to a given thermal conditions or a programmed series of conditions for a given period.

Thermal conductivity – The rate of heat transfer through a material.

Thermal cycling – Same as **Temperature cycling.**

Thermal endurance – Minimum time required at a selected temperature for a material or system of materials to deteriorate to some predetermined minimum acceptable level of electrical, mechanical, or chemical performance under prescribed test conditions.

Thermal equilibrium – Item temperature condition in which the rate of change is less than 2°C/hr.

Thermal fatigue – Mechanical failure mode of materials subjected to the stress of alternating cycles of heating and cooling, including loss of temper, off-tolerance states, and work hardening.

Thermal gradient stresses – Cyclically cumulative or one-shot mechanical stresses within an integral assembly caused by an unmatched distribution of either heat or thermal expansion between adjacent parts of an assembly, such as printed circuit board warping and integrated circuit walk-out from sockets.

Thermal rating – The temperature range within which a material will function without unacceptable degradation.

Thermal resistance – The opposition to heat flow through a material.

Thermal shock – Subjecting a material to rapid and wide range changes in temperature to ascertain its ability to withstand those rapid changes.

Thermal survey – The measurement of thermal response characteristics, such as time to equilibrium, etc, at points of interest within an equipment when temperature extremes are applied to the equipment.

Tolerance – The allowable variation in a quality characteristic within which an item is judged to be acceptable, and equal to the difference between the upper limit and lower limit.

Tolerance limits – Minimum and/or maximum permissible parameter values for an item as determined by nominal and basic values and allowances.

Transducer – A device which changes the format of a sensor output in order to facilitate using it, but does not perform analog-digital conversions.

Transparent failure – A fault tolerant failure causing no obvious performance degradation.

Treatment – A combination of levels of different factors involved in an experiment.

Treatment combination – One set of levels of all factors involved in an experiment run.

Troubleshooting time – The period required to isolate the fault at the system, unit, module or part level.

Ultrasonic vibration – Potentially stressful mechanical vibration waveforms in the frequency band above 20 kHz. Contrast with **Sonic vibration**.

Unilateral tolerance – Tolerance which allows variations in one positive or negative direction from the specified dimension.

Unit – A self-contained collection of parts and/or assemblies within one package performing a specific function or group of functions, and removable as a single package from an operating system, such as an autopilot computer, VHF communications transmitter, voltmeter, etc.

Variable data – Data which contains the measured value as opposed to go/no-go indication.

Vibration – Periodic movement of an item.

Vibration survey – The measurement of vibration response characteristics at points of interest within an equipment when vibration excitation is applied to the equipment.

Warranty – Explicit or implied guarantee from the seller to the purchaser that an item will perform as specified or implied for a minimum specified period of time or usage.

Warranty claim – Action started by the equipment user for authorized warranty repair, replacement, or reimbursement made from the local dealer or manufacturer.

Warranty maintenance – Any corrective maintenance during the warranty period of the product or during a promised free maintenance period after repair.

Warranty period – Calendar time during which the warranty is in effect.

Wearout – The process of attrition which results in an increasing failure rate.

Wearout failure – Malfunction due to equipment deterioration caused by environmental stresses such as abrasion, radiation, fatigue and creep; or corrosion and other chemical reactions.

Wearout life/Wearout period/Wearout failure period – The total most appropriate life units after useful life during which equipment failure rate increases above the failure rate during useful life.

White-box testing – Testing, in contrast to **Black-box testing** and **Gray-box testing**, that is conducted to examine the internal functions which must collectively result in the specified external output.

Workmanship defects – Defects caused by human error during fabrication and assembly.

Yield – The probability that an equipment is free of screenable latent defects when offered for acceptance.

The following is a collection of some acronyms, arranged in alphabetic order, and their corresponding descriptions, based on [1; 2; 4; 7; 8; 9; 10], which may be encountered in *ESS* literature and documents:

AC – Alternating current.

AFC – Amortized facility cost.

Ag – Silver.

AI – Airborne inhabited.

Al – Aluminium.

AIM – Avalanche induced migration.

ALC – Air Force Logistics Center.

ALSTTL – Advanced low-power Schottky transistor-transistor logic.

Am – Americium.

AOQ – Average outgoing quality.

AOQL – Average outgoing quality limit (or level).

AQL – Acceptance quality levels.

AQR – Arrival quality reporting.

ARL – Acceptance reliability levels.

As – Arsenic.

ASICs – Application-specific integrated circuits.

ASTTL – Advanced Schottky transistor-transistor logic.

ATE – Automatic test equipment.

Au – Gold.

AU – Airborne uninhabited.

bit – Binary digit.

BIT – Built-in test.

BITE – Built-in test equipment.

BOC – Best operational capability.

B.S. – British Standards.

CCCs – Ceramic chip carriers.

CCD – Charge-coupled device.

CDE – Chance defective exponential.

CERDIPs – Ceramic dual-in-line packages.

CIM – Computer integrated manufacturing.

Cl – Chlorine.

CML – Current mode logic.

CMOS – Complementary metal oxide semiconductor.

CMOS/SOS – CMOS fabricated on silicon or sapphire.

CND – Cannot duplicate.

COB – Chip on board.

COE – Coefficient of thermal expansion.

COQ – Cost of quality.

COS – Chip on substrate.

Cr – Chromium.

Cu – Copper.

CVD – Chemical vapor deposition.

DC – Direct current.

DID – Diffusion induced dislocations.

DIPs – Dual-in-line packages.

DMOS – Double-diffused MOS.

DOA – Dead on arrival.

DOE – Design of experiments.

DR – Discrimination ratio.

DRAM – Dynamic random access memory.

DTC – Direct test cost.

DTL – Diode-transistor logic.

EAPROM – Electrically alterable, programmable read-only memory.

EBIC – Electron-beam-induced current.

ECL – Emitter coupled logic.

EDAX – Energy dispersive analysis of X rays.

EED – Emitter edge dislocation.

EEPROMs – Electrically erasable PROM's (E^2PROM's).

EMI – Electromagnetic interference.

EOS – Electrical overstress.

EPROMs – Erasable PROM's.

ESD – Electrostatic discharge.

ESS – Environmental stress screening.

ESSEH – Environmental stress screening of electronic hardware.

ETM – Electrothermomigration.

eV – Electron volts.

FAMECA/FMECA – Failure modes, effects and criticality analysis.

FAMOS – Floating-gate avalanche MOS.

FBT – Functional board tester.

FET – Field effect transistor.

FFAT – Failure-free acceptance tests.

FFT – Failure-free tests.

FIT – Failures in unit time (1 failure per 10^9 device-hours).

FL – Fault location.

FLOTOX – Floating oxide.

FMA – Failure modes analysis.

FMEA – Failure modes and effects analysis.

FOM – Figures of merit.

FPHP – Flat plate heat pipe.

FRACAS – Failure reporting and corrective action system.

FRB – Failure review board.

FRU – Field replaceable unit.

FTA – Fault tree analysis.

FTCs – Flip tab carriers.

FTTL – Fast transistor-transistor logic.

Ga – Gallium.

GaAs – Gallium arsenide.

GB – Ground benign.

GF – Ground fixed.

GIDEP – Government Industry Data Exchange Program.

GM – Ground mobile.

GRB – Ground benign environment.

GRF – Ground fixed environment.

GRM – Ground mobile environment.

HAST – Highly accelerated stress tests.

HC – Heat concentration.

HCMOS – High-speed CMOS.

HMOS – High-speed MOS.

HD – Heat dissipation.

HDSCs – High-density signal carriers.

HEMT – High-electron-mobility transistors (also known as MODFET's).

HMOS – Highly scaled MOS.

HTCMOS – High-speed TTL-compatible CMOS.

HTOT – High-temperature operating tests.

HTRB – High-temperature reverse bias.

HTTL – High-speed transistor-transistor logic.

IC – Integrated circuit.

ICA – In-circuit analyzer.

ICs – Integrated circuits.

ICT – In-circuit tester.

IEEE – Institute of Electrical and Electronic Engineers.

IES – Institute of Environmental Sciences. A professional organization of engineers and scientists involved in the study and application of climatic testing. The IES is a valuable source for ESS studies and cost analysis data.

IIL – Integrated injection logic.

IMPATT – Impact avalanche and transit time.

I/O – Input/output.

ISL – Integrated Schottky logic.

I^{3}L – Isoplanar integrated injection logic.

I-V – Current-voltage (characteristic).

LBS – Loaded board shorts.

LCC – Life-cycle cost.

LCCCs – Leadless ceramic chip carriers.

LCMs – Liquid cooled modules.

LCRU – Least-cost replaceable unit.

LED – Light emitting diode.

LQL – Limiting quality level.

LRU – Line replaceable unit.

LSC – Logistic support cost.

LSTTL – Low-power Schottky transistor-transistor logic.

LTTL – Low-power transistor-transistor logic.

LSCS – Logistic support cost savings.

LSI – Large-scale integration.

LTPD – Lot tolerance percent defective.

MCMs – Multichip modules.

MIPS – Million instructions per second.

MLE – Maximum likelihood estimate.

MNOS – Metal nitride oxide semiconductor.

MOS – Metal oxide semiconductor.

MSI – Medium-scale integration.

MTBF – Mean time between failures.

MTBR – Mean time between removals.

MTBUR – Mean time between unscheduled removals.

NFF – No fault found.

NMOS – N-channel metal oxide semiconductor.

NPL – Natural process limit.

NTF – No trouble found.

OC – Operating characteristic.

OEM – Original equipment manufacturer.

PA – Procuring activity.

PCB – Printed circuit board.

PCC – Plastic chip carriers.

PED – Plastic encapsulated device.

PEP – Production engineering phase.

PFMEA – Process failure modes and effects analysis.

PGA – Pin grid array.

PICs – Power integrated circuits.

PIN – P intrinsic N.

PIND – Particle impact noise detector.

PLCCs – Plastic leaded chip carriers.

PM – Performance monitoring.

PM – Preventive maintenance.

PMOS/pMOS – MOS that will conduct through a p-channel.

PPM – Parts per million.

PQFPs – Plastic quad flat packs.

PROMs – Programmable read-only memory.

ps – Picosecond (10^{-12} seconds).

PSG – Phosphosilicate glass.

PSR – Power stress ratio.

Pt – Platinum.

PTHs – Plated through holes.

PWA – Printed wiring assembly.

PWB – Printed wiring board.

QA – Quality assurance.

QC – Quality control.

QE – Quality engineering.

QFD – Quality function deployment.

QFP – Quad flat pack.

QPL – Qualified parts list.

QRV – Quasi-random vibration.

RAM – Random access memory.

RAM – Reliability, availability and maintainability.

RAS – Reliability, availability and serviceability.

RCM – Reliability centered maintenance.

RDGT – Reliability development/growth testing.

REDR – Recombination enhanced defect reactions.

RFI – Radio frequency interference.

RGM – Reliability growth management.

RH – Relative humidity.

RISE – Reliability improvement of selected equipment.

ROI – Return on investment.

ROM – Read-only memory.

RQL – Rejectable quality level.

RQT – Reliability qualification testing.

R&R – Remove and replace.

RSER – Residual SER.

RTD – Resistance temperature detector.

RTOK – Retest OK.

RTV – Room temperature vulcanized.

SAW – Surface acoustic wave.

SCPs – Single-chip packages.

SEM – Scanning electron microscope.

SEMs – Standard electronic modules.

SER – Soft error rates.

Si – Silicon.

SIMs – Superintegration modules.

Si$_3$N$_4$ – Silicon nitride.

SiO$_2$ – Silicon dioxide.

SIP – Single in-line package.

SOP – Small outline (integrated circuit) package.

SOR – System operational reliability.

Sn – Tin.

SQL – Specified quality level.

SRAM – Static RAM.

SRD – Standard reporting designation.

SRU – Shop replaceable unit.

SS – Screening strength.

SSI – Small-scale integration.

STTL – Schottky transistor-transistor logic.

TAAF – Test, analyze and fix.

TAB – Tape automated bonding.

TAC – Thermal accomodation coefficient.

TCM – Thermal conduction module.

TDDB – Time-dependent dielectric breakdown.

TEC – Thermal expansion coefficient.

TEM – Transmission electron microscope.

THB – Temperature humidity bias.

Ti – Titanium.

TS – Test strength.

TTL – Transistor-transistor logic.

TTL-LS – Transistor-transistor logic (low-power Schottky).

TWT – Traveling wave tube.

ULSI – Ultralarge-scale integration.

UUT – Unit under test.

UV – Ultraviolet.

VHF – Very high frequency.

VLSI – Very-large-scale integration.

VMOS – V-groove MOS (or vertical MOS if stated).

VSR – Voltage-stress ratio.

WUC – Work unit code.

3.3 COMMON MISCONCEPTIONS ABOUT *ESS*

Like any specialized technology, *ESS* is subject to misunderstanding. The following list [2; 5] includes the more common misconceptions:

1. *ESS* **IS A TEST.**

   ESS **IS NOT A TEST.** It is not intended to validate design, which requires a failure-free simulation process. Rather, *ESS* is a screening process that requires stimulation to force latent defects in products to failure that would otherwise occur in the field.

2. *ESS* **IS THE SAME AS BURN-IN OR AGING AND THEREFORE EQUIPMENT WITH PREVIOUSLY APPROVED BURN-IN PROCEDURES SHOULD NOT BE SUBJECT TO** *ESS*.

   ESS **EVOLVED FROM BURN-IN TECHNIQUES BUT IS A CONSIDERABLY ADVANCED PROCESS.** Burn-in is a generally lengthy process of powering a product at a specified constant temperature. *ESS* is an accelerated process of stressing a product in continuous cycles between predetermined environmental extremes, primarily temperature cycling plus random vibration.

   This misconception is based on the assumption that historical "burn-in" procedures conducted on the electronic equipment, currently in the inventory, are as cost effective as the *ESS* temperature cycling and random vibration screens. Table 3.1 illustrates the differences between burn-in and *ESS* procedures.

   The findings and recommendations of preference [8] refute the assumptions and arguments that historical burn-in procedures are as effective as *ESS*. Applicable findings and recommendations are as follows:

TABLE 3.1– Comparison of burn-in test and *ESS* procedure.

Criteria	Burn-in	ESS
Temperature	Operating or accelerated	Cycled from high to low operating
Vibration	Sinusoidal (if used)	Random, normally 20–2000 Hz
Temperature rate of change	Usually constant, but sometimes cycled	5°C per minute minimum
Length of time	Normally 168 hr or less	• 5 or 10 minutes perpendicular to each axis of orientation for vibration, and • 10 to 20 cycles for temperature cycling.

(a) Temperature cycling and random vibration have been found to be the most generally useful environmental stresses for screening.

(b) Temperature cycling precipitates 2/3 of all defects and random vibration the remaining 1/3.

(c) Random vibration is replacing sine wave vibration because it is more effective than either fixed frequency or swept sine vibration as an environmental screen.

(d) Defect precipitation is accelerated by random vibration.

(e) Effective module level screening can be achieved when performed either with a module operating or nonoperating. However, the additional defects precipitated by the application of power may not always justify the added costs.

The finding that "temperature cycling precipitates 2/3 of all defects and random vibration the remaining 1/3" requires further comment. These results were based on circumstantial evidence at the time of failure and do not reflect the findings of subsequent failure analyses, which often show that a defect induced by one environment was detected during the other. The failure physics is not so simply described. To do so can lead to a decision to delete one environment or the other from a screen based on superficial economic grounds without considering the actual failure mechanism.

The statement that "effective module level screening can be achieved when performed with a module operating or nonoperating" is certainly a generality. Most of the literature shows that there are distinct advantages to be gained by monitoring equipment performance while environmental stresses are being applied, especially when dynamic environments are being applied such as random vibration with rapid temperature cycling. These added expenses of monitoring during screening should be assessed against the projected failure precipitation to be gained from this monitoring effort.

3. *ESS* STRESSES THE EQUIPMENT BEYOND ITS DESIGN LIMITS.

THE SCREENING LEVELS SHOULD NOT EXCEED DESIGN LIMITS, BUT THEY MUST BE OF SUFFICIENT STRENGTH TO PRECIPITATE FAILURES DUE TO WEAK PARTS AND MANUFACTURING DEFECTS AT THE EARLIEST TIME SUCH THAT CORRECTIONS ARE MOST COST EFFECTIVE. An objection to *ESS* is that it stresses the equipment beyond its design limits and, therefore, the temperature and vibration stresses should be limited to the equipment's projected operational environment. However, the projected operational environment is usually less severe than the design limit. Effective screening requires stresses of sufficient magnitude and time duration to precipitate failures from latent defects without accumulating significant damage (i.e., exceeding design limits and consuming useful life) to the remaining nondefective structural elements. *ESS* should not stress the equipment such that fatigue failures are precipitated. Subcontractors should then conduct temperature and vibration surveys to identify hot spots and vibration resonances and conduct step screens to determine the most effective screening stresses for precipitating the early failures due to manufacturing defects. Screens that do not precipitate these early defects are not accomplishing the ESS objectives and are not cost effective. Screens that are too severe will damage good parts and detract from the useful life of the equipment.

4. *ESS* CAN BE USED TO VALIDATE DESIGN.

WHILE *ESS* MAY OCCASIONALLY EXPOSE A DESIGN INCONSISTENCY, ITS INTENT AND METHODOLOGIES ARE DIFFERENT. Design validation uses environmental simulation and laboratory analysis to determine whether a product is viable, and whether it will work within the environment for which it was designed, usually called its mission profile. *ESS*, on the other hand, applies maximum permissible stimulation to products for the purpose of exposing latent defects related to parts, manufacture, assembly technique or process. While *ESS* produces data that can be analyzed for pattern part failures, process defects, or design errors, for example, its principal intent is to force failures. The mission profile, in this sense, is irrelevant; what is important is that the maximum degree of stress be applied to precipitate

TABLE 3.2– *ESS–RDGT* comparisons.

Criteria	*ESS*	*RDGT**
Purpose	To precipitate and remove early failures due to manufacturing defects	To identify and correct reliability problems due to design
When conducted	During production	Prior to production
Length	Normally – Ten temperature cycles and 10 minutes of random vibration	Normally a fixed-length test of three times the equipment *MTBF*
Number of units	100% of all production units	Minimum of two units (if possible)
Pass/fail criteria	None – Screening should precipitate a maximum of early failures	*MTBF* growth must correlate with the selected growth model

* Reliability Development/Growth Testing.

failure without detracting from the product's useful life.

5. *ESS* **UNCOVERS ALL DESIGN PROBLEMS.**

ESS **DOES NOT UNCOVER ALL DESIGN PROBLEMS.** The misconception, that since *ESS* may uncover design problems, Reliability Development/Growth Testing (*RDGT*) is the same as *ESS* and, therefore, there is no need for *ESS* if *RDGT* has been conducted. While it is true that *ESS* uncovers some design problems (even after *RDGT*), this is not the objective of *ESS*. As defined previously, the objective of *ESS* is to accelerate early failures due to manufacturing defects such that repair is accomplished during the most cost-effective period of the program. A comparison of *ESS* and *RDGT* shows the fallacy of the misconception that *RDGT* testing replaces *ESS*. Table 3.2 compares *ESS* and *RDGT* in several criteria.

As shown by Table 3.2, *ESS* and *RDGT* are entirely different procedures. *ESS* is a screening procedure tailored to precipitate the maximum amount of manufacturing defects possible. *RDGT* is a "test, analyze and fix" procedure structured to find and correct design problems. *RDGT* is normally conducted in conjunction with first article testing on a one-time basis while *ESS* is a production screening process conducted on 100% of the production units.

6. *ESS* **IS APPLICABLE ONLY TO NEW DESIGN.**

GRANTED, *ESS* WILL PRECIPITATE MORE DESIGN RELATED
FAILURES ON NEW OR MAJOR MODIFIED EQUIPMENT THAN
ON EQUIPMENT ALREADY APPROVED AND IN THE INVEN-
TORY, BUT THIS DOES NOT RESTRICT *ESS* TO ONLY NEW
OR MAJOR MODIFIED EQUIPMENT. Properly conducted, *ESS* is
a cost-effective procedure for eliminating early failures resulting from
part defect tolerance allowances or manufacturing processes on all elec-
tronic equipment.

This misconception is the assumption that *ESS* is a reliability improvement "lessons learned" effort and is, therefore, applicable only to
a new design or major modified equipment. To be correct, this assumption would also have to assume that only new or major modified
equipment have defective parts and manufacturing defects. However,
military part specifications such as MIL-STD-38510 MIL-STD-883 allow part manufacturers to accept and ship electronic parts with up to
5% defectives, based on a lot tolerance percent defective (LTPD) of
5%. This translates into the possibility of fifty (50) defective parts per
thousand that can be precipitated by *ESS*, regardless of whether new,
modified, or remanufactured existing equipment are involved. Likewise,
manufacturing processes can degrade as a result of new procedures, new
personnel and out-of-tolerance automation such that defects can be introduced by the manufacturing process on any equipment — not just
new or major modified.

7. *ESS* IS NOT NECESSARY FOR EQUIPMENT EXHIBITING A
   SATISFACTORY *MTBF*.

THE NUMBER OF FAILURES CONTRIBUTING TO THE DEFI-
CIENT *MTBF* DURING THE EARLY FAILURE PERIOD ARE
ATTRIBUTED PRIMARILY TO MANUFACTURING DEFECTS
THAT COULD HAVE BEEN ELIMINATED BY AN EFFECTIVE
ESS PROGRAM. The argument that *ESS* is not necessary for equipment currently in the inventory and exhibiting a satisfactory *MTBF*
may or may not be true. Certainly an equipment exhibiting a satisfactory *MTBF* has normally accumulated sufficient operating hours such
that the early manufacturing defects have been eliminated. The question to be addressed is "would *ESS* have been cost effective for this
particular equipment, or would it be effective for additional buys of the
same design?" To answer this question, historical failure data should
be analyzed (if available). Only after plotting the *MTBF* growth from
historical data, calculating the number of failures attributed to manufacturing defects and deriving the cost benefits from a cost model, can
the decision as to the effectiveness of an *ESS* program be determined.
Therefore, additional buys of hardware that is in service with a proven
MTBF should have *ESS* considered.

8. **THE FEWER THE FAILURES PRECIPITATED BY** *ESS*, **THE MORE SUCCESSFUL THE SCREENING.**

   THE FEWER THE FAILURES, THE LESS SUCCESSFUL THE SCREENING. The *ESS* objective is to eliminate as many of the weak parts and manufacturing defects as possible by the tailored application of environmental stresses. These stresses must be tailored to eliminate as many of these defects as possible while assuring that wearout failures are not induced and then precipitated by overstress. Unless screening effectively precipitates a large portion of the manufacturing defects, screening is not cost effective.

   Some subcontractors have the mistaken theory that they will be penalized in proportion to the amount of failures precipitated by an *ESS* program. It is true that *ESS* failures are costly to subcontractors, driving them to improve the manufacturing process so that the cost impact of failures is reduced. Therefore, successful *ESS* does indirectly reduce the failure rate. However, this is not the reason for fewer failures being equated with successful testing. This misconception may be the result of a false correlation of *ESS* with Reliability Development/Growth Testing (*RDGT*) and Reliability Qualification Testing (*RQT*).

9. *ESS* **INVOLVES RANDOM SAMPLING.**

   ESS **REQUIRES THAT ALL PRODUCTS (100%) BE EXPOSED.** Because latent product defects are random by nature, screening all products is the only way to assure the effectiveness of the program.

10. *ESS* **FORCES NEEDLESS FAILURES.**

    ESS **IS DESIGNED TO FORCE ONLY THOSE FAILURES THAT WOULD NORMALLY OCCUR IN THE FIELD.** ESS programs are based on a product profile that determines the maximum allowable stress the product can absorb without affecting its useful life. This determination is made prior to implementing the program by analyzing product function, tolerances and failure rates. Ideally, there is never a loss of useful life; if there is, it will be negligible, particularly in light of the failures precipitated in house.

11. **ALL** *ESS* **SYSTEMS ARE ALIKE.**

    ESS **IS AS DIFFERENT AS EACH PRODUCT THAT IS MADE.** It is, by design, a product-specific program that must be tailored according to predetermined product variables. The type of screen, profile variations, fixturing, power supplies, inputs/outputs and other considerations must be resolved before the program is implemented.

12. *ESS* **IS EXPENSIVE.**

    ESS **SAVES MONEY AND CAN DELIVER A FAST PAYBACK.** An *ESS* program is a cost that can and should be measured for its

ROI potential. By implementing the program properly, dramatic cost savings can be realized through reduced field warranty repair expense, lower unit costs, improved product value perception in the marketplace, and, ultimately, increased profitability. *ESS* is ideally implemented at a point in production where the cost to repair is the lowest.

REFERENCES

1. DOD-HDBK-344 (USAF), *Environmental Stress Screening of Electronic Equipment*, Department of Defense, Washington, DC, October 20, 1986.

2. Thermotron Industries, *The Environmental Stress Screening Handbook*, 291 Kollen Park Drive, Holland, MI 49423, 36 pp., 1988.

3. Mandel, C. E., "Environmental Stress Screening (ESS)," *Proceedings of the Institute of Environmental Sciences*, pp. 294-302, May 1984.

4. Saari, A. E. et al, *Environmental Stress Screening* , RADC-TR-86-149, September 1986.

5. Pennington, Duane R., "Environmental Stress Screening – Some Misconceptions," *The Journal of Environmental Sciences*, pp. 26-29, May/June 1986.

6. Tustin, W., "Recipe for Reliability: Shake and Bake," *IEEE Spectrum*, pp. 37-42, December 1986.

7. Pellicone, V. H. and Popolo, J., *Improved Operational Readiness Through Environmental Stress Screening*, RADC-TR-87-225, Final Technical Report, November 1987.

8. Institute of Environmental Sciences, *Environmental Stress Screening Guidelines for Assemblies*, 940 East Northwest Highway, Mount Prospect, IL 60056, 1984.

9. Omdahl, T. P., *Reliability, Availability, and Maintainability (RAM) Dictionary*, ASQC Quality Press, 361 pp., 1988.

10. Amerasekera, E. A. and Campbell, D. S., *Failure Mechanisms in Semiconductor Devices*, John Wiley, New York, 205 pp., 1987.

Chapter 4

ESS TYPES AND THEIR EFFECTIVENESS

4.1   FLAW-STIMULUS RELATIONSHIPS

Before discussing the flaw-stimulus relationships, two defect definitions need to be clarified, namely, *patent defect* and *latent defect*.

A patent defect is a flaw which has advanced to the point where an anomaly actually exists or it can be defined in terms of an out-of-tolerance, or a specification, condition which can be readily detected by an inspection or a test procedure [1].

Patent defects represent the majority of the defect population in an equipment and are readily detected without the need for stress screens. A small percentage of defects however, cannot be detected by conventional means. Such defects are termed *latent defects*.

A latent defect or a flaw is some irregularity due to manufacturing processes or materials which will advance to a patent defect when exposed to environmental or other stimuli [1].

A latent defect is characterized as an inherent or induced weakness or flaw in a material which will manifest itself as a failure in the operational environment. A subset of this is a design defect which should be corrected during maturation.

Both patent and latent defects are introduced into the product during fabrication, assembly, handling and test operations. The patent defects pass through various assembly stages until they are detected by a test or inspection of sufficient thoroughness and are subsequently eliminated from the product. When good quality control test and inspection procedures are applied, all but the most subtle patent defects should be detected and eliminated prior to shipment. Some examples of patent defects are the following:

61

1. Parts

   (a) Broken or damaged in handling.

   (b) Wrong part installed.

   (c) Correct part installed incorrectly.

   (d) Failure due to electrical overstress or electrostatic discharge.

   (e) Missing parts.

2. Interconnections

   (a) Incorrect wire termination.

   (b) Open wire due to handling damage.

   (c) Wire shorted to ground due to misrouting or insulation damage.

   (d) Missing wire.

   (e) Open etch on printed wiring board.

   (f) Open plated through-hole.

   (g) Shorted etch.

   (h) Solder bridge.

   (i) Loose wire strand.

Latent defects cannot be detected until they are transformed to patent defects by environmental stress applied over time. Stress screening is the vehicle by which latent defects are transformed into detectable failures. Some examples of latent defects are the following:

1. Parts

   (a) Partial damage through electrical overstress or electrostatic discharge.

   (b) Partial physical damage during handling.

   (c) Material or process induced hidden flaws.

   (d) Damage inflicted during soldering operations (excessive heat).

2. Interconnections

   (a) Cold solder joint.

   (b) Inadequate/excessive solder.

   (c) Broken wire strands.

   (d) Insulation damage.

   (e) Loose screw termination.

   (f) Improper crimp.

(g) Unseated connector contact.

(h) Cracked etch.

(i) Poor contact termination.

(j) Inadequate wire stress relief.

For example, a cold solder joint represents a flaw. After vibration and/ or thermal cycling, the joint will (we assume) crack. The joint would now have a detectable (patent) defect.

Note, however, that not all latent defects are screenable; i.e., capable of being eliminated from the equipment in the factory by use of stress screens. It is only those latent defects, whose failure threshold can be accelerated by the stresses imposed by the screens, which are screenable. Such screenable defects, if not eliminated from the product in the factory, will result in premature or early-life failures in the field. It is the screenable early-life failure which the stress screening program must be designed to remove.

Some flaws or latent defects can be stimulated into patent defects by thermal cycling, some by vibration and some by voltage cycling (or some other stimulus). Not all flaws respond to all stimuli. This effect can be shown in table form, as has been done in many publications. One such table is given in Table 4.1, which is taken from MIL-STD-883, for integrated circuits.

The black dot indicates that the test listed in the column on the left can detect or precipitate the flaw listed in the row across the top. The table shows, for example, that burn-in is not effective in finding particle contaminants, seal defects, package defects, external lead defects and thermal mismatch problems. A further screen using vibration, thermal cycling and visual exams would be capable of detecting these defects if the stimuli levels were correctly selected and the product is screenable; i.e., capable of withstanding, without meaningful degradation, a level of stimulation which will expose the flaw. Note also that the favorite of many, thermal cycling, will not stimulate silicon defects, particle contamination and electrical stability problems.

Since bulk silicon defects and particulate contamination are among the most numerous problems in solid state components, selecting thermal cycling only as a screen is a serious mistake.

Consider another table, Table 4.2, which is taken from the 1985 IES Guidelines [2] and addresses the burn-in types which are effective in the screening of IC's.

This table shows that the selection of the burn-in type must match the flaws sought, or the flaws will (probably) not be found. An example here is that dynamic burn-in will (probably) not precipitate oxide defects, a defect type common in IC's.

The concept of flaw-stimulus relationships can also be shown in a Venn diagram form, as in Fig. 4.1, for a hypothetical but specific product. It would be different for a different product. For clarity, not all stimuli are shown.

TABLE 4.1– Screening processes for IC failure mechanisms.

Screening test	Failure mechanism									
	Substrate mounting defects	Bulk silicon defects	Substrate surface defects	Bonding and wire	Particle contamination + extraneous material	Seal defects	Package defects	External lead defects	Thermal mismatch	Electrical stability
Internal visual exam	•		•	•	•					
External visual exam						•	•	•		
Stabilization bake		•	•	•						•
Thermal cycling	•		•	•		•	•		•	
Thermal shock	•		•	•		•	•		•	
Centrifuge	•			•	•		•			
Shock	•			•	•		•			
Vibration	•			•	•		•			
X ray	•	•	•	•	•	•	•			
Burn-in	•									•
Leakage tests						•				

TABLE 4.2– Burn-in screening effectivity as a function of failure mechanisms detected.

Failure mechanism	Dynamic burn-in	Steady-state burn-in	High voltage cell stress test
Surface defects	●	●	
Oxide defects		●	●
Metallization defects	●	●	
Contamination/Corrosion -intermetallic		●	
Junction anomalies (breakdown)	●		●
Wire bond	●	●	●

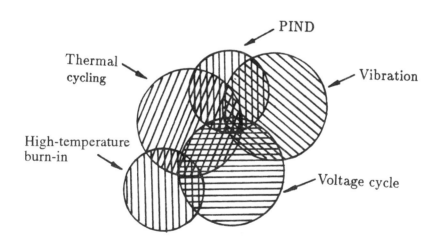

Fig. 4.1– Flaw precipitation–stimulus relationships.

Note that, for the hypothetical example given, there are many latent defects that will not be transformed into patent defects by any one stimulus. For example, a solder splash which is just barely clinging to a circuit board would probably not be broken loose by high-temperature burn-in or voltage cycling, but vibration or thermal cycling would probably break the particle loose. Also, to find that the defect exists, the particle must be seen by eye, heard by sound [as in Particle Impact Noise Detection (PIND)], or by the electrical or mechanical function of the device being screened showing some sort of a change. Note that the defect may only be observable during a stimulation and not be observable during a bench test.

Another example would be that of a latent defect of a chemical nature where a high-temperature bake would cause a reaction to proceed and a defect to show up. Applying vibration, mechanical shock or centrifuging would be completely ineffective in precipitating such a defect. The stimulus must be chosen to precipitate flaws into defects which can be found during or after the screen.

ESS consists of exposing a product to one or more stressful environments or stimuli, such as thermal cycling, vibration, high temperature, electrical stimuli, thermal shock and others. Profile characteristics, such as stress extremes, rates of change, amplitude, frequency and duration, must be tailored for each product. A description of the more commonly used *ESS* environments, that is partially adopted from [3] by permission, follows.

4.2   *ESS* TYPES

1. TEMPERATURE CYCLING

Temperature cycling consists of multiple cycles of changing temperature between predetermined extremes. A typical temperature profile is shown in Fig. 4.2.

As all variables in a screen are product dependent, temperature extremes must stop short of damaging the product but must be far enough apart to allow for optimum stressful ranges of temperature change between the extremes. Cycle times must follow this same rule of thumb, as it is the constant rate of change that provides the expansion and contraction necessary to sufficiently stress the product.

The product temperature change rate is dependent on specific thermal properties of the product, the difference between product and air temperatures, and the surface conduction factors involving air velocity and direction. Assuming proper air flow, the temperature of a typical circuit board would look like Fig. 4.3 when subjected to a typical temperature profile.

Since the dwell period at the temperature extremes does not significantly contribute to the stress, many manufacturers allow the product

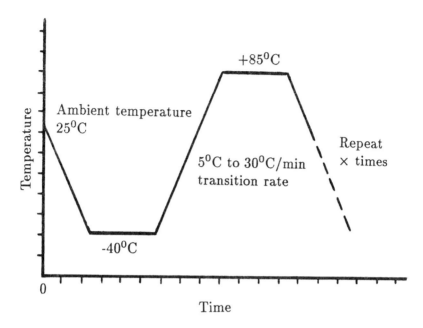

Fig. 4.2– A typical temperature profile for *ESS* [3].

to remain at temperature extremes only long enough to allow a functional test of the product as shown in Fig. 4.4, where the dwell periods are shorter than those in Fig. 4.3.

Another step some manufacturers have taken, to maximize the stress during thermal cycling, is to adjust the chamber air temperature so that the high and the low temperatures are close to the extremes the product can withstand. However, in this stress profile, it must be determined that multiple cycles between these two extremes do not erode the product's useful life. Figure 4.5 illustrates such a profile.

Air flow in a thermal cycling chamber is critical as it directly affects the delay time in which product temperature reaches the chamber air temperature. Figure 4.6 illustrates product temperature during three separate runs of a profile that had identical air temperature change rates. The only difference between the runs is the air velocity. Clearly, the figure shows that the lower the air velocity, the slower the product temperature reaches the chamber air temperature, and the less stress is brought to bear on the product. The higher air velocities cause a more rapid product temperature change and thus subject the product to a higher degree of stress.

The optimum air flow depends on the product. There is an air veloc-

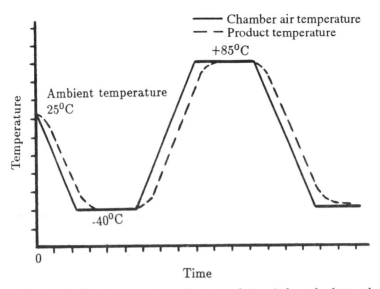

Fig. 4.3– The temperature of a typical circuit board when subjected to a typical temperature profile [3].

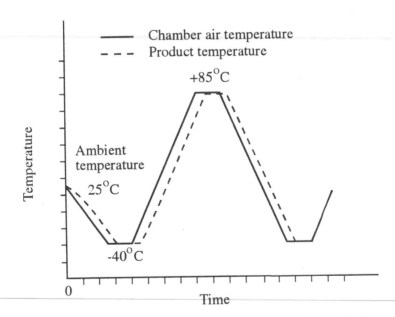

Fig. 4.4– Dwell period at the temperature extremes of a typical temperature profile [3].

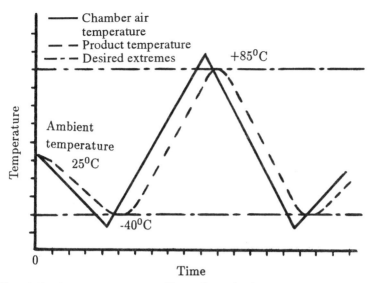

Fig. 4.5– A temperature profile with its high and low temperatures close to the product's design extremes [3].

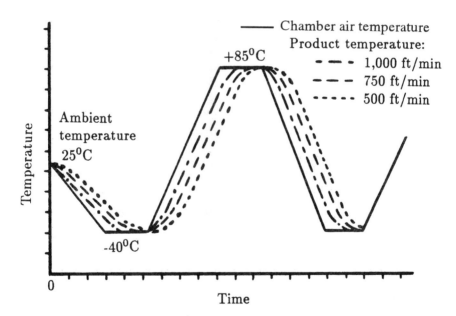

Fig. 4.6– Influence of air velocity on the product temperature
approaching the chamber air temperature [3].

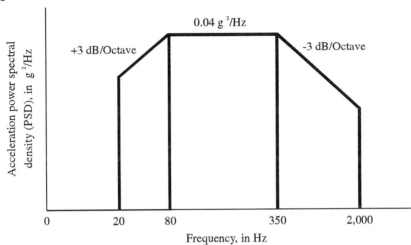

Fig. 4.7– A typical input power spectral density, specified by
[4], which many manufacturers used as a starting profile for the further tailoring to their individual product.

ity at which maximum heat transfer is obtained. Exceeding that air
velocity can be counterproductive. The air velocities listed on Fig. 4.6
may or may not be the most appropriate for your product. The correct air velocities and air direction for a particular product can only be
determined through experimentation.

The general consensus is that thermal cycling is the most effective
screen. Like any screen, it must be properly implemented. Failure
rates must be analyzed to determine which latent defects are causing
the failures, and experimentation must be performed to determine the
screen profile best suited to trigger those particular latent defects into
failures.

2. RANDOM VIBRATION

Random vibration is considered to be the most effective of three principal types of vibration; namely: sine wave, fixed frequency; sine wave,
swept frequency; random vibration. Random vibration involves the
excitation of a product with a predetermined profile over a wide frequency range, usually from 20 to 2,000 Hz. Figure 4.7 is a typical input
power spectral density, recommended by the Naval Material Command
($NAVMAT$) Document P-9492 [4], which many manufacturers used as
a starting profile for the further tailoring to their individual product.
Product stress is created through simultaneous excitation of all resonant frequencies within the profile's range.

Random vibration is applied by mounting or attaching a product to
an electrodynamic shaker, controlled by a closed-loop digital system

and dedicated fixturing. Fixturing must be extremely rigid to ensure that stress is transmitted directly to the product and that the process is repeatable with reasonable accuracy. Products may be vibrated on a single axis or on multiple axes, concurrently or consecutively. Currently, there are varying opinions as to which provides the more stressful environment.

Random vibration screens generally require less time to run than other *ESS* programs and are considered particularly effective in exposing mechanical defects, such as loose solder, improper bonds and printed circuit board shorts.

The primary drawbacks are equipment cost and lack of screen uniformity. Regarding the former, electrodynamic shakers may be expensive to install, control and maintain. Random vibration also is, by nature, less effective than temperature cycling in providing a uniform stress environment. To avoid overstress, random vibration screens are tailored to exert maximum stress on parts at joints located at the center of a product and are gradually reduced for parts closer to the edge, a factor that may result in failure to expose all potential defects.

Nonetheless, random vibration is considered superior to other forms of vibration and is often used, where financial objectives allow, in conjunction with thermal cycling screens.

3. HIGH-TEMPERATURE BURN-IN

This process is generally a static one in which a product is subjected to elevated temperatures over a predetermined period. This screen evolved from the idea that continuous operation of a product would force infancy failures. It was also believed that providing additional heat would escalate the rate of failure. Research now shows, however, that increases in screen effectiveness are related to changing temperatures during heat-up and final cool-down, rather than powering at a constant high temperature.

4. ELECTRICAL STRESS

Electrical stress is a process used to exercise circuitry and/or simulate junction temperatures on semiconductors. There are two basic types: (a) power cycling, which consists of turning product power on and off at specified intervals, and (b) voltage margining, which involves varying input power above and below nominal product power requirements.

Electrical stress does not, according to research, expose the number of defects commonly found through thermal or random vibration screens; consequently, it is considered much less effective. It can, however, be relatively inexpensive to implement electrical stress with another screen stress to increase overall screen effectiveness. It may also be necessary to power products in order to find soft failures.

5. THERMAL SHOCK

Thermal shock is a process that exposes products to severe temperature extremes, usually in rapid fashion; i.e., a product is continually transferred–either mechanically or manually–from an extremely hot environment to an extremely cold environment and back. Thermal shock is generally considered a cost-effective way to screen for defects at the component level, particularly IC's, which require a high degree of stress to experience the rates of change needed to force latent defects into failure.

Thermal shock may also be useful at other levels of assembly, as long as the severity of its rates of change do not cause needless damage. This is a particular risk with more complex assemblies, such as those containing components other than IC's. Of equal consideration is the cost efficiency of a thermal shock screen. Generally, the equipment needed to provide an effective screen is expensive, because products must be shifted from one environment to another; consequently, there will always be an unused chamber. Also, if manual transfer is involved, the risk of accidental product damage increases. Finally, a thermal shock screen is difficult to monitor. This limits opportunities for accumulating field failure analysis data.

6. SINE WAVE VIBRATION, FIXED FREQUENCY

Sine wave, fixed frequency vibration is a form of vibration that operates on a fixed-sine wave or single operational frequency. This method usually requires a mechanical shaker with a frequency range of up to 60 Hz. While less expensive and easier to control than a random vibration screen, a fixed frequency vibration screen is generally viewed as not providing an effective level of stress.

7. LOW TEMPERATURE

This concept is similar to that for a high-temperature screen, but it is based on the principle that failures will be forced by the contrast between the heat generated by powering a product and the cold environment.

8. SINE WAVE VIBRATION, SWEPT FREQUENCY

Sine wave, swept frequency vibration is a form of vibration that operates on a swept-sine wave or multiple operational frequency. Usually it requires a hydraulic shaker with a frequency range of up to 500 Hz. It is viewed to be similar to fixed frequency vibration in terms of overall effectiveness.

9. COMBINED ENVIRONMENT

Depending on product complexity, cost and/or reliability specifications, environmental screens may be used in concert with each other. For example, thermal cycling and random vibration are often combined in an *ESS* program. The primary considerations should be whether an additional stress-applied simultaneously or consecutively-will expose a significant number of additional defects, and whether the cost of the additional stress is justifiable.

In general, the types of stress used and how the stress is applied is entirely dependent on what will precipitate the greatest number of failures in a specific product in the shortest period of time.

As with all screens, the most effective profile is product-dependent. Some recommendations are available, however, in the IES ESSEH Guidelines [2], the RADC studies [5; 6] and the Navy Manufacturing Screening Program, NAVMAT P-9492 [4].

4.3 SCREEN PARAMETERS

Screening strength and the failure rate of defects are a function of specific screen stresses (parameters) and the time duration of the stress application. Temperature cycling, high-temperature burn-in, random and swept sine wave vibration screening parameters are defined as follows:

4.3.1 TEMPERATURE CYCLING SCREEN PARAMETERS

1. Maximum temperature (T^*_{max}) – The maximum temperature to which the screened assembly will be exposed. This should not exceed the lowest of the maximum ratings of all the parts and materials comprising the assembly. Note that nonoperating temperature ratings for parts are higher than operating ratings.

2. Minimum temperature (T^*_{min}) – The minimum temperature to which the screened item will be exposed. This should not exceed the highest of the minimum ratings of all the parts and materials comprising the assembly.

3. Range (R) – The range is the difference between the maximum and the minimum applied external (chamber) temperature $(T^*_{max} - T^*_{min})$. Temperatures are expressed in °C.

4. Temperature rate of change $(\dot{T}^*)$ – This parameter is the average rate of change of the temperature of the item to be screened as it transitions between T^*_{max} and T^*_{min}, and is given by

$$\dot{T}^* = \frac{1}{2}\left[\left(\frac{T^*_{max} - T^*_{min}}{t_1}\right) + \left(\frac{T^*_{max} - T^*_{min}}{t_2}\right)\right], \qquad (4.1)$$

where

$$t_1 = \text{transition time from } T^*_{min} \text{ to } T^*_{max} \text{ in minutes,}$$

and

$$t_2 = \text{transition time from } T^*_{max} \text{ to } T^*_{min} \text{ in minutes.}$$

5. Dwell – Maintaining the chamber temperature constant, once it has reached the maximum (or minimum) temperature, is referred to as dwell. Dwell at the temperature extremes may be required to allow the item being screened to achieve the chamber temperature at the extremes. The duration of the dwell is a function of the thermal mass of the item being screened. For assemblies which have low thermal mass, part case temperatures will track chamber temperatures closely thereby eliminating the need for dwell. Units and systems may have a greater thermal lag, and achieving high rates of temperature change may be difficult. Dwell at temperature extremes is required in such instances.

6. Number of cycles – The number of transitions between temperature extremes (T^*_{max} or T^*_{min}) divided by two.

4.3.2 HIGH-TEMPERATURE BURN-IN SCREEN PARAMETERS

1. Temperature delta (ΔT^*) – The absolute value of the difference between the chamber temperature at which the equipment is being screened and 25°C, or

$$\Delta T^* = |T^* - 25°C|, \tag{4.2}$$

where

$$T^* = \text{chamber temperature.}$$

2. Duration – The time period over which the temperature is applied to the item being screened, in hours.

4.3.3 VIBRATION SCREEN PARAMETERS

1. *grms* level for random vibration – The *rms* value of the applied power spectral density over the vibration frequency spectrum.

2. *g*-level for swept sine wave vibration – The constant *rms* acceleration applied to the equipment being screened throughout the frequency range above 40 Hz. The *g*–level below 40 Hz may be less.

3. Duration – The time period over which the vibration excitation is applied to the item being screened, in minutes.

4. Axes of vibration – This can be single axis or multiple axes, depending on the sensitivity of defects to particular axial inputs.

4.4 POPULARITY AND EFFECTIVENESS OF *ESS* TYPES

In the past several years, a number of independent studies have been conducted analyzing the effectiveness of various types of *ESS* processes. In 1981 the IES guidelines for stress screening [7] presented Fig. 4.8 as a result of a poll to determine what was being done in stress screening. The statement requiring an answer in that poll was Question 7, page A-3 of [7] "RANK THE OVERALL EFFECTIVENESS OF SCREENING ENVIRONMENTS, WITH 1 BEING THE MOST EFFECTIVE, 2 NEXT EFFECTIVE, ETC. INCLUDE ONLY THOSE SCREENS YOU HAVE USED."

Figure 4.8 gives valuable information in the context of the survey, but taken out of context, it is not correct at all. Unfortunately, many novices have simply found the figure and proceeded to draw conclusions from it with no other facts as to assembly levels or defect types precipitated by the screens being considered. To further compound the potential confusion, the French published in [8] a similar result, Fig. 4.9.

Again, the figure is valuable when considered in its context, but very misleading if taken alone, at face value. Note that the order of "effectiveness" is different in the USA than in Europe. This, of course, is due to the question asked which required the respondent only to include those stimuli that he had used. In the USA, it is widely accepted that room temperature power on is very ineffective in screening, yet it is "rated" number three in "effectiveness" in Europe! Also, compare the relative position of random vibration as number two in the USA and as number eight in Europe! This is probably because of the military requirement for random vibration in the qualification, acceptance and screen testing on military programs in the USA. Hence, random vibration equipment is found in many companies and is readily available. This may not be the case in Europe.

Figures 4.8 and 4.9 can be named "Popularity Contests" since this title really seems to be more fitting than "Effectiveness." Many very serious mistakes have been made by a simplistic interpretation of the two "Effectiveness Rankings." For example, one company tried thermal cycling on 100% of their production for nearly a year and found hardly any improvement in their field failure rate. The troublesome defects were not susceptible to thermal cycling, but probably would have been found by vibration. The company concluded that it was much cheaper to just accept field failures and fix them when they occurred, than to attempt stress screening. Another company replaced 100% burn-in with a 100% vibration screen and, of course, found different defect

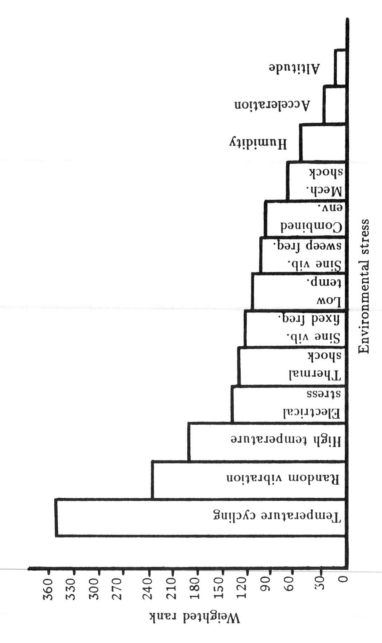

Fig. 4.8– Effectiveness comparison of environmental stress screens [6].

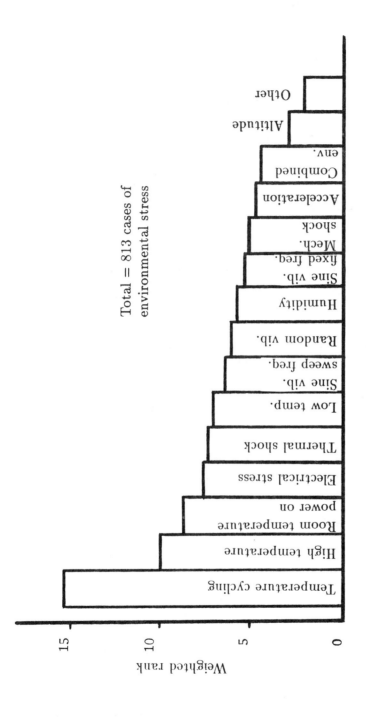

Fig. 4.9– Effectiveness comparison of environmental stress screens by the French *ESS* Task Team [7].

types. A correlation of the cost of screening and the effectiveness in decreasing field failures was inconclusive, as would be expected, considering the gross change in the screen and the attendant flaw types precipitated. These failures of stress screening, by not being effective, have led many to conclude that stress screening was simply not worth doing. But if the stimulus-flaw precipitation relationships had been taken into account, and corrective action emphasized, then the cost effectiveness may have been proven.

4.5   ADVANTAGES OF TEMPERATURE CYCLING

Organizations such as the Institute of Environmental Sciences (IES) [7] and the Rome Air Development Center (RADC) (now Rome Laboratory) [5; 6] have compiled in-use data from which they concluded that thermal cycling is the most effective type of screen. Both IES [7] and the French *ESS* Task Team [8] reached the same conclusion at this point, as shown in Figs. 4.8 and 4.9.

While it has been shown that the most effective screening program may involve more than one screen, thermal cycling is considered the single most effective screen in terms of identifying latent defects over a period of time. According to the ESSEH guidelines published in 1981 [7] and 1984 [9], thermal cycling, when compared with random vibration (ranked second most effective), regularly detected an average of two-thirds more latent product defects.

Thermal cycling provides the additional advantage of a uniform stress environment, one in which all areas of the product are subjected to an equal amount of stress throughout the screen profile. Some of the other environmental stresses cannot, by their nature, provide uniformity and have a higher inherent risk factor. Stress uniformity also makes thermal cycling easier to control, permitting greater flexibility in implementation, revision or refinement.

Thermal cycling takes less time to perform than most other types of stress. It is less time consuming than any other form in terms of its ability to force the greatest number of defects into failure; random vibration, which has a shorter cycle time, does not provide the same overall efficiency. The net result is a better product throughput with minimal impact on production through thermal cycling.

The basic concept of thermal cycling involves three uncomplicated principles. First, the temperature extremes must be as far apart as possible–regardless of the product's intended operational limits–as long as they don't cause needless damage. Second, the rate of change between the extremes must be as rapid as possible–usually equal to or greater than 5°C per minute–to create the optimum level of stress, again, without damaging the product. Fig-

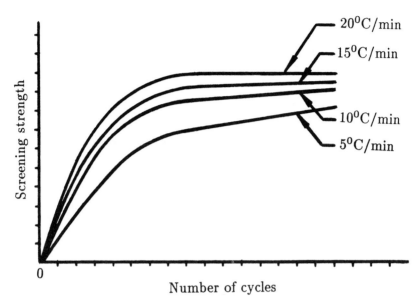

Fig. 4.10– A comparison of various temperature change rates
and their impact on screen effectiveness [3].

ure 4.10 shows a comparison of various temperature change rates and their
impact on screen effectiveness [3].

Finally, the product temperature must follow, as closely as possible, the
temperature inside the chamber. This is achieved by assuring that high velocity air is properly routed over the product. Figure 4.7 illustrates how
fast the product temperature changes at various air velocities. Clearly, the
temperature of the product changes faster as the air velocity across it is increased. Faster temperature change creates greater stress on the product;
consequently, it can more quickly force latent defects into failure.

A good starting point to determine an effective thermal cycling stress
profile would be to start with the following:

1. Temperature extremes equal to or greater than a 100°C temperature
   difference between hot and cold.

2. Product temperature rate of change equal to or greater than 5°C per
   minute.

3. Air velocity of approximately 750 fpm (feet per minute) at the product.

4. Enough cycles to effectively cause latent defects to surface as product
   failures.

REFERENCES

1. Hobbs, G. K., "Development of Stress Screens," *Proceedings of the Annual Reliability and Maintainability Symposium*, pp. 115-118, 1987.

2. Institute of Environmental Sciences, *Environmental Stress Screening Guidelines for Parts,* 940 East Northwest Highway, Mount Prospect, IL 60056, 1985.

3. Thermotron Industries, *The Environmental Stress Screening Handbook,* 291 Kollen Park Drive, Holland, MI 49423, 36 pp., 1988.

4. Fiorentino, E., *RADC Guide to Environmental Stress Screening, RADC TR-86-138,* August 1986.

5. Saari, A. E., "On the Implications of R & M 2000 Environmental Stress Screening," *IEEE Transactions on Reliability*, Vol. R-36, No. 3, pp. 342-345, 1987.

6. Willoughby, W. J., "Navy Manufacturing Screening Program," *Department of the Navy, NAVMAT P-9492,* May 1979.

7. Institute of Environmental Sciences, *Environmental Stress Screening Guidelines,* 940 East Northwest Highway, Mount Prospect, IL 60056, 1981.

8. Alis, B., "The French Environmental Stress Screening Program," *Proceedings of the Annual Reliability and Maintainability Symposium, 31st Annual Technical Meeting*, Philadelphia, Pennsylvania, pp. 439-442, 1985.

9. Institute of Environmental Sciences, *Environmental Stress Screening Guidelines,* 940 East Northwest Highway, Mount Prospect, IL 60056, 1984.

Chapter 5

ESS STANDARDS
AND DOCUMENTS
FOR ELECTRONIC ASSEMBLIES

5.1   BACKGROUND

The July/August 1989 issue of the *Journal of Environmental Sciences* included an overview [1] of military standards, directives, pamphlets, and other guidance documents describing *ESS*. This overview summarized the salient features of nine different *ESS* documents issued by the Department of Defense, Army, Air Force, Navy, and the Institute of Environmental Sciences.

However, in the two and a half years since this overview was published, some of these documents have been revised. In addition, a few new *ESS* documents have been added to those reviewed previously. Furthermore, the proliferation of *ESS* documents continues and the administrative confusion over which documents to use and how to apply *ESS* most effectively has not diminished. This is especially unfortunate since all of the documents describe *ESS* as essentially *the same core technical process*. It is the dozens of variations in details that create the seeming chaos.

The following overview [2] updates the previous summary by a brief commentary. The commentaries are intended to establish a perspective for assessing the relationships or conflicts between the various documents and the potential implications of using them as a guidance.

One potential shortcoming of nearly all of the documents reviewed is that they seem to apply primarily to the procurement of new hardware. When applied to spares, retroactive screening of previously unscreened (or differently screened) hardware, repaired hardware in service, or second-source procurements, some of the guidelines and requirements in these documents may not be practical, sensible or cost effective. Therefore, the user should be very cautious about adopting traditional or so-called "standard" *ESS* procedures, since they could significantly alter existing business arrangements and obligations.

The readers are strongly cautioned not to rely solely on these summaries for a complete understanding of these documents and their implications. Your particular business environment and *ESS* needs will be critical factors affecting the relevance of these documents for you, since they represent a wide range of influence, from regulations and standards to manuals, pamphlets and handbooks. It is essential that you become familiar with the content *and implications* of the complete and most recent versions of each applicable document before you make decisions involving the *ESS* program or facility development for your products.

5.2 AN UPDATED OVERVIEW

The following overview is quoted, with permission, directly from [2]:

1. **AFP 800-7, "USAF R & M 2000 Process"** (1 January 1989).

   TAILORABILITY INDEX: Moderate to High.

   PURPOSE: Information and Guidance.

   COMMENTARY: Although issued more than a year after DOD 4245.7-M and NAVSO P-6071, AFP 800-7 has become considerably more influential than either of its two predecessors. *ESS* is given special visibility in a detailed appendix, with specific parameters for temperature cycling and random vibration, included in a chart titled "R&M 2000 Baseline Regimen."

   It is *very important* to know that the notes accompanying the chart provide considerably more opportunity for tailoring than the chart by itself might otherwise imply. In particular, the notes provide the flexibility to tailor temperature change rates and vibration profiles based on response surveys.

   In general, this guidance document encourages tailoring and discourages a pass/fail mind-set toward *ESS*. Unfortunately, what the small print hides, the large print seems to take away: "Contractors offering alternative screens must demonstrate that their alternatives are AS EFFECTIVE AS R&M 2000 IN PRECIPITATING LATENT...DEFECTS TO FAILURE." How to quantitatively demonstrate comparable effectiveness without first applying the R&M 2000 baseline regimen in actual production is not explained. Analytical "proofs" would be extremely difficult to derive and even harder to gain approval for. Therefore, corporate best practices for *ESS* could be viewed as "noncompliant" even though proven effective on other programs.

2. **AMC-R 702-25, "AMC Environmental Stress Screening Program"** (29 May 1987).

TAILORABILITY INDEX: Potentially High (Depending on Contract Implementation).

PURPOSE: Information and Contract Administration.

COMMENTARY: This Army regulation potentially gives the contractor more responsibility and technical latitude than any of the other *ESS requirements* documents. The contractor has the responsibility for optimizing the *ESS* process. Toward this end, Task 301 of MIL-STD-785B^6 and a FRACAS (Failure Review and Corrective Action System) must be implemented. Appendix A of this document contains a baseline Statement of Work for *ESS* to be used in Invitations for Bids (IFB's), Requests for Proposals (RFP's), and awarded contracts.

Confusion can result from the apparent contradiction of defining a *minimum ESS* baseline (Table 1, p. A-4) that is not to exceed the hardware design capability (which implies that the minimum could be reduced). The statement that the recommended *ESS* values "are not hard inflexible requirements" (Paragraph 3.3, p. A-2) that can be modified if they exceed hardware design capability, is seemingly contradicted by the following cryptic note: "In most cases, it will be necessary to exceed the design specifications of an item to reach a level that will precipitate out infant and latent defects inherent in the item" (Paragraph 3.3, p. A-2).

3. **DOD 4245.7-M, "Transition from Development to Production"** (September 1985)/ **NAVSO P-6071, "Best Practices"** (March 1986).

TAILORABILITY INDEX: Moderate to High.

PURPOSE: Information and Guidance.

COMMENTARY: DOD 4245.7-M, "Transition from Development to Production," and its companion document, NAVSO P-6071, "Best Practices," are valuable policy guidance documents that provide excellent philosophical overviews of the proper way to use *ESS* in the manufacturing screening template. In particular, Best Practices offers a very useful executive-style summary of the important issues associated with successfully using *ESS*. This summary is accompanied by a unique chart that contrasts the traps and consequences of some current approaches with the potential benefits of applying the Best Practices strategies. Perhaps their most valuable contribution is the strong emphasis placed on keeping *ESS* dynamic and flexible through intelligent tailoring.

Unfortunately, both documents include recommended *ESS* starting conditions that are not supported by technical references or otherwise explained. In particular, the issue of input versus response-defined environmental control strategies is not addressed, even though the choice of control strategy can have a very large impact on *ESS* effectiveness. (In fact, neither document defines whether the *ESS* conditions represent *ESS* facility inputs or responses of the assembly being screened, although the numbers suggest input conditions.) Also, the level of assembly to which the recommended starting conditions apply is not defined, but it appears to be intended for LRUs and "black boxes."

4. **DOD-HDBK-344 (USAF), "Environmental Stress Screening (ESS) of Electronic Equipment"** (20 October 1986).

TAILORABILITY INDEX: High.

PURPOSE: Information and Guidance.

COMMENTARY: DOD-HDBK-344 (USAF) combines enlightened administrative philosophy for *ESS* goals and implementation, with obscure and questionable technical baselines that subvert the ambitious attempts of this document to model the ESS process. On the positive side, this very detailed handbook covers a wide variety of important issues, including contractual aspects, planning for development and production phase *ESS*, and incorporating results of different program test phases. Especially valuable are the questions to be asked regarding these subjects.

Unfortunately, the technical details included in this handbook are of questionable validity. Vibration and temperature cycling parameters are defined only in terms of *ESS* facility inputs; the response of the hardware being screened is not considered. Defining *ESS* screening strengths (for vibration and temperature cycling) with three- and four-place decimal values (Tables 5.14 through 5.18) is particularly suspect for general application since actual hardware can have widely varying mechanical and thermal characteristics that cannot be generically characterized with such precision. No references or rationale are cited to validate the very precise data presented in these tables. This handbook can be useful, but must be applied with caution to avoid inappropriate assumptions and unjustifiable expectations that might come from trying to apply the seemingly rigorous, but unpedigreed, mathematical models.

5. **MIL-HDBK-338-1A, "Electronic Reliability Design Handbook,"** Volume I of II (12 October 1988).

TAILORABILITY INDEX: High.

PURPOSE: Information and Guidance.

COMMENTARY: This document consists of more than 1,050 pages dealing with all aspects of reliability program planning and execution. Although only 21 pages long, the section on assembly-level *ESS* is nevertheless noteworthy for its generally realistic approach to determining appropriate screens based on thermal and vibration response surveys. The need for tailoring, continuous reevaluation of screen cost-effectiveness, understanding the root causes of failures, and the imprecise prior knowledge of latent defect quantities and screen strengths are constantly emphasized. The need to involve competent engineering personnel" (Paragraph 11.2.3.2.2) in determining appropriate screen parameters and surveys is also highlighted.

The most significant weakness in this document's *ESS* discussion is the graphic presentation of relatively precise screen strengths for various screen types and conditions (Figures 11.2.3.2-2 through 11.2.2.2-6). These are presented without description of how they were derived and under what conditions they apply (or don't apply). The influence of fixture characteristics, equipment structure, and power-on effects are not considered in the discussion. As stated in the document (see above), it is not possible to calculate screen strengths accurately. Unfortunately, without reasonably accurate screen strength values, the Stress Screening Model presented (Paragraph 11.2.3.3.1) is of more academic interest than practical utility and, therefore, of questionable value. (This same shortcoming handicaps MIL-HDBK-344.)

6. **MIL-STD-781D, "Reliability Testing for Engineering Development, Qualification, and Production"** (17 October 1986)/**MIL-HDBK-781, "Reliability Test Methods, Plans, and Environments for Engineering Development, Qualification, and Production"** (14 July 1987).

TAILORABILITY INDEX: Low to Nil (Systematic Optimization of the Production Process is Not Addressed).

PURPOSE: Contract Administration and Compliance.

COMMENTARY: In view of its pre-eminence as a reliability standard, it is unfortunate that MIL-STD-781D treats *ESS* more as a qualification test than as a process improvement tool. Indeed, the contract-like language (i.e., "shall be as specified in") throughout Task 401 is oriented more toward keeping the production process unchanged than encouraging systematic improvements. This document has a distinctly different emphasis than DOD 4254.7-M/NAVSO P-6071 and AFP 800-7, which emphasize tailoring, not compliance.

Perhaps its most serious flaw is its foundation of a "standard *ESS*," implying that there is only one right way to screen. The standard also instructs the *procuring activity* (not the equipment supplier, who should understand the assemblies best) to *specify* the *ESS* conditions. References to the conceptually different NAVMAT P-9492 and the even more rigid MIL-STD-2164 only enhance the confusion since none of these documents are integrated or related in philosophy or content.

The handbook describes three methods for "monitoring" *ESS* (Paragraph 4.4) although one is actually the so-called "Standard *ESS*." The Computed *ESS* Time Interval Method uses conceptually cryptic "prespecified probabilities" and is derived from a reference that is not published in the literature and therefore not available to the user for clarification of the concept. The Graphical Method artificially smooths *ESS* data in a way that could obscure important trends in defect population disclosure.

7. **MIL-STD-2164 (EC), "Environmental Stress Screening Process for Electronic Equipment"** (5 April 1985).

TAILORABILITY INDEX: Nil.

PURPOSE: Contract Administration and Compliance.

COMMENTARY: MIL-STD-2164 is the *only ESS* document which the IES ESSEH Working Group has asked the preparing activity to withdraw from use. (The request was declined.) About the only part of this document that relates to *ESS* as generally accepted is the title. Otherwise this document is the antithesis of the majority of published literature on *ESS*. It directly contradicts the thoughtful guidance of DOD 4245.7-M/NAVSO P-6071 and AFP-800-7 about flexible, dynamic *ESS* processes by establishing *ESS* as a standard *test*. Conspicuously absent are any words about improving the production process or hardware quality. Instead, "the standard provides for a uniform *ESS*" (Paragraph 1.1).

Very detailed requirements (similar to those found in MIL-STD-810) for facility and fixture characteristics, control tolerances, and test profiles make it difficult to innovate different approaches and could drive *ESS* facility costs upward. Restrictions on sinusoidal crosstalk (Paragraph 4.5.1) would disallow many effective screening facilities currently in use. The temperature cycle described in Figure 1 could be as long as 120 hr (certainly not oriented toward an efficient production flow!). Extensive criteria for *scoring failures* during the various *test* phases (i.e., predefect-free test, environmental test, defect-free test) are defined.

An appendix relates test durations to *MTBF*, probability of rejecting or accepting defective hardware, and specified failure rates. Doing so seems to contradict MIL-STD-785B, which defines a basic conceptual difference between reliability *engineering* tests, such as *ESS*, and reliability/accounting tests for supplying data to determine compliance (Appendix A, Paragraph 50.3.1.1). MIL-HDBK-785B also cautions that "Reliability values measured during ESS ⋯ cannot be expected to correlate with reliability values in service" (Appendix A, Paragraph 50.3.1.4.2).

It is ironic that such a rigidly detailed *requirements* document would caution the procuring activity not to tell the contractor how to achieve the required *ESS* results. It is also unfortunate that this document is sometimes invoked only because it is referenced in MIL-HDBK-781.

8. **NAVMAT P-9492, "Navy Manufacturing Screening Program"** (May 1979)/**NAVMAT P-4855-1A (NAVSO P-3641, "Navy Power Supply Reliability"** (January 1989).

TAILORABILITY INDEX: Low to Moderate.

PURPOSE: Guidance and Program Development.

COMMENTARY: NAVMAT P-9492 is the grandfather of all of the current *ESS* standards and documents. Although often imposed as a contract requirement, it was never intended to be used as such (see Overview). The often cited vibration spectrum in this document was originally presented as a facility capability, not a *required* input spectrum (Paragraph 4.2). This document was also one of the only ones to define temperature change rates in terms of equipment response instead of chamber air conditions (Paragraph 4.1).

NAVMAT P-9492 was to have been rewritten, but the revised draft is currently in limbo. However, the basic process introduced in NAVMAT P-9492 has been extended in NAVMAT P-4855-1A, which references its predecessor. Therefore, there are situations in which these documents must be used together.

It is also worth noting that both documents deal with *ESS* at the Line Replaceable Unit (LRU), or "black-box", level only and not with Shop Replaceable Units (SRU's) such as printed wiring assemblies, although this is never clearly stated. NAVMAT P-9492 has also been misapplied (with disastrous results) to large items such as torpedoes and equipment cabinets.

9. **Sacramento Air Logistics Center, "Environmental Stress Screening Handbook"** (15 June 1988).

TAILORABILITY INDEX: High.

PURPOSE: Information and Guidance; Program Administration.

COMMENTARY: It's unfortunate that this document does not have more visibility within the *ESS* community because, in many ways, it is one of the most enlightened documents published by a government agency. Of particular note is the scope of program activities to which this document applies. Whereas many *ESS* documents focus only on technical minutia, this handbook considers often-overlooked administrative as well as technical concerns. For example, such issues as previous contractor experience, decision criteria for *ESS* applicability, cost effectiveness in the production process, and contractor development of an appropriate *ESS* methodology are addressed. Statement of Work examples are also included.

In the technical area, there are a few inconsistencies. For example, the R&M 2000 Guidelines cited define a 6-*grms* spectrum from 50 to 1,000 Hz. However, in Statement of Work examples included in the handbook, 6-g (rms) spectra are also defined for frequency ranges from 30 to 200 (2,000?) Hz and 100 to 1,000 Hz. These are three very different spectra. Unfortunately, the value of 6g (rms) seems to be given a "magical" status. Also, even though "failure-free tests" are stated to be "not appropriate for *ESS*," the same Paragraph (4.1.7) requires that at least the last 25% of the temperature cycles be failure free.

Overall, however, this document has much to commend it. It should be given serious consideration by other organizations involved with *ESS*.

10. **TE000-AB-GTP-020A, "Environmental Stress Screening Requirements and Application Manual for Navy Electronic Equipment"** (January 1992) (*Supersedeas NAVSEA Notice 3900*).

TAILORABILITY INDEX: High to Low (Depending on Contract Implementation).

PURPOSE: Contract Administration and Information.

COMMENTARY: The TE000 document is very difficult to evaluate. It is a curious mixture of contract requirements, technical advice, and undocumented pronouncements. Unfortunately, it is often difficult to separate these elements from each other. No references are cited for the data, surveys, and experiences used as justification for requirements.

Perhaps the most disturbing aspect of TE000 is that it seems to be based on the basic premise that there is no other way to implement a valid *ESS* program. It tries to anticipate all possible *ESS* situations and define an approved action for each.

There are also surprising instances where the guidance in TE000 runs counter to published *ESS* studies and consensus in the *ESS* community. For example, published studies and other *ESS* guidelines support applying vibration before thermal cycling instead of vice versa (Paragraph 5.1). TE000 does not require power on/off, or performance monitoring, during higher-assembly-level vibration, nor does it require performance monitoring during temperature cycling changes from cold to hot (Paragraphs 2.3.2 and 2.4.1). Yet these are the situations in which intermittent defects are most frequently detectable. The proof test approach to validating random vibration screens (Paragraph 2.4.2) could be unnecessarily expensive and time consuming. Measuring input vibration levels on the hardware being vibrated (Paragraph 2.4.1) is generally not a recommended practice.

TE000 *does* contain a good deal of useful technical information [such as the Printed Wiring Assembly (PWA) natural frequency and displacement equations] and valid *ESS* guidance (such as the importance of understanding the equipment's vibration and thermal *responses*. However, the user should (if not already) become familiar with the large body of *ESS* literature that has been published and weigh the contents of TE000 carefully against the literature base and his/her own experiences.

11. **Warner Robins Air Logistics Center (WR-ALC), Environmental Stress Screening (ESS) Handbook** (February 1990).

    TAILORABILITY INDEX: Potentially High (depending on contract implementation).

    PURPOSE: Contract Administration and Guidance.

    COMMENTARY: The WR-ALC *ESS* Handbook is based on AFP 800-7. To its credit, the handbook focuses on *ESS* as an equipment production process (Paragraph 3.1) and clearly states that *ESS* is *not* a test (Paragraph 3.3.2). It emphasizes the need to tailor screens based on vibration and thermal response surveys (Section 5 and Paragraph 6), using "these guidelines [as] a starting point for designing an *ESS* regimen" (Paragraphs 5.3.2 and 6.1).

    However, portions of this handbook could interfere with implementing an effective, tailored *ESS* program:

    (a) Paragraph 1.1 of the handbook defines its purpose as providing both requirements and guidelines, a potentially confusing dichotomy.

    (b) The screen strength model method (Paragraph 5.3.3) is based on DOD-HDBK-344 and therefore suffers the same handicaps described elsewhere in this overview.

    (c) The strong recommendation to specify MIL-STD-781D, Task 401, in all statements of work (Paragraph 6.3.2 and Appendices B and C) could destroy all of the tailorability and common sense that is otherwise present in this handbook. As detailed elsewhere in this review, MIL-STD-781D describes *ESS* in very different terms than the WR-ALC handbook and invokes conflicting criteria.

    (d) Perhaps the most important potential conflict is between Items 1 and 3 of the Statement of Work example in Appendix B, which *requires* that two conflicting documents, MIL-STD-781D and the R&M 2000 *ESS* Guidelines be applied. It is indeed unfortunate that a reference to the WR-ALC Handbook was not included.

12. **IES, "Environmental Stress Screening Guidelines for Assemblies"** (March 1990).

TAILORABILITY INDEX: High.

PURPOSE: Information and Guidance.

COMMENTARY: The Institute of Environmental Sciences has been an integral part of the *ESS* development process since its first ESSEH (Environmental Stress Screening of Electronic Hardware) Conference in 1979 (the year that NAVMAT P-9492 was released). Since that time, the IES has continually worked to ensure that a useful information base related to recent *ESS* experiences, philosophies, and trends are available to the *ESS* community and other involved organizations.

The IES "Environmental Stress Screening Guidelines for Assemblies" [3; 4; 5] has been a significant part of this effort. These guidelines have evolved considerably from their first appearance in 1981. They started as a survey of industry experiences, but have grown since then from a survey report into a technical and administrative handbook. What is most significant is that these guidelines are the *only ESS* documents in existence that have been *continually updated by recognized experts* to incorporate the most current information available.

Although initially prompted by the requirements of defense programs, these guidelines have been drafted so that they *can be applied to commercial as well as military programs*. Even in situations where a procuring organization already has its own *ESS* document, the IES Guidelines help provide a general philosophical and technical perspective which can assist in the effective application of other *ESS* documents.

The Guidelines include program management guidance, cost-effectiveness analysis techniques, descriptions of vibration and thermal survey methodologies, and *ESS* tailoring principles. Although not perfect, this document is a uniquely valuable resource in that it represents the closest the *ESS* community has been able to come to a consensus on this subject and it continues to evolve.

REFERENCES

1. Caruso, H., "An Overview of Environmental Stress Screening Standards and Documents for Electronic Assemblies," *Journal of Environmental Sciences*, Vol. 32, No. 4, pp. 15-25, July/August 1989.

2. Caruso, H., "An Updated Overview of Environmental Stress Screening Standards and Documents for Electronic Assemblies," *Journal of Environmental Sciences*, Vol. 34, No. 4, pp. 49-61, March/April 1992.

3. Institute of Environmental Sciences, *Environmental Stress Screening Guidelines*, 940 East Northwest Highway, Mount Prospect, IL 60056, 1981.

4. Institute of Environmental Sciences, *Environmental Stress Screening Guidelines,* 940 East Northwest Highway, Mount Prospect, IL 60056, 1984.

5. Institute of Environmental Sciences, *Environmental Stress Screening Guidelines for Parts,* 940 East Northwest Highway, Mount Prospect, IL 60056, 1985.

Chapter 6

THE STATISTICAL QUANTIFICATION AND OPTIMIZATION OF *ESS*

6.1 INTRODUCTION

ESS planning objectives require a quantitative specification of what the *ESS* program is to achieve. It is not enough to say "precipitate early failures" without specifying the reliability of the process. The *ESS* objective is thus to meet some reliability requirement at the time of product delivery. Given that the product was designed right to start with, *ESS* cannot improve the product beyond its designed-in reliability, but *ESS* can bring it close to it and do so at minimum cost. Cost may involve the screening process, its duration, cost of early failures (repairs and replacements) and penalties paid for incomplete screening, as well as the cost of failing "good" parts in the assembly that may occur during screening because of the elevated stress levels.

Two parameters appear to be important in the economical design of *ESS* programs:

1. The degree of contamination; i.e., the proportion of parts with latent defects contained initially in an otherwise well-designed system.

2. The predictive life distribution of parts in the assembly which exhibit early failures.

Both are statistical concepts. They govern the *ESS* process and determine the rate at which the purification proceeds. Both can be estimated from prior experience supplemented with current data.

Systems consist of components. Component populations contain latent defects. These combine into system latent defects that cause early failures (so-called infant mortality failures). In addition to component latent defects, production flaws are introduced into the assemblies at each phase of the manufacturing process. These production flaws manifest themselves in the same way as parts' latent defects – they cause early system failures if not detected and corrected. We will start with the quantification and optimization of parts level screening, and proceed to system level screening, part-system screening and finally the Bayesian approach to *ESS*.

6.2   ASSUMPTIONS

The following assumtions are very important in understanding and applying the models and results presented in this chapter. They are vital to the validity of the developed models.

6.2.1   BIMODAL MIXED-EXPONENTIAL LIFE DISTRIBUTION ASSUMPTION

We are considering a lot of N parts in the assembly having a proportion p_b of substandard parts. These substandard parts in the assembly have hidden latent defects which will very probably provoke the "early failures." They cannot be detected by quality control (even if applied 100%) because they appear to be normal "good" parts.

Beginning with the Hughes proposed Military Standard [1] for stress screening of electronic equipment, it has been conventional to think of two distinct subpopulations of parts in the assembly: a main subpopulation of "good" parts and a much smaller subpopulation of defectives. But for a mature product which has been already burned-in, broken-in and debugged, it will be assumed that the so-called "good" products will have a constant failure rate of their own, λ_g, and the so-called "bad" products will have a constant failure rate of their own, λ_b, which would be several orders of magnitude greater than the failure rate, λ_g. Therefore, a mixed-exponential distribution may be used to represent the life distribution of the product's population or of the whole system.

There are two points which should be carefully considered relative to the applicability of this model.

First, the constant failure rate assumption relative to the subpopulations plays a crucial role in *ESS*. The assessment that failure times can be predicted using an exponential distribution, with a specified failure rate, means that we would use the same exponential distribution for predicting the life of a new part, as for predicting the remaining life of a part given that the part had been used for some specific time period. If, as is the case, the failure rate cannot be specified in advance, then the predictive life distribution will not be the exponential.

Second, the quantification of stress levels needs to be carefully considered. In [2] there are reminders of the fact that the stress level should be kept under a "threshold which could precipitate failure modes that would never occur in normal operation, or which could damage the part." In addition to this assumption of "nondamaging" stress levels, it is assumed that the stress level used can be translated into a multiplicative time factor. That is, the failure rate of either subpopulation, say, λ, under stress level with a stress acceleration factor A_F can be related to the failure rate, λ', under normal operating conditions by

$$\lambda = A_F \, \lambda', \text{ for } A_F \geq 1.$$

6.2.2   UNIFORM-STRESS ENVIRONMENT ASSUMPTION

It is assumed for all models presented in this chapter that the screen is conducted under a uniform-stress environment. Under uniform-stress environment the specified stress types and corresponding profiles, such as temperature profile and/or random vibration spectrum, etc., are consistently applied without any variation throughout the whole screening process. A uniform-stress environment may correspond to a single-stress environment, such as thermal cycling only or random vibration only, etc. It may also correspond to a simultaneously-combined-stress environment, such as combined-temperature-vibration, etc., during which two or more stress types, each with a fixed stress profile, are applied for the same duration. Sequential stress, profile-varying and other nonuniform-stress environments are excluded here.

6.3   PART LEVEL SCREENING [2]

6.3.1   THE FAILURE RATE

The mathematics of the screening process shows that the residue of latent defects, left over after screening, is a direct function of screen duration for any prescribed stress level. Assume that in an assembly there is a proportion, p_b, of substandard parts with latent defects which will eventually cause early failures. Obviously, if the mean time to failure of good parts in the assembly is in the millions of hours, and of the bad parts in the assembly failing is in a few thousand, hundred or ten hours, then the latter are failing due to early causes. If in an assembly of N parts there are N_b weak parts, the proportion of latent defectives is

$$p_b = N_b/N.$$

The proportion of good parts in the assembly is then

$$p_g = N_g/N,$$

so that

$$N_b + N_g = N,$$

and

$$p_b + p_g = 1.$$

The quantities p_b and p_g are also the probabilities of picking an early failing part or picking a good part, respectively, when drawing a part at random from the assembly. Of course, one cannot know whether the drawn part comes from the p_b or the p_g subpopulation in the assembly.

Now let the weak parts in the assembly have a failure rate of λ_b and the good parts in the assembly a failure rate of λ_g, with the respective reliabilities of

$$R_b(t) = e^{-\lambda_b t},$$

and

$$R_g(t) = e^{-\lambda_g t}.$$

Since with a probability of p_b the part drawn at random from the assembly will have a reliability of $R_b(t)$, and with a probability of p_g it will have a reliability of $R_g(t)$, any part drawn at random from the assembly will have a statistical probability of survival (reliability) of

$$R(t) = p_b R_b(t) + p_g R_g(t),$$

or

$$R(t) = p_b e^{-\lambda_b t} + p_g e^{-\lambda_g t}. \tag{6.1}$$

Equation (6.1) is based on the principle of uncertainty because we do not know what part in the assembly we have drawn.

The probability density function of such a randomly drawn part in the assembly is then

$$f(t) = -d[R(t)]/dt,$$

which results in

$$f(t) = p_b \lambda_b e^{-\lambda_b t} + p_g \lambda_g e^{-\lambda_g t}, \tag{6.2}$$

and the part's failure rate is

$$\lambda(t) = f(t)/R(t). \tag{6.3}$$

Substituting Eqs. (6.1) and (6.2) into Eq. (6.3) yields the failure rate equation

$$\lambda(t) = \frac{p_b \lambda_b e^{-\lambda_b t} + p_g \lambda_g e^{-\lambda_g t}}{p_b e^{-\lambda_b t} + p_g e^{-\lambda_g t}}. \tag{6.4}$$

This function decreases with time. Its initial value, λ_i, at $t = 0$ is

$$\lambda_i = \lambda(0) = p_b \, \lambda_b + p_g \, \lambda_g, \qquad (6.5)$$

while the limiting final value, λ_f, is that of a good part in the assembly, λ_g, as t approaches infinity, and it thus becomes

$$\lambda_f = \lim_{t \to \infty} \lambda(t) = \lambda_g. \qquad (6.6)$$

This may be proven if we rewrite Eq. (6.4) in the form

$$\lambda(t) = \lambda_g + \frac{p_b \, (\lambda_b - \lambda_g) \, e^{-(\lambda_b - \lambda_g) \, t}}{p_g + p_b \, e^{-(\lambda_b - \lambda_g) \, t}}, \qquad (6.7)$$

and see that the numerator of the second term on the right vanishes as t goes to infinity.

Figure 6.1 is a plot of the failure rate Eq. (6.7) for an assumed parts population in the assembly which contains 1% of weak parts ($p_b = 0.01$) and 99% of good parts ($p_g = 0.99$). Let the good parts in the assembly, under a "ground benign" environment, have a failure rate of 10^{-7} failures per hour (MIL-HDBK-217) while the weak parts in the assembly have a failure rate of 10^{-3} failures per hour, under the same environment. During ESS let an acceleration factor of 100 be applied (without damaging parts in the assembly or subtracting from their life span). By failing the weak parts in the assembly during ESS the initial failure rate of the part's population will drop from 10^{-5} failures per hour [see Eq. (6.5)] to a value very close to the failure rate of the good parts in the assembly (10^{-7} fr/hr). This process of elimination of the weak parts in the assembly may take thousands of hours under the benign environment while it will take much less time under ESS. After ESS we have a highly reliable parts population in the assembly, while in the first several thousand hours – without ESS – the parts in the assembly population would be very unreliable. This can be ascertained by looking at the ordinate of Fig. 6.1.

The proper physical interpretation of the decaying failure rate phenomenon is that the longer a part in the assembly survives, the more likely it is that it came from the good population p_g. One should remember that no single physical part in the assembly ever has the failure rate of Eq. (6.7). It is either good with a failure rate of λ_g, or it is weak with a failure rate of λ_b. Yet, from Eq. (6.7) we can draw very important conclusions.

6.3.2 SCREEN DURATION

Screen duration, T_s, is a function of the stress level(s) applied. However, it is important to keep the stress level below a threshold which could precipitate failure modes that would never occur in normal operation or which could damage the part in the assembly. On the other hand, for ESS to be successful,

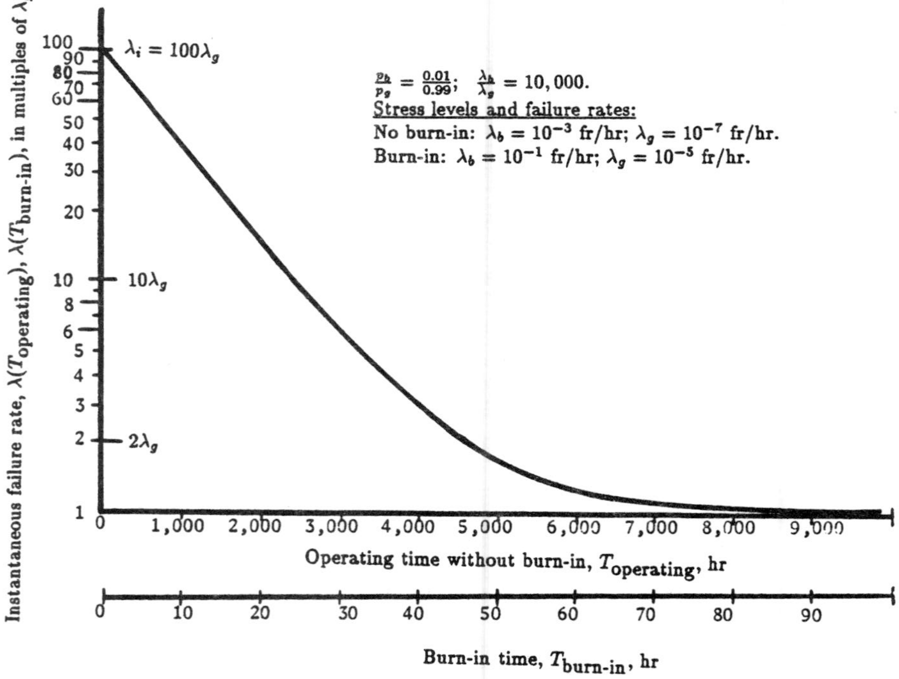

$\lambda_i = 100\lambda_g$
$10\lambda_g$
$2\lambda_g$
$\frac{p_h}{p_g} = \frac{0.01}{0.99}$; $\frac{\lambda_h}{\lambda_g} = 10,000$.
Stress levels and failure rates:
No burn-in: $\lambda_b = 10^{-3}$ fr/hr; $\lambda_g = 10^{-7}$ fr/hr.
Burn-in: $\lambda_b = 10^{-1}$ fr/hr; $\lambda_g = 10^{-5}$ fr/hr.
Instantaneous failure rate, $\lambda(T_{operating})$, $\lambda(T_{burn\text{-}in})$, in multiples of λ_g
1 2 3 4 5 6 8 10 20 30 40 50 60 70 80 90 100
0 1,000 2,000 3,000 4,000 5,000 6,000 7,000 8,000 9,000
Operating time without burn-in, $T_{operating}$, hr
0 10 20 30 40 50 60 70 80 90
Burn-in time, $T_{burn\text{-}in}$, hr

Fig. 6.1– Early failures at the assembly level versus stress level.

it is necessary to accelerate the occurrence of failure modes which normally can occur. Acceleration is achieved by increasing the stress levels (without going into excess) and it "compresses" the time failures occur.

Equation (6.7) lends itself to calculating the *ESS* time, T_s, under various stress conditions. To perform this calculation we must extract the time factor t from the equation. First, we put the equation into the form

$$\lambda(T_s) = \lambda_g \ (1 + \varepsilon), \tag{6.8}$$

where

$$\lambda(T_s) = \text{failure rate of the assembly at the end of } ESS,$$

and

$$\varepsilon = \frac{\lambda(T_s) - \lambda_g}{\lambda_g},$$

which is called the *Screen Residue*, and correspondingly $(1 - \varepsilon)$ is called the *Screen Efficiency*, and is a function of the *ESS* time T_s. The reason for selecting this form of the failure rate equation is that we can make ε arbitrarily small, and this provides us with the possibility of controlling how close we want the failure rate to get to λ_g in the *ESS* process. Thus, in accordance with Eq. (6.8), we may preselect a desired residue value ε and obtain parts in the assembly of a failure rate $\lambda(T_s)$ corresponding to the preselected ε in a screening process of T_s hours duration.

We now further rewrite Eq. (6.7) as

$$\lambda(T_s) = \lambda_g \left[1 + \left(\frac{1}{\lambda_g} \right) \frac{D \ p_b \ e^{-D \ T_s}}{p_g + p_b \ e^{-D \ T_s}} \right], \tag{6.9}$$

where

$$D = \lambda_b - \lambda_g$$

is the difference of the failure rates between the weak and the good parts in the assembly. From Eqs. (6.8) and (6.9) we get the *Screen Residue* as

$$\varepsilon = \left(\frac{1}{\lambda_g} \right) \frac{D \ p_b \ e^{-D \ T_s}}{p_g + p_b \ e^{-D \ T_s}}. \tag{6.10}$$

For the sake of simplicity we may define the ratio of weak parts to good parts in the initial assembly as

$$\alpha = p_b / p_g,$$

and the ratio of the early failure rate to the good parts' failure rate in the assembly as

$$\beta = \lambda_b / \lambda_g,$$

and we may write Eq. (6.10) in the form

$$\varepsilon = \frac{\alpha \ (\beta - 1) \ e^{-D \ T_s}}{\alpha \ e^{-D \ T_s} + 1}. \tag{6.11}$$

Rearranging Eq. (6.11) yields

$$e^{-D \ T_s} \left[\frac{\alpha \ (\beta - 1 - \varepsilon)}{\varepsilon} \right] = 1, \tag{6.12}$$

and, finally, taking the logarithm of both sides yields the screen time T_s as

$$T_s = \frac{1}{D} \log_e \left[\frac{\alpha \ (\beta - 1 - \varepsilon)}{\varepsilon} \right], \tag{6.13}$$

which is a function of the preselected *Screen Residue*, ε, and of the difference D between the weak parts and good parts failure rates in the assembly. Thus, for a preselected value of ε in a screen duration of time T_s we get parts in the assembly with a failure rate given by Eq. (6.8); i.e.,

$$\lambda(T_s) = \lambda_g \ (1 + \varepsilon).$$

Under normal field operating conditions D will be a comparatively small number which makes T_s large. But under the accelerated burn-in, or stress screen environment, T_s can be drastically reduced. In other words, self-elimination of early failures in the field, without a preceding *ESS*, may be a very protracted process of low equipment reliability, while in a properly planned *ESS* process the elimination of early failures can occur very fast. Also, it is important to know that replacement parts or components used in field maintenance come from burned-in populations. This will affect the final equations.

Present military screening (burning-in) specifications, such as R&M 2000, require the use of prescreened items for field maintenance.

6.3.3 THE NUMBER AND COST OF FAILURES

While in a Poisson process, with an exponential underlying life distribution where the failure rate λ is constant, the expected number of failures in time t is given by

$$H(t) = \lambda \ t, \tag{6.14}$$

the case is different when the failure rate is not constant but is a function of time, such as given by Eq. (6.7). To calculate the expected number of failures such a case we need to apply Renewal Theory [4]. We proceed by first obtaining the renewal rate, $h(t)$, of the *Renewal Process of the Screen*, and by integration of $h(t)$ we obtain the number of failures in t; i.e., $H(t)$.

The *pdf* for the times-to-failure distribution is given by Eq. (6.2), or

$$f(t) = p_b\,\lambda_b\,e^{-\lambda_b\,t} + p_g\,\lambda_g\,e^{-\lambda_g\,t}.$$

The Laplace transform of this *pdf*, $f^*(s)$, is

$$f^*(s) = \frac{p_b\,\lambda_b}{s + \lambda_b} + \frac{p_g\,\lambda_g}{s + \lambda_g}. \tag{6.15}$$

According to Renewal Theory [4], the Laplace transform of the renewal rate function, $h^*(s)$, can be obtained by

$$h^*(s) = \frac{f^*(s)}{1 - f^*(s)}. \tag{6.16}$$

Substituting Eq. (6.15) into Eq. (6.16) and simplifying yields

$$h^*(s) = \frac{s\,(p_b\,\lambda_b + p_g\,\lambda_g) + \lambda_b\,\lambda_g}{s^2 + s\,(p_b\,\lambda_g + p_g\,\lambda_b)}, \tag{6.17}$$

which can be rearranged as

$$h^*(s) = \frac{\lambda_b\,\lambda_g}{A}\left(\frac{1}{s} - \frac{1}{s + A}\right) + (p_b\,\lambda_b + p_g\,\lambda_g)\frac{1}{s + A},$$

or

$$h^*(s) = \frac{\lambda_b\,\lambda_g}{A}\frac{1}{s} + \left(p_b\,\lambda_b + p_g\,\lambda_g - \frac{\lambda_b\,\lambda_g}{A}\right)\frac{1}{s + A}, \tag{6.18}$$

where

$$A = p_b\,\lambda_g + p_g\,\lambda_b. \tag{6.19}$$

The inverse Laplace transform of Eq. (6.18) yields

$$h(t) = \frac{\lambda_b\,\lambda_g}{A} + \left(p_g\,\lambda_g + p_b\,\lambda_b - \frac{\lambda_b\,\lambda_g}{A}\right)e^{-A\,t}. \tag{6.20}$$

The initial value of $h(t)$, h_i, is

$$h_i = h(0) = p_g\,\lambda_g + p_b\,\lambda_b, \tag{6.21}$$

while the limiting final value is

$$h_f = h(\infty) = \frac{\lambda_b\,\lambda_g}{A}, \tag{6.22}$$

which occurs as t approaches infinity.

Using this notation we may rewrite Eq. (6.20) in the form

$$h(t) = h_f + (h_i - h_f)\,e^{-A\,t}, \tag{6.23}$$

and the integral of the renewal rate function becomes the expected number of renewals which equals the expected number of failures in t; i.e.,

$$H(t) = h_f\, t + \frac{1}{A}(h_i - h_f)(1 - e^{-A\,t}). \tag{6.24}$$

At $t = 0$ the number of failures is zero, and grows continuously with t.

As to the cost of failures, $C(t)$, in time t, counted from $t = 0$, this is a linear function of the number of failures $H(t)$; i.e.,

$$C(t) = H(t) \cdot c, \tag{6.25}$$

where c is the average cost of a single failure. In general, the cost of a single failure, c, depends on the screening phase at which the failure occurs. Thus, for instance, a part failure in the assembly during ESS may cost \$5.00 (loss of the part), but the same failure occurring during operation in the field may cost \$15,000.00 or more. Figure 6.2 illustrates the comparative cost of early failures without and with ESS.

6.3.4   THE POWER FUNCTION OF A SCREEN

In Eq. (6.8) we have seen that in a screen of duration T_s the failure rate of a population of items drops from an initial value of λ_i to a value of

$$\lambda(T_s) = \lambda_g(1 + \varepsilon).$$

Thus, the difference

$$\lambda_i - \lambda_g(1 + \varepsilon)$$

is a measure of how good the screen is. To formalize this as a *Power Function*, we form the ratio of the actual screen result

$$\lambda_i - \lambda_g(1 + \varepsilon)$$

to the ideal case of

$$\lambda_i - \lambda_g,$$

in which all failures are eliminated and we call this ratio the *Power Function of the Screen*, PS; i.e.,

$$PS = \frac{\lambda_i - \lambda_g\,(1 + \varepsilon)}{\lambda_i - \lambda_g}. \tag{6.26}$$

Using the preceding results for λ_i and ε, we may write the PS equation in the explicit form of

$$PS = \frac{p_g\,(1 - e^{-D\,T_s})}{p_g\,(1 - e^{-D\,T_s}) + e^{-D\,T_s}}. \tag{6.27}$$

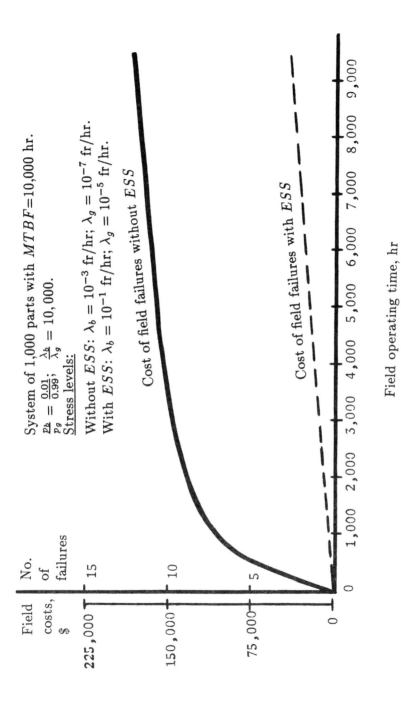

System of 1,000 parts with $MTBF = 10,000$ hr.

$\frac{p_b}{p_g} = \frac{0.01}{0.99}; \quad \frac{\lambda_b}{\lambda_g} = 10,000.$

Stress levels:

Without ESS: $\lambda_b = 10^{-3}$ fr/hr; $\lambda_g = 10^{-7}$ fr/hr.

With ESS: $\lambda_b = 10^{-1}$ fr/hr; $\lambda_g = 10^{-5}$ fr/hr.

Cost of field failures without ESS

Cost of field failures with ESS

Field operating time, hr

Fig. 6.2– Cost of early failures with and without ESS.

The PS function supplies the answer to the question of "How good is a screen?". This function, as given by Eq. (6.26), becomes unity, or 100%, only if

$$\varepsilon = 0,$$

which is not achievable practically.

To make use of the methods outlined in this section, the practicing reliability engineer must first obtain basic information on part quality levels and failure rates. MIL-HDBK-217 provides the expected failure rates of "good" parts in various environments, including the maximum permissible electrical and thermal stresses. The proportions and failure rates of "early" failing parts are not so readily available and the user will have to establish a data collection system (based on experience) for such information. In the absence of such data, one could initially use the part manufacturers' claimed quality levels and assume (as a ballpark estimate) that the defective parts fail at least 10,000 times faster than good parts.

Recommended starting regimes for ESS have been published by the military and by the Institute for Environmental Sciences $ESSEH$ Committee.

The use of the equations generated in the previous paragraphs is illustrated with the following example, in which only the failures due to component parts in the assembly are considered.

EXAMPLE 6–1

An equipment consisting of approximately 3,750 electronic parts is subjected to an alternating sequence of screening environments including thermal shock, random vibration, thermal cycling, and extensive electrical testing at ambient temperatures to measure performance and detect failures.

In this example we will lump all the different component part types in the assembly together and consider average failure and defect rates across all parts in the assembly. Also no attempt will be made to evaluate the relative stress levels of the different screen environments. Instead some average stress level will be assumed across the entire screening process.

From data obtained on this and similar equipment, the following are estimated:

The initial part defect rate, at the start of equipment screening is

$$p_b = 250 \text{ ppm (parts per million)}.$$

The failure rate of an average defective part in the assembly is

$$\lambda_b = 15,000 \text{ fr}/10^6 \text{ hr}.$$

The failure rate of an average good part in the assembly is

$$\lambda_g = 0.5 \text{ fr}/10^6 \text{ hr}.$$

Determine the screen duration, T_s, for a specified *Power Function* of 96%.

SOLUTION TO EXAMPLE 6–1

The proportion of good parts in the assembly is

$$p_g = 1 - p_b = 1 - 250 \times 10^{-6} = 999,750 \times 10^{-6}.$$

The initial failure rate, from Eq. (6.5), will be

$$
\begin{aligned}
\lambda_i &= p_b\,\lambda_b + p_g\,\lambda_g, \\
&= (250 \times 10^{-6}) \times 0.015 + (999,750 \times 10^{-6}) \times (0.5 \times 10^{-6}), \\
&= 4.25 \times 10^{-6},
\end{aligned}
$$

which is 8.5 times the failure rate, λ_g, of good parts in the assembly; i.e.,

$$\frac{\lambda_i}{\lambda_g} = \frac{4.25 \times 10^{-6}}{0.5 \times 10^{-6}} = 8.5.$$

Substitution into Eq. (6.26) gives the *Power Function of the Screen*, or

$$
\begin{aligned}
PS &= \frac{\lambda_i - \lambda_g\,(1+\varepsilon)}{\lambda_i - \lambda_g}, \\
&= \frac{\lambda_i/\lambda_g - (1+\varepsilon)}{\lambda_i/\lambda_g - 1}, \\
&= \frac{8.5 - (1+\varepsilon)}{8.5 - 1}, \\
&= \frac{7.5 - \varepsilon}{7.5},
\end{aligned}
$$

which when rearranged becomes

$$\varepsilon = 7.5\,(1 - PS).$$

If a *Power Function* of 96% is desired then

$$\varepsilon = 7.5\,(1 - 0.96) = 7.5 \times 0.04 = 0.3.$$

The required screening time, T_s, can now be obtained from Eq. (6.13) by substituting the values of

$$
\begin{aligned}
\alpha &= p_b/p_g = 0.000250/0.999750 = 0.00025, \\
\beta &= \lambda_b/\lambda_g = 15,000/0.5 = 30,000,
\end{aligned}
$$

and

$$D = \lambda_b - \lambda_g = (15,000 - 0.5) \times 10^{-6} = 0.015.$$

This gives the required screening time of

$$T_s = \frac{1}{0.015} \log_e \left[0.00025 \, \frac{(30,000 - 1 - 0.3)}{0.3} \right],$$

$$= \frac{1}{0.015} \log_e \, 25,$$

$$= 3.219/0.015,$$

or

$$T_s = 215 \text{ hr.}$$

If these values are applied to the equipment containing 3,750 parts the following estimates are obtained:

Initial defective parts in the assembly $= 3,750 \times 250 \times 10^{-6} = 0.9375$.

Expected failures of defective parts during equipment screening, N_{fb}, is

$$N_{fb} = 0.9375[1 - R_b(215 \text{ hr})],$$

$$= 0.9375 \left(1 - e^{-\lambda_b \, t} \right),$$

$$= 0.9375 \left[1 - e^{-(15,000 \times 10^{-6}) \times 215} \right],$$

or

$$N_{fb} = 0.9002.$$

The number of defective parts in the assembly escaping the screen, N_{eb}, is

$$N_{eb} = 0.9375 - 0.9002 = 0.0373,$$

which is 4% of 0.9375, verifying the desired power function of 96%.

The expected number of failures of good parts in the assembly during screening, N_{fg}, is

$$N_{fg} = (3,750 - 0.9375)[1 - R_g(215)],$$

$$= (3,750 - 0.9375) \left[1 - e^{-(0.5 \times 10^{-6}) \times 215} \right],$$

or

$$N_{fg} = 0.403,$$

which is consistent with the expected equipment $MTBF$ of about 533.5 hr under the screening environment. This $MTBF_{equip} = 533.5$ hr is obtained from the following:

$$N_{fg} = \lambda_{equip} \times T_s,$$

$$= \frac{T_s}{MTBF_{equip}},$$

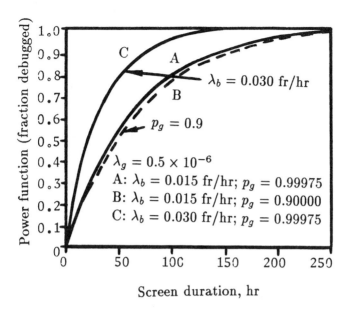

Fig. 6.3– The power function of a screen for Example 6–1.

or

$$MTBF_{\text{equip}} = \frac{T_s}{N_{f_g}},$$
$$= \frac{215}{0.403},$$

or

$$MTBF_{\text{equip}} = 533.499 \text{ hr or } 533.5 \text{ hr.}$$

Figure 6.3 depicts the power function for this example and shows its sensitivity to variations in the part quality level, p_g, and the defective parts failure rate, λ_b.

6.4   SYSTEM LEVEL SCREENING

The *ESS* of a system composed of a mixture of parts is different from that of a homogeneous-part population [3].

In a parts population, each failure of a part constitutes the end of life for that part. Thus, when evaluating the distribution of the times to failure of parts we are really dealing with the times to the first failure.

In a repairable system of mixed parts, a failure of a module or assembly does not terminate its life, because upon failure it is repaired (renewed) and continues to operate, sometimes to a second, or to a third failure or more. The *pdf's* of these times to the first, second, etc., failure, add up to form the renewal rate, which we denote by $h(t)$ to distinguish it from the instantaneous failure rate, $\lambda(t)$, of a part.

The $\lambda(t)$ of a part is a measure of its proneness to fail in the next short interval of time. This failure rate may be constant (CFR), increasing with time due to aging (IFR) or decreasing with time (DFR). However, for parts with a DFR it means that the parts population as a whole improves as weak parts are eliminated by some purification process such as burn-in. Therefore, we assign to it a theoretical failure rate, which is a proportionate mixture of the failure rates of the good and the weak parts, and exhibits a DFR property as early failing parts die out.

While a single part may not have a DFR, assemblies and systems do exhibit a DFR as they improve through repair or replacement of the early failures. In this case a DFR becomes a physical reality and the failure rate decreases initially. Later it usually becomes constant, until wearout reverses the trend and the system exhibits an IFR, in accordance with the "bathtub" curve concept.

As to the renewal rate concept, also referred to as the average long-term failure rate, $h(t)$, it is a measure of the frequency with which parts fail and are replaced with identical new ones in a sequence of failures and replacements. In the case of parts, the renewal rate is based strictly on the rate at which replacements occur because of failures. On the other hand, assemblies and systems are usually repaired when they fail and an actual "renewal" occurs. This is a different concept from the failure rate. The latter applies to the proneness of an operating item to fail and the former to the sequential repair or replacement of failed items.

From Eq. (6.23), which is the renewal rate of a part, we see that the renewal rate of a system of N identical parts may be written as

$$h_s(t) = N\, h(t) = N\, h_f + N\, (h_i - h_f)\, e^{-A\, t}, \tag{6.28}$$

and when the parts are different, we obtain

$$h_s(t) = \sum_{j=1}^{N} h_{f_j} + \sum_{j=1}^{N} (h_{i_j} - h_{f_j})\, e^{-A_j\, t}, \tag{6.29}$$

where the subscript j refers to the jth part in the system. Equations (6.28) and (6.29) follow from the theorem of the *"Superposition of Renewal Processes"* discussed by Cox [4].

Finally, by rewriting Eq. (6.28), we develop the system's ESS duration time, T_s, as follows:

$$h_s(T_s) = N\, h_f \left[1 + \frac{1}{N\, h_f} N\, (h_i - h_f)\, e^{-A\, T_s} \right],$$

or

$$h_s(T_s) = N \, h_f \left[1 + \frac{(h_i - h_f)}{h_f} \, e^{-A \, T_s} \right].$$ (6.30)

The second term in the bracket is the screen residue of the renewal rate, denoted by ζ, which is analogous to ε, the part failure rate developed in the previous section, then

$$\zeta = \frac{(h_i - h_f)}{h_f} \, e^{-A \, T_s},$$ (6.31)

and the renewal rate of Eq. (6.30) may be written as

$$h_s(T_s) = N \, h_f \, (1 + \zeta).$$ (6.32)

By taking the logarithm of Eq. (6.31) we can calculate the duration of *ESS* for a pre-selected ζ from

$$T_s = \frac{1}{A} \log_e \left[\frac{(h_i - h_f)}{\zeta \, h_f} \right],$$ (6.33)

which, interestingly, is independent of N.

For the usual case where a system consists of mixed parts, the screen duration cannot be derived in a closed form. Each part will have its own exponential decay factor and the renewal rate equation becomes

$$h_s(T_s) = \sum_{j=1}^{m} N_j \, h_{f_j} + \sum_{j=1}^{m} N_j \, (h_{i_j} - h_{f_j}) \, e^{-A_j \, T_s},$$ (6.34)

where

$m = $ total number of all part types in the system,

$N_j = $ number of parts of the *jth* part type in the system,

and

$A_j = $ exponential decay factor for the *jth* part type,

or

$$A_j = p_{bj} \, \lambda_{bj} + p_{gj} \, \lambda_{gj}.$$

Applying Eq. (6.31), Eq. (6.34) can thus be rewritten as

$$h_s(T_s) = \sum_{j=1}^{m} N_j \, h_{f_j} \, (1 + \zeta) = (1 + \zeta) \sum_{j=1}^{m} N_j \, h_{f_j},$$ (6.35)

where

$$\zeta = \zeta_j = \frac{h_{i_j} - h_{f_j}}{h_{f_j}} e^{-A_j T_s},$$

which is assumed to be identical for all j, and $\sum\limits_{j=1}^{m} N_j h_{f_j}$ is the final renewal rate of the system. Rearranging Eq. (6.35) yields

$$h_s(T_s) = \sum_{j=1}^{m} N_j h_{f_j} + \zeta \sum_{j=1}^{m} N_j h_{f_j},$$

or

$$\zeta = \frac{h_s(T_s) - \sum\limits_{j=1}^{m} N_j h_{f_j}}{\sum\limits_{j=1}^{m} N_j h_{f_j}}.$$

But from Eq. (6.34) we know that

$$h_s(T_s) - \sum_{j=1}^{m} N_j h_{f_j} = \sum_{j=1}^{m} N_j (h_{i_j} - h_{f_j}) e^{-A_j T_s}.$$

Therefore,

$$\zeta = \frac{\sum\limits_{j=1}^{m} N_j (h_{i_j} - h_{f_j}) e^{-A_j T_s}}{\sum\limits_{j=1}^{m} N_j h_{f_j}}. \tag{6.36}$$

The screen duration, T_s, can be derived for a selected value of ζ from Eq. (6.36) by iteration using a computer program. This program is presented in Appendix 6A, written in FORTRAN language, which is compatible with most personal computers.

In practice a value for ζ may be selected based upon the specified $MTBF$. From Eq. (6.34) the final renewal rate of the system, as $t \rightarrow \infty$, is

$$h_{fs} = h_s(\infty) = \sum_{j=1}^{m} N_j h_{f_j}. \tag{6.37}$$

Correspondingly, Eq. (6.34) can be rewritten as

$$h_s(T_s) = h_{fs} + \sum_{j=1}^{m} N_j (h_{i_j} - h_{f_j}) e^{-A_j T_s}, \tag{6.38}$$

which compares with Eq. (6.23) for the case of part level screening.

This is the reciprocal of the system's final $MTBF, M_{fs}$, in the ESS environment. Then,

$$M_{fs} = \frac{1}{h_{fs}}. \tag{6.39}$$

If we denote M_G to be the specified operational $MTBF$, then the renewal rate at the end of ESS should be

$$h_s(T_s) = \frac{K}{M_G}, \tag{6.40}$$

where K is the ratio of the final renewal rate under the ESS to that under operational stress levels; i.e.,

$$K = \frac{h_{fs}}{h'_{fs}}, \tag{6.41}$$

where h'_{fs} is the final renewal rate during operation.

Therefore, from Eqs. (6.35) and (6.37)

$$h_s(T_s) = (1 + \zeta) \, h_{fs}. \tag{6.42}$$

Substitution of Eqs. (6.39) and (6.40) into Eq. (6.42), and rearranging, yields the screen residue factor as

$$\zeta = \frac{h_s(T_s)}{h_{fs}} - 1 = K \left(\frac{M_{fs}}{M_G} \right) - 1, \tag{6.43}$$

or

$$\zeta = \frac{M'_{fs}}{M_G} - 1, \tag{6.44}$$

where

$$M'_{fs} = K \cdot M_{fs} \tag{6.45}$$

is the final achievable $MTBF$ in long-term operation.

The expected number of failures during an ESS duration of T_s, $H_s(T_s)$, can be obtained by integrating Eq. (6.38) with respect to T_s; i.e.,

$$H_s(T_s) = \int_0^{T_s} h_s(t) \, dt = \int_0^{T_s} \left[h_{fs} + \sum_{j=1}^{m} N_j \, (h_{i_j} - h_{f_j}) \, e^{-A_j \, t} \right] dt,$$

or

$$H_s(T_s) = h_{fs} \, T_s + \sum_{j=1}^{m} \frac{N_j}{A_j} \, (h_{i_j} - h_{f_j}) \, (1 - e^{-A_j \, T_s}), \tag{6.46}$$

which compares with Eq. (6.24) for the case of part level screening.

Table 6.1 provides a simplified procedure for applying the technique described in this section. It is necessary to estimate the defect proportions and the failure rates of the "good" and of the substandard parts under the *ESS* environment.

Presently there is no generally available source for failure rates of substandard parts, although *MIL-HDBK-217* provides the failure rates of "good" parts under various environments.

The following simplified example, based on the parts-count method, is presented for illustrative purposes only. It was derived from the analysis of the system-level test data of an inertial navigation system [3], which is subjected to an extensive *ESS* regimen.

EXAMPLE 6–2

An inertial navigation system was subjected to an extensive *ESS* regimen. Table 6.2 was obtained by performing Steps 1 through 4 of the procedure in Table 6.1, where the failure rates λ_b and λ_g are estimates in two environments for an "average" part and an "average" connection.

In practice, it is recommended that each of the two groups be broken down further, at least to the type of part and type of connection, or even to individual parts, if practical. The specified operational $MTBF$ goal is $M_G = 357$ hr. Determine the optimum ESS duration following the steps listed in Table 6.1.

SOLUTION TO EXAMPLE 6–2

Step 5 of the procedure yields the values tabulated in Table 6.3. Then, the final system renewal rate in the ESS environment, h_{fs}, is calculated using Eq. (6.37); i.e.,

$$h_{fs} = \sum_{j=1}^{2} N_j\, h_{f_j},$$
$$= N_{\text{part}}\, (h_f)_{\text{part}} + N_{\text{connection}}\, (h_f)_{\text{connection}},$$
$$= 3,750\, (1.100 \times 10^{-6}) + 22,000\, (2.250 \times 10^{-7}),$$

or

$$h_{fs} = 0.009075 \text{ fr/hr}.$$

Similarly, the final system renewal rate under the operational environment, h'_{fs}, can be calculated also using Eq. (6.37); i.e.,

$$h'_{fs} = \sum_{j=1}^{2} N_j\, h'_{f_j},$$
$$= N_{\text{part}}\, (h'_f)_{\text{part}} + N_{\text{connection}}\, (h'_f)_{\text{connection}},$$
$$= 3,750\, (3.000 \times 10^{-7}) + 22,000\, (5.500 \times 10^{-8}),$$

TABLE 6.1– Step-by-step procedure to determine the optimum *ESS* duration for a specified screen residue ζ [3]. "Reprinted with permission from the Proceedings of the IES, 1988."

Step	Description	Symbols	Equation
1	Group like elements (component parts electrical connections, etc.) into categories.	j	
2	Count the number of items in each category.	N_j	
3	Estimate the proportion of substandard items in each category.	p_{b_j}	
4	Estimate the failure rates of the standard and substandard items in each category for both the *ESS* and operational environments.	$\lambda_{g_j}, \lambda_{b_j}$ $\lambda'_{g_j}, \lambda'_{b_j}$	
5	Calculate the initial and final renewal rates for each category and their exponential decay factors.	h_{i_j}, h_{f_j} A_j	(6.21), (6.22) (6.19)
6	Calculate the system's final renewal rate for both environments and their ratio.	h_{f_Λ}, K	(6.37)
7	Identify the specified *MTBF*.	M_G	
8	Calculate the system's final *MTBF*.	M_{fs}	(6.39)
9	Determine the required value of ζ to meet the specified *MTBF*.	ζ	(6.44)
10	Iterate to determine the screen duration for the selected value of ζ.	T_s	(6.33)
11	If this screen duration is not practical, repeat calculation with a relaxed ζ and determine the achieved *MTBF*.	M_G	(6.42), (6.40)

TABLE 6.2– Immediate results from performing Steps 1 through 4 of the procedure in Table 6.1, for Example 6–2.

Parameter	Part	Connection
N	3,750	22,000
p_b	1.75×10^{-4}	2.30×10^{-5}
λ_b	1.50×10^{-2}	4.50×10^{-2}
λ_b'	4.00×10^{-3}	1.00×10^{-2}
λ_g	1.10×10^{-6}	2.25×10^{-7}
λ_g'	3.00×10^{-7}	5.50×10^{-8}

TABLE 6.3– Immediate results from performing Step 5 of the procedure in Table 6.1, for Example 6–2.

Environment	Parameter	Parts	Connections
	A	1.500×10^{-2}	4.500×10^{-2}
ESS	h_i	3.725×10^{-6}	1.260×10^{-6}
	h_f	1.100×10^{-6}	2.250×10^{-7}
	A'	4.000×10^{-3}	1.000×10^{-2}
Operational	h_i'	1.000×10^{-6}	2.850×10^{-7}
	h_f'	3.000×10^{-7}	5.500×10^{-8}

or

$$h_{f_s}' = 0.0023350 \text{ fr/hr.}$$

Step 6 is completed using Eq. (6.41), or taking the ratio of the above two values, to obtain the K factor as

$$K = \frac{h_{f_s}}{h_{f_s}'} = \frac{0.009075}{0.002335} = 3.8865.$$

Since the specified $MTBF$ under operational environment is

$$M_G' = 357 \text{ hr,}$$

then the $MTBF$ in the ESS environment will be

$$M_G = \frac{M_G'}{K} = \frac{357}{3.8865} = 91.9 \text{ hr.}$$

TABLE 6.4– Optimum *ESS* durations for three ζ values, for Example 6–2.

ζ	System T_s, hr	$MTBF$ at $t = T_s$	
		ESS	Operational
0.30	94	85	330
0.20	117	92	357
0.15	135	96	372

From Eq. (6.39)

$$M_{fs} = \frac{1}{h_{fs}} = \frac{1}{0.009075} = 110.2 \text{ hr,}$$

which is the maximum *MTBF* achievable under *ESS*. This completes Steps 7 and 8.

To perform Step 9 and select a value of ζ, we proceed according to Eq. (6.44) as follows:

$$\zeta = 3.8865 \left(\frac{110.2}{357} \right) - 1,$$

or

$$\zeta \cong 0.20.$$

The optimum screen duration, T_s^*, for the system, is computed from Eq. (6.36) by iteration, according to Step 10, using the computer program presented in Appendix 6A and setting ζ to the selected value 0.20. The optimum screen duration for other ζ values can be computed similarly. Appendix 6B is the sample output of this computer program where the optimum *ESS* durations of $T_s^* = 94$ hr, 117 hr and 135 hr are obtained for three specified screen residue values of $\zeta = 0.30$, 0.20 and 0.15, respectively.

These results for T_s^* together with the corresponding achieved *MTBF's* are listed in Table 6.4. A sample calculation of the achieved *MTBF's* is given next in the explanation of Fig. 6.5.

The system's renewal rate equation, Eq. (6.35), can be rewritten by substituting Eq. (6.37) into it; i.e.,

$$h_s(T_s) = (1 + \zeta) \, h_{fs}.$$

This equation is plotted graphically in Fig. 6.4 which shows the three values of ζ and the corresponding *ESS* durations. For example, when $\zeta = 0.15$

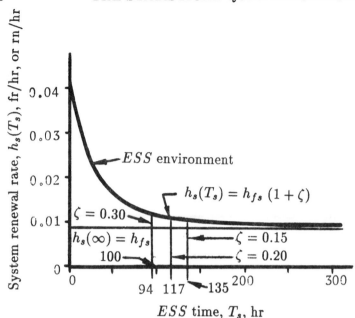

Fig. 6.4– The system's renewal rate as a function of screen duration for Example 6–2.

the required ESS duration is 135 hr according to Table 6.4, and the system renewal rate, $h_s(T_s)$, is

$$h_s(T_s) = h_s(135) = (1 + 0.15)\ (0.009075) = 0.010436 \text{ fr/hr}.$$

Similarly, when $\zeta = 0.20$ the required ESS duration is 117 hr according to Table 6.4, and the system renewal rate, $h_s(T_s)$, is

$$h_s(T_s) = h_s(135) = (1 + 0.20)\ (0.009075) = 0.010890 \text{ fr/hr}.$$

Similarly, when $\zeta = 0.30$ the required ESS duration is 94 hr according to Table 6.4 and the system renewal rate, $h_s(T_s)$, is

$$h_s(T_s) = h_s(135) = (1 + 0.30)\ (0.009075) = 0.011798 \text{ fr/hr}.$$

The system's $MTBF$, under both ESS and operational environments, can be obtained by taking the reciprocal of the renewal rate under each environment; i.e., the $MTBF$ under ESS environment is

$$MTBF(T_s) = \frac{1}{h_s(T_s)},$$

and the $MTBF$ under operational environment is

$$MTBF'(T_s) = \frac{1}{h'_s(T_s)} = K\ [MTBF(T_s)].$$

These two equations are plotted in Fig. 6.5, respectively. For example, when the *ESS* duration is

$$T_s = 117 \text{ hr}$$

for a specified screen residue factor of $\zeta = 0.20$, which corresponds to a time period under operational environment of

$$T_{\text{operation}} = K \ (T_s) = 3.8865 \ (117) = 454.72 \text{ hr or } 455 \text{ hr,}$$

the system renewal rate under *ESS* environment is 0.010890 fr/hr according to the sample caculation for Fig. 6.4. Then, the system's *MTBF* under the *ESS* environment is

$$MTBF(T_s = 117 \text{ hr}) = \frac{1}{h_s'(T_s = 117 \text{ hr})} = \frac{1}{0.010890},$$

or

$$M_G = MTBF(T_s = 117 \text{ hr}) = 91.83 \text{ hr or } 92 \text{ hr,}$$

which corresponds to an *MTBF* value under the operational environment, $MTBF'(T_s)$, of

$$MTBF'(T_s = 117 \text{ hr}) = K \ [MTBF(T_s = 117 \text{ hr})] = 3.8865 \ (91.83),$$

or

$$M_G' = MTBF'(T_s = 117 \text{ hr}) = 356.89 \text{ hr or } 357 \text{ hr.}$$

Consider a limiting case when the screening duration approaches infinity, then the system renewal rate under the *ESS* environment is

$$h_s(\infty) = h_{fs} = 0.009075 \text{ fr/hr.}$$

The corresponding system *MTBF* under the *ESS* environment is

$$MTBF(\infty) = M_{fs} = \frac{1}{h_{fs}} = \frac{1}{0.009075} = 110.19 \text{ hr or } 110 \text{ hr,}$$

which corresponds to a system *MTBF* under the operational environment of

$$MTBF'(\infty) = M_{fs}' = K \ (M_{fs}) = 3.8865 \ (110.19) = 428.25 \text{ hr,}$$

or

$$MTBF'(\infty) = M_{fs}' = K \ (M_{fs}) \cong 428 \text{ hr.}$$

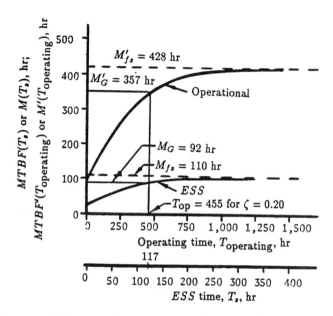

Fig. 6.5– The system's $MTBF$ as a function of screen duration for Example 6–2.

6.5 TWO-LEVEL SCREENING: THE LIFE-CYCLE COST MODEL [5]

In this section, ESS for both part and system levels will be studied together. A general modeling concept for a two-level screening program is discussed with no particular reference to any product. The two levels of screening are for the part level and the system level. These are two different types of screens. At the part level all the parts are screened and only those passing the screen are accepted. The screened part population thus contains a higher fraction of good parts than the original population. The system is then assembled with parts from this population. Upon failure, the parts are replaced by new ones from the screened population. System level screening is performed to detect defects induced during the manufacturing process, like connections as discussed in the previous section. Unlike part level screening, all connections have to be screened at once. Upon failure, only the failed connections are fixed and screening is continued for the rest of the screen duration. Optimal screen durations are obtained by minimizing the life-cycle costs. Again, the bimodal exponential distributions are used to represent both part life and connection life distributions, and then the life-cycle cost model is developed.

6.5.1 PART LEVEL SCREENING

As stated earlier, the objective of part level screens is to screen out the sub-standard parts by subjecting them to stresses that induce failures in these parts. The concept of bimodal distributions is used to model the failure distribution of these parts. The main population is assumed to consist of a fraction of good parts and a fraction of substandard parts. The substandard parts fail at a much higher failure rate than the good ones. Parts are screened for a specific duration and only those passing without failure are accepted. The system is then assembled from the screened population. As the screen duration increases, the probability of substandard parts passing the screen decreases, thereby reducing the fraction of substandard parts in the screened population. The reduction in the fraction of substandard parts depends on the screen duration. The longer the screen duration the larger the reduction. However, as the screen duration increases beyond a certain point, the savings in field repair costs may not be commensurate with the increasing screening costs. The optimal screen durations are obtained by minimizing the life-cycle costs.

Let the screen duration be T_s. The probability of a part passing the screen is

$$R(T_s) = p_b \ R_b(T_s) + p_g \ R_g(T_s). \tag{6.47}$$

The probability that the part passing the screen is substandard is

$$p_b' = \frac{p_b \ R_b(T_s)}{p_b \ R_b(T_s) + p_g \ R_g(T_s)}. \tag{6.48}$$

The probability that the part passing the screen is standard is

$$p_g' = 1 - p_b' = \frac{p_g \ R_g(T_s)}{p_b \ R_b(T_s) + p_g \ R_g(T_s)}. \tag{6.49}$$

Therefore, the field reliability, during mission time t, of a part screened for T_s hr is

$$R'(T_s, t) = p_b' \frac{R_b'(t + T_s')}{R_b(T_s)} + p_g' \frac{R_g'(t + T_s')}{R_g(T_s)}, \tag{6.50}$$

where

$R_b' = $ field reliability function of bad parts which escaped the screen,

$R_g' = $ field reliability function of good parts which passed the screen,

and

$T_s' = $ equivalent screen time under use stress level such that $R(T_s) = R'(T_s')$.

Substituting Eqs. (6.48) and (6.49) into Eq. (6.50) yields

$$R'(T_s, t) = \frac{p_b\, R_b'(t + T_s') + p_g\, R_g'(t + T_s')}{p_b\, R_b(T_s) + p_g\, R_g(T_s)}.$$ (6.51)

The probability density function of the screened part in the field is given by

$$f'(T_s, t) = -\frac{d[R'(T_s, t)]}{dt},$$

or

$$f'(T_s, t) = \frac{p_b\, f_b'(t + T_s') + p_g\, f_g'(t + T_s')}{p_b\, R_b(T_s) + p_g\, R_g(T_s)},$$ (6.52)

where

R' = reliability function of screened parts under use stress level,

f' = pdf of screened parts under use stress level,

f_b' = pdf of screened bad parts under use stress level,

and

f_g' = pdf of screened good parts under use stress level.

If the screen stress level is the same as the use stress level, then the parts' life distribution parameters during screen and field operation are identical. Therefore, Eqs. (6.51) and (6.52) become

$$R'(T_s, t) = R(T_s, t) = \frac{p_b\, R_b(t + T_s) + p_g\, R_g(t + T_s)}{p_b\, R_b(T_s) + p_g\, R_g(T_s)},$$ (6.53)

and

$$f'(T_s, t) = f(T_s, t) = \frac{p_b\, f_b(t + T_s) + p_g\, f_g(t + T_s)}{p_b\, R_b(T_s) + p_g\, R_g(T_s)}.$$ (6.54)

If a mixed-exponential distribution with parameters of λ_b and λ_g is used, then Eqs. (6.47), (6.48), (6.49), (6.53) and (6.54) become

$$R(T_s) = p_b\, e^{-\lambda_b\, T_s} + p_g\, e^{-\lambda_g\, T_s},$$ (6.55)

$$p_b' = \frac{p_b\, e^{-\lambda_b\, T_s}}{p_b\, e^{-\lambda_b\, T_s} + p_g\, e^{-\lambda_g\, T_s}},$$ (6.56)

$$p_g' = \frac{p_g\, e^{-\lambda_g\, T_s}}{p_b\, e^{-\lambda_b\, T_s} + p_g\, e^{-\lambda_g\, T_s}},$$ (6.57)

$$R'(T_s, t) = \frac{p_b \, e^{-\lambda_b \, (t+T_s)} + p_g \, e^{-\lambda_g \, (t+T_s)}}{p_b \, e^{-\lambda_b \, T_s} + p_g \, e^{-\lambda_g \, T_s}}, \tag{6.58}$$

and

$$f'(T_s, t) = \frac{p_b \, \lambda_b \, e^{-\lambda_b \, (t+T_s)} + p_g \, \lambda_g \, e^{-\lambda_g \, (t+T_s)}}{p_b \, e^{-\lambda_b \, T_s} + p_g \, e^{-\lambda_g \, T_s}}, \tag{6.59}$$

respectively.

Upon failure during system level screening and field use, parts are replaced by new ones from the screened population. The number of renewals of a screened part during field operation can then be obtained using renewal theory as given by Eq. (6.24), or

$$H_p(T_s, t) = \int_0^t h'(\tau) \, d\tau = h'_f \, t + \frac{1}{A'}(h'_i - h'_f)(1 - e^{-A' \, t}), \tag{6.60}$$

where

$$
\begin{aligned}
H_p(T_s, t) &= \text{number of renewals of a part in time } t \text{ given that} \\
&\quad \text{it has already been screened for } T_s \text{ hr,} \\
h'(t) &= \text{renewal rate function for a screened part} \\
&\quad \text{as given by Eq. (6.23),} \\
&= h'_f + (h'_i - h'_f) \, e^{-A' \, t}, \\
h'_i &= \text{initial value of renewal rate } h'(t), \, h'(0), \text{ as given} \\
&\quad \text{by Eq. (6.21),} \\
&= p'_g \, \lambda_g + p'_b \, \lambda_b, \\
h'_f &= \text{limiting final value of renewal rate } h'(t), \, h'(\infty), \\
&\quad \text{as given by Eq. (6.22),} \\
&= \frac{\lambda_b \, \lambda_g}{A'},
\end{aligned}
$$

and

$$
\begin{aligned}
A' &= \text{a constant as given by Eq. (6.19) for the} \\
&\quad \text{screened parts,} \\
&= p'_b \, \lambda_g + p'_g \, \lambda_b.
\end{aligned}
$$

Equation (6.60) will be used for the number of field part failures in the life-cycle cost model.

As the screen duration tends to infinity, the renewal density tends to the failure rate of good parts and the number of renewals becomes equal to the expected number of renewals for an exponential process; i.e., from Eq. (6.48)

$$\lim_{T_s \to \infty} p_b' = \lim_{T_s \to \infty} \frac{p_b\, e^{-\lambda_b\, T_s}}{p_b\, e^{-\lambda_b\, T_s} + p_g\, e^{-\lambda_g\, T_s}},$$

$$= \lim_{T_s \to \infty} \frac{p_b\, e^{-(\lambda_b - \lambda_g)\, T_s}}{p_b\, e^{-(\lambda_b - \lambda_g)\, T_s} + p_g},$$

$$= \frac{0}{0 + p_g},$$

or

$$\lim_{T_s \to \infty} p_b' = 0.$$

Correspondingly,

$$\lim_{T_s \to \infty} p_g' = \lim_{T_s \to \infty} (1 - p_b') = 1,$$

$$\lim_{T_s \to \infty} h_i' = \lim_{T_s \to \infty} (p_g'\, \lambda_g + p_b'\, \lambda_b),$$

$$= 1 \times \lambda_g + 0 \times \lambda_b,$$

or

$$\lim_{T_s \to \infty} h_i' = \lambda_g.$$

$$\lim_{T_s \to \infty} A' = \lim_{T_s \to \infty} (p_b'\, \lambda_g + p_g'\, \lambda_b),$$

$$= 0 \times \lambda_g + 1 \times \lambda_b,$$

or

$$\lim_{T_s \to \infty} A' = \lambda_b,$$

and

$$\lim_{T_s \to \infty} h_f' = \lim_{T_s \to \infty} \left(\frac{\lambda_b\, \lambda_g}{A'} \right),$$

$$= \frac{\lambda_b\, \lambda_g}{\lambda_b},$$

or

$$\lim_{T_s \to \infty} h_f' = \lambda_g.$$

Therefore,

$$\lim_{T_s \to \infty} H_p(T_s, t) = \lim_{T_s \to \infty} [h_f'\, t + \frac{1}{A'}(h_i' - h_f')(1 - e^{-A'\, t})],$$

$$= \lambda_g\, t + \frac{1}{\lambda_b}(\lambda_g - \lambda_g)(1 - e^{-\lambda_b\, t}),$$

or

$$\lim_{T_s \to \infty} H_p(T_s, t) = \lambda_g\ t. \tag{6.61}$$

In other words, as the screen duration increases, substandard parts are eliminated and the population consists of good parts only. The field savings asymptotically approach that of an exponential renewal process with the failure rate of the good parts. However, after a certain duration, the decrease in field expenses may not be commensurate with the increasing screening costs. The optimal duration is obtained by considering the trade-offs between savings and screening costs.

Often screening is done at higher stress levels (accelerated stress levels). The failure rates of substandard and good parts will be different during and after screening. In such a case, the fraction of substandard parts after screening is computed with the accelerated failure rates, whereas the field failure rates are used for the computation of the number of failures in the field. In other words, Eq. (6.56) is computed with accelerated failure rates.

6.5.2   SYSTEM LEVEL SCREENING

Defects induced in the assembly process are precipitated during system level screening. The connections are possible sources of these defects. Unlike part level screening, these connections cannot be screened separately. The assembled system is screened for a specified duration and only those connections that fail during screening are fixed. In other words, the system is minimally repaired. The connections in the system are assumed to form a reliability-wise series type configuration. The number of failures during the screening process is given by the integral of the failure rate function of the system due to connections only. In a series system, the system's failure rate is the sum of the failure rates of the individual connections. It is reasonable to assume that all connections have the same times-to-failure distribution, consequently the same failure rate. The total failure rate of the connections in the system is given by

$$\lambda_{cs}(t) = n_c\ \lambda_c(t), \tag{6.62}$$

where

n_c = total number of connections in the system.

The number of connection failures for one system in time t is then given by

$$H_{cs}(t) = \int_0^t \lambda_{cs}(\tau)\ d\tau,$$

$$= \int_0^t n_c\ \lambda_c(\tau)\ d\tau,$$

or

$$H_{cs}(t) = n_c \int_0^t \lambda_c(\tau)\, d\tau. \tag{6.63}$$

Note that the general expression for the reliability of one connection is

$$R_c(t) = e^{-\int_0^t \lambda_c(\tau)\, d\tau},$$

or

$$\int_0^t \lambda_c(\tau)\, d\tau = -\log_e R_c(t). \tag{6.64}$$

Substituting Eq. (6.64) into Eq. (6.63) yields

$$H_{cs}(t) = -n_c \log_e R_c(t). \tag{6.65}$$

With sufficient screen duration all early life failures can be moved into the shop floor thereby reducing field repair costs. If the connections also come from a mixed population of substandard and good ones, Eq. (6.65) becomes

$$H_{cs}(t) = -n_c \log_e[p_{cb}\, R_{cb}(t) + p_{cg}\, R_{cg}(t)], \tag{6.66}$$

where

$$p_{cb} = \text{proportion of bad connections in the system,}$$
$$p_{cg} = \text{proportion of good connections in the system,}$$
$$R_{cb}(t) = \text{reliability of bad connections in } t,$$

and

$$R_{cg}(t) = \text{reliability of good connections in } t.$$

If a mixed-exponential distribution with parameters of λ_{cb} and λ_{cg}, denoting the failure rates of bad and good connections, respectively, is used, then Eq. (6.66) becomes

$$H_{cs}(t) = -n_c \log_e \left(p_{cb}\, e^{-\lambda_{cb}\, t} + p_{cg}\, e^{-\lambda_{cg}\, t} \right). \tag{6.67}$$

6.5.3   COST MODEL FORMULATION

The life-cycle cost model is developed based on the following assumptions:

1. Parts and connections form a series configuration.

2. The parts are replaced upon failure from the screened population.

3. The connections are minimally repaired upon failure.

Parts may fail during system level screening. The replacement costs of parts during screening are much lower than those of field replacements. The same is true for the repair costs of connections. As the duration of the screens increases, field expenses decrease but the screening costs increase. Optimal screen durations are those which minimize the life-cycle cost. The expected life-cycle cost is

$$C(T_{sp}, T_{ss}) = n_p\{C_{SP} + C_{TP}T_{sp} + H_p(T_{sp}, T_{ss})C_{HP}$$
$$+ [H_p(T_{sp}, T_w + T_{ss}) - H_p(T_{sp}, T_{ss})]C_{FP}\}$$
$$+ \{C_{SS} + C_{TS}T_{ss} + H_{cs}(T_{ss})C_{HS}$$
$$+ [H_{cs}(T_w + T_{ss}) - H_{cs}(T_{ss})]C_{FS}\},$$

where

n_p = number of parts in the system,

C_{SP} = fixed cost of part level screening per part,

C_{TP} = cost of part level screening per part per hour,

C_{HP} = cost of part replacement in house,

C_{FP} = cost of part replacement in field operation,

C_{SS} = fixed cost of system level screening,

C_{TS} = cost of system level screening per hour,

C_{HS} = cost of in-house repair for the system due to connections alone,

C_{FS} = cost of field repair for the system due to connections alone,

T_{sp}, T_{ss} = screen durations for part and system level screens, respectively,

$H_p(T_{sp}, t)$ = number of renewals of a screened part in time t as given by Eq. (6.60),

$H_{cs}(t)$ = number of renewals of all connections in the system in time t as given by Eq. (6.67),

and

T_w = warranty period.

The first term in Eq. (6.68) corresponds to part level screening and the second to the screening at the system level. The first term contains the fixed screening cost per part, C_{SP}, cost of screening per part, $C_{TP}T_{sp}$, cost of part replacements during system level screening and warranty period, $H_p(T_{sp}, T_{ss})C_{HP} + [H_p(T_{sp}, T_w + T_{ss}) - H_p(T_{sp}, T_{ss})]C_{FP}$. The second term contains fixed system level screening cost, C_{SS}, cost of screening, $C_{TS}T_{ss}$, cost of repairs during screening, $H_{cs}(T_{ss})C_{HS}$, and field repair costs, $[H_{cs}(T_w + T_{ss}) - H_{cs}(T_{ss})]C_{FS}$. The field repair costs are computed for the product warranty period.

The optimum screening times for minimum life-cycle cost, T_{sp}^* and T_{ss}^*, can be obtained by taking the partial derivatives of Eq. (6.68) with respect to T_{sp} and T_{ss}, respectively, setting them equal to zero, and solving the two equations for T_{sp} and T_{ss}; i.e.,

$$
\begin{cases}
\frac{\partial C(T_{sp},T_{ss})}{\partial T_{sp}} = 0, \\
\frac{\partial C(T_{sp},T_{ss})}{\partial T_{ss}} = 0,
\end{cases}
$$

or

$$
\begin{cases}
n_p \left\{ C_{TP} + \frac{\partial H_p(T_{sp},T_{ss})}{\partial T_{sp}} C_{HP} \right. \\
\left. + \left[\frac{\partial H_p(T_{sp},T_w+T_{ss})}{\partial T_{sp}} - \frac{\partial H_p(T_{sp},T_{ss})}{\partial T_{sp}} \right] C_{FP} \right\} = 0, \\
n_p \left\{ \frac{\partial H_p(T_{sp},T_{ss})}{\partial T_{ss}} C_{HP} + \left[\frac{\partial H_p(T_{sp},T_w+T_{ss})}{\partial T_{ss}} - \frac{\partial H_p(T_{sp},T_{ss})}{\partial T_{ss}} \right] C_{FP} \right\} \\
+ \left\{ C_{TS} + \frac{\partial H_{cs}(T_{ss})}{\partial T_{ss}} C_{HS} + \left[\frac{\partial H_{cs}(T_w+T_{ss})}{\partial T_{ss}} - \frac{\partial H_{cs}(T_{ss})}{\partial T_{ss}} \right] C_{FS} \right\} = 0.
\end{cases}
$$

$$(6.68)$$

In this equation the associated partial derivatives are given as follows:

$$
\begin{aligned}
\frac{\partial H_p(T_{sp},T_{ss})}{\partial T_{sp}} &= \frac{\partial \left[h'_f T_{ss} + \frac{1}{A'}(h'_i - h'_f)(1 - e^{-A'\,T_{ss}}) \right]}{\partial T_{sp}}, \\
&= \left[\frac{1}{(A')^2} \right] (h'_i - h'_f) \frac{\partial A'}{\partial T_{sp}} (1 - e^{-A'\,T_{ss}}) \\
&\quad + \frac{1}{A'}(h'_i - h'_f)(-e^{-A'\,T_{ss}})(-T_{ss} \frac{\partial A'}{\partial T_{sp}}), \\
&= \left[\frac{h'_i - h'_f}{(A')^2} \right] (h'_i - h'_f) \frac{\partial A'}{\partial T_{sp}} \\
&\quad \cdot \left[-(1 - e^{-A'\,T_{ss}}) + A'\,T_{ss}\,e^{-A'\,T_{ss}} \right],
\end{aligned}
$$

or

$$
\frac{\partial H_p(T_{sp},T_{ss})}{\partial T_{sp}} = \left[\frac{h'_i - h'_f}{(A')^2} \right] \frac{\partial A'}{\partial T_{sp}} \left[(A'\,T_{ss} + 1)\,e^{-A'\,T_{ss}} - 1 \right].
$$

$$(6.69)$$

$$\frac{\partial A'}{\partial T_{sp}} = \frac{\partial(p_b' \, \lambda_g + p_g' \, \lambda_b)}{\partial T_{sp}},$$

$$= \lambda_g \, \frac{\partial p_b'}{\partial T_{sp}} + \lambda_b \, \frac{\partial p_g'}{\partial T_{sp}},$$

or

$$\frac{\partial A'}{\partial T_{sp}} = \lambda_g \, \frac{\partial}{\partial T_{sp}} \left[\frac{p_b \, e^{-\lambda_b \, T_{Ap}}}{p_b \, e^{-\lambda_b \, T_{Ap}} + p_g \, e^{-\lambda_g \, T_{Ap}}} \right]$$

$$+ \lambda_b \, \frac{\partial}{\partial T_{sp}} \left[\frac{p_g \, e^{-\lambda_g \, T_{Ap}}}{p_b \, e^{-\lambda_b \, T_{Ap}} + p_g \, e^{-\lambda_g \, T_{Ap}}} \right].$$

Differentiating and rearranging yields

$$\frac{\partial A'}{\partial T_{sp}} = \frac{p_b \, p_g \, (\lambda_b - \lambda_g)^2 \, e^{-(\lambda_b - \lambda_g) \, T_{Ap}}}{(p_b \, e^{-\lambda_b \, T_{Ap}} + p_g \, e^{-\lambda_g \, T_{Ap}})^2}. \tag{6.70}$$

Then,

$$\frac{\partial H_p(T_{sp}, T_{ss})}{\partial T_{sp}} = \left[\frac{h_i' - h_f'}{(A')^2} \right] \frac{p_b \, p_g \, (\lambda_b - \lambda_g)^2 \, e^{-(\lambda_b - \lambda_g) \, T_{Ap}}}{(p_b \, e^{-\lambda_b \, T_{Ap}} + p_g \, e^{-\lambda_g \, T_{Ap}})^2}$$

$$\cdot \left[(A' \, T_{ss} + 1) \, e^{-A' \, T_{AA}} - 1 \right]. \tag{6.71}$$

Similarly,

$$\frac{\partial H_p(T_{sp}, T_w + T_{ss})}{\partial T_{sp}} = \left[\frac{h_i' - h_f'}{(A')^2} \right] \frac{p_b \, p_g \, (\lambda_b - \lambda_g)^2 \, e^{-(\lambda_b - \lambda_g) \, T_{Ap}}}{(p_b \, e^{-\lambda_b \, T_{Ap}} + p_g \, e^{-\lambda_g \, T_{Ap}})^2}$$

$$\cdot \left[(A' \, T_w + A' \, T_{ss} + 1) \, e^{-A' \, (T_w + T_{AA})} - 1 \right]. \tag{6.72}$$

$$\frac{\partial H_p(T_{sp}, T_{ss})}{\partial T_{ss}} = \frac{\partial \left[h_f' \, T_{ss} + \frac{1}{A'} \, (h_i' - h_f') \, (1 - e^{-A' \, T_{AA}}) \right]}{\partial T_{ss}},$$

$$= h_f' + \frac{1}{A'} \, (h_i' - h_f') \, A' \, e^{-A' \, T_{AA}},$$

or

$$\frac{\partial H_p(T_{sp}, T_{ss})}{\partial T_{ss}} = h_f' + (h_i' - h_f') \, e^{-A' \, T_{AA}}. \tag{6.73}$$

Now

$$\frac{\partial H_p(T_{sp}, T_w + T_{ss})}{\partial T_{ss}}$$

$$= \frac{\partial \left[h'_f (T_w + T_{ss}) + \frac{1}{A'} (h'_i - h'_f)(1 - e^{-A' (T_w + T_{AA})}) \right]}{\partial T_{ss}},$$

$$= h'_f + \frac{1}{A'} (h'_i - h'_f) A' e^{-A' (T_w + T_{AA})},$$

or

$$\frac{\partial H_p(T_{sp}, T_w + T_{ss})}{\partial T_{ss}} = h'_f + (h'_i - h'_f) e^{-A' (T_w + T_{AA})}. \tag{6.74}$$

$$\frac{\partial H_{cs}(T_{ss})}{\partial T_{ss}} = \lambda_{cs}(T_{ss}) = n_c \, \lambda_c(T_{ss}),$$

or

$$\frac{\partial H_{cs}(T_{ss})}{\partial T_{ss}} = n_c \left[\frac{p_{cb} \, \lambda_{cb} \, e^{-\lambda_{cb} \, T_{AA}} + p_{cg} \, \lambda_{cg} \, e^{-\lambda_{cg} \, T_{AA}}}{p_{cb} \, e^{-\lambda_{cb} \, T_{AA}} + p_{cg} \, e^{-\lambda_{cg} \, T_{AA}}} \right]. \tag{6.75}$$

Similarly

$$\frac{\partial H_{cs}(T_w + T_{ss})}{\partial T_{ss}} = n_c \left[\frac{p_{cb} \, \lambda_{cb} \, e^{-\lambda_{cb} \, (T_w + T_{AA})} + p_{cg} \, \lambda_{cg} \, e^{-\lambda_{cg} \, (T_w + T_{AA})}}{p_{cb} \, e^{-\lambda_{cb} \, (T_w + T_{AA})} + p_{cg} \, e^{-\lambda_{cg} \, (T_w + T_{AA})}} \right].$$

$$\tag{6.76}$$

It may be observed that T_{sp}^* and T_{ss}^* cannot be obtained analytically by simply substituting the preceding derivatives into Eq. (6.68) and solving for T_{sp} and T_{ss}. However, the optimum screening times, T_{sp}^* and T_{ss}^*, can be obtained by solving the following two-variate, unconstrained optimization problem:

$$Min \; C(T_{sp}, T_{ss}). \tag{6.77}$$

The quasi-Newton method [5] can be applied to deal with this minimization problem. Note that if zero or negative solutions are obtained, then no screening will be necessary economically.

During ESS, stresses applied at different screens cater to specific effect types. In such a case it may be reasonable to assume that the parts do not age during the system level screen. Then, the optimization problem reduces to that of two separate problems; i.e., one for the part level screen, and one for the system level screen.

EXAMPLE 6-3

A 1,000 part–1,000 connection system is subjected to a two-level *ESS*. Assume that the corresponding costs for the parts and connections are the same. The mixed-exponential distribution parameters under operational stress level are the following:

$$p_b = 0.01; \ p_g = 0.99; \ \lambda_e = 10^{-3} \text{ fr/hr}; \ \lambda_g = 10^{-7} \text{ fr/hr};$$
$$p_{ce} = 0.01; \ p_{cg} = 0.99; \ \lambda_{ce} = 10^{-3} \text{ fr/hr}; \ \lambda_{cg} = 10^{-7} \text{ fr/hr}.$$

The associated costs are the following:

$$C_{SP} = \$0.200; \ C_{TP} = \$0.010; \ C_{HP} = \$500; \ C_{FP} = \$5,000;$$
$$C_{SS} = \$200.0; \ C_{TS} = \$10.00; \ C_{HS} = \$500; \ C_{FS} = \$5,000.$$

The warranty period is $T_w = 20,000$ hr. Do the following:

1. Find the optimum screening durations and the associated life-cycle costs for the parts and the system, T^*_{sp}, T^*_{ss}, and $C(T^*_{sp}, T^*_{ss})$, respectively, for the following cases:

   Case 1: The stress levels at part level screening are the same as operational stresses.

   Case 2: The stress levels at part level screening are twice the operational stresses.

   Case 3: The stress levels at part level screening are three times the operational stresses.

   Compare and discuss the results obtained for these three cases. Note that the failure rates under higher stress levels can be determined using the Inverse Power Law assuming that the power, n, is equal to one; i.e.,

   $$\frac{\lambda'}{\lambda} = \left(\frac{S'}{S}\right)^n = \frac{S'}{S},$$

   or

   $$\lambda' = \left(\frac{S'}{S}\right)\lambda,$$

   where

   S = operational stress level,

   S' = accelerated stress level,

   λ = failure rate at operational stress level,

   λ' = failure rate at accelerated stress level,

   n = product specific constant,

   and in this case

   $n = 1.$

2. Assume that system level screening stimulates only the connection failures and does not age the parts. Then both the screens have to be designed independently of each other. This is equivalent to solving two minimization problems, one for the parts alone and the other for the connections alone. Redo the problem for the preceding three cases using these considerations.

SOLUTIONS TO EXAMPLE 6–3

1. The cost function given by Eq. (6.68) is minimized using the quasi-Newton method [5].

Case 1: Since the stress levels at part level screening are the same as the operational stresses, the failure rates of bad and good parts during the screen are

$$\lambda_b' = \lambda_b = 10^{-3} \ \text{fr/hr},$$

and

$$\lambda_g' = \lambda_g = 10^{-7} \ \text{fr/hr}.$$

The optimal solutions for the data given are

$$T_{sp}^* = 50.498 \ \text{hr or 51 hr},$$
$$T_{ss}^* = 2,177.178 \ \text{hr or 2,177 hr},$$

and

$$C(T_{sp}^*, T_{ss}^*) = \$62,969.06 \ \text{or} \ \$62,969.$$

Case 2: Since the stress levels at part level screening are twice the operational stresses, then according to the Inverse Power Law the failure rates of bad and good parts during the screen are twice those under field operating conditions; i.e.,

$$\lambda_b' = 2 \times \lambda_b = 2 \times 10^{-3} \ \text{fr/hr},$$

and

$$\lambda_g' = 2 \times \lambda_g = 2 \times 10^{-7} \ \text{fr/hr}.$$

The optimal solutions for the data given, using the same procedure as before, are

$$T_{sp}^* = 456.839 \ \text{hr or 457 hr},$$
$$T_{ss}^* = 1,841.235 \ \text{hr or 1,841 hr},$$

and

$$C(T_{sp}^*, T_{ss}^*) = \$60,774.61 \ \text{or} \ \$60,775.$$

Case 3: Since the stress levels at part level screening are three times the operational stresses, then according to the Inverse Power Law the failure rates of bad and good parts during the screen are three times those under the field operating conditions; i.e.,

$$\lambda_b' = 3 \times \lambda_b = 3 \times 10^{-3} \ \text{fr/hr},$$

and

$$\lambda_g' = 3 \times \lambda_g = 3 \times 10^{-7} \ \text{fr/hr}.$$

The optimal solutions for the data given previously are

$$T_{sp}^* = 461.411 \ \text{hr or } 461 \ \text{hr},$$
$$T_{ss}^* = 1,726.695 \ \text{hr or } 1,727 \ \text{hr},$$

and

$$C(T_{sp}^*, T_{ss}^*) = \$58,882.61 \text{ or } \$58,883.$$

In **Case 1**, where the stress levels are the same as the operational stresses, the part level screen duration is almost negligible as compared with the system level screen duration. This implies that screening at the system level alone is sufficient to minimize life-cycle costs. On the other hand, when part level screening is done at higher levels, such as in **Cases 2 and 3**, both screens need to be implemented for the minimum life-cycle costs. The total life-cycle costs decrease with the increase in the stress levels. This is so because the screening costs are assumed to be the same at all stress levels, and no penalty is assumed for the increased stress levels.

2. In all the three cases studied previously it was assumed that the failure mechanism in the parts is stimulated by system level screens. This may not be the case in general. In such a case, it is reasonable to assume that system level screening stimulates only the connection failures and does not age the parts. Then, both screens have to be designed independently of each other. Now the life-cycle cost model becomes

$$\begin{aligned} C(T_{sp}, T_{ss}) = \ & n_p[C_{SP} + C_{TP}T_{sp} + H_p(T_{sp}, T_w)C_{FP}] \\ & + \{C_{SS} + C_{TS}T_{ss} + H_{cs}(T_{ss})C_{HS} \\ & + [H_{cs}(T_w + T_{ss}) - H_{cs}(T_{ss})]C_{FS}\}. \end{aligned}$$

$$(6.78)$$

Then

$$Min \ C(T_{sp}, T_{ss}) \iff [Min \ C_p(T_{sp}) \ \cap \ Min \ C_{cs}(T_{ss})],$$

where

$$C_p(T_{sp}) = \text{total cost of parts during screen and}$$
$$\text{warranty},$$

or

$$C_p(T_{s_r}) = n_p[C_{SP} + C_{TP}T_{sp} + H_p(T_{sp}, T_w)C_{FP}],$$

and

$$C_{cs}(T_{ss}) = \text{total cost of connections during screen}$$
$$\text{and warranty},$$

or

$$C_{cs}(T_{ss}) = C_{SS} + C_{TS}T_{ss} + H_{cs}(T_{ss})C_{HS}$$
$$+ [H_{cs}(T_w + T_{ss}) - H_{cs}(T_{ss})]C_{FS}.$$

This is equivalent to solving two minimization problems, one for the parts alone and the other for the connections alone; i.e.,

$$Min \; C_p(T_{sp}), \tag{6.79}$$

and

$$Min \; C_{cs}(T_{ss}). \tag{6.80}$$

These two minimization problems are also solved by using the quasi-Newton method with the data from the three cases. In the model presented, it is assumed that system level screening is done at operational stress levels. Hence, the screen duration and the cost for the system level screen are the same for all the three cases; i.e.,

$$T_{ss}^* = 1,506.954 \text{ hr or } 1,507 \text{ hr},$$

and

$$C(T_{ss}^*) = \$40,432.31 \text{ or } \$40,432.$$

The screen duration and the cost for the part level screen, and the total two-level screening costs are the following:

Case 1:

$$T_{sp}^* = 1,621.348 \text{ hr or } 1,621 \text{ hr},$$
$$C(T_{sp}^*) = \$36,414.48 \text{ or } \$36,414,$$

and

$$C(T_{sp}^*, T_{ss}^*) = C(T_{sp}^*) + C(T_{ss}^*) = \$76,846.79 \text{ or } \$76,847.$$

Case 2:

$$T_{sp}^* = 1,157.280 \text{ hr or } 1,157 \text{ hr},$$
$$C(T_{sp}^*) = \$26,773.32 \text{ or } \$26,773,$$

and

$$C(T_{sp}^*, T_{ss}^*) = C(T_{sp}^*) + C(T_{ss}^*) = \$67,205.63 \text{ or } \$67,206.$$

Case 3:

$$T_{sp}^* = 906.690 \text{ hr or } 907 \text{ hr,}$$
$$C(T_{sp}^*) = \$22,600.59 \text{ or } \$22,601,$$

and

$$C(T_{sp}^*, T_{ss}^*) = C(T_{sp}^*) + C(T_{ss}^*) = \$63,032.29 \text{ or } \$63,032.$$

In all previous three cases, the costs are greater than the corresponding costs when the parts do age with ESS; i.e.,

Case 1: $76,847 > $62,969.

Case 2: $67,206 > $60,775.

Case 3: $63,032 > $58,883.

This is so because in this case it was assumed that the parts do not age during system level screening. Therefore, more time is needed for part level screening which costs more.

6.6 BAYESIAN APPROACH TO ESS

6.6.1 WHY THE BAYESIAN APPROACH?

The failure rate function, $\lambda(t)$, of mixed-exponential distributions given by Eq. (6.4) yields a decreasing failure rate with increasing t which approaches λ_g as t tends to infinity. The classical approach presented in Section 6.3 was to specify some small value for *Screen Residue*, ε, and choose the screen duration, T_s, sufficiently large so that the difference between $\lambda(t)$ and λ_g is less than ε. To accomplish this, the failure rates λ_b and λ_g have to be specified in advance of testing. The difficulty with this approach is that it requires specifying exactly unknown quantities such as λ_b and λ_g, and ignores costs. This may lead to unreasonable ESS duration times. To overcome these shortcomings of the classical approach, Barlow et al [6] proposed a Bayesian approach which assumes a joint prior distribution for the unknown parameters.

The Bayesian approach to decision analysis requires the specification of a loss, or utility, as a function of decision variables and logically possible outcomes. Since outcomes cannot be known in advance, a probability assessment must be made for possible outcomes. The expected value of the loss function is then computed relative to this probability assessment and that decision is made which minimizes the expected loss. Let

$T_s^* = $ optimal duration for the screen under a given stress level l.

We will now introduce the notation and discuss the costs involved in stressing and stopping the screen. Let

C_b = cost of having a "bad" part escape the screen,

and

C_g = cost of having a "good" part destroyed by the screen.

The costs C_b and C_g are "decision" costs in the sense that they describe the cost of wrong "decisions" regarding the part. Since the major concern in ESS is to purify the population of parts in the assembly, the cost of C_b is usually much larger than C_g. In any event, it will only be necessary to specify the ratio C_b/C_g. We will not consider the cost of ESS as a function of time t.

6.6.2 THE JOINT PRIOR DISTRIBUTION FOR PARAMETERS OF THE BIMODAL MIXED-EXPONENTIAL LIFE DISTRIBUTION

Since in the mixture pdf given by Eq. (6.2), p, λ_b and λ_g are in general unknown, we will need to specify a joint prior distribution for them. The design problem to be solved is the determination of $T_s = T_s^*$ in such a way that *the expected total cost with respect to the joint prior distribution for p, λ_b and λ_g is minimized.* We will consider families of prior distributions which are convenient and large enough to accommodate different shades of opinion. The prior joint $pdf's$ for λ_b and λ_g are conditional to the stress level l.

Assume that p is independent of λ_b and λ_g. The beta family of prior distributions with parameters of α and β can be used for p. Note that the uniform pdf on $(0,1)$ is a special case of the beta pdf. An exponential pdf with parameter θ is used as the prior distribution for λ_g and another exponential pdf with parameter τ, shifted by λ_g, is used as the conditional distribution for λ_b given λ_g. Then, the joint prior distribution for p, λ_b and λ_g, given $l = 1$, is

$$f(p, \lambda_b, \lambda_g) = f(\theta)\, f(\lambda_b, \lambda_g),$$
$$= f(\theta)\, f(\lambda_g)\, f(\lambda_b|\lambda_g),$$
$$= \left[\frac{\Gamma(\alpha + \beta)}{\Gamma(\alpha)\, \Gamma(\beta)}\, p^{\alpha-1}\, (1 - p)^{\beta-1}\right] \left(\theta\, e^{-\theta\, \lambda_g}\right) \left(\frac{\tau\, e^{-\tau\, \lambda_b}}{e^{-\tau\, \lambda_g}}\right),$$

or

$$f(p, \lambda_b, \lambda_g) = \theta\, \tau\, \frac{\Gamma(\alpha + \beta)}{\Gamma(\alpha)\Gamma(\beta)} p^{\alpha-1}\, (1 - p)^{\beta-1}\, e^{-(\theta-\tau)\, \lambda_g}\, e^{-\tau\, \lambda_b}, \quad (6.81)$$

for $0 < p < 1$ and $0 < \lambda_g < \lambda_b$, where

α, β = parameters of the beta pdf for p,

θ = parameter of the exponential prior *pdf* for λ_g such that $1/\theta$ is the prior expected value of λ_g,

and

τ = parameter of the exponential prior *pdf* for λ_b such that $1/\tau$ is the prior expected value of λ_b.

It follows that

$$E(\lambda_g) \quad = 1/\theta, \tag{6.82}$$

$$E(\lambda_b|\lambda_g) = \lambda_g + 1/\tau, \tag{6.83}$$

and

$$E(\lambda_b) = \frac{\theta + \tau}{\theta \, \tau}. \tag{6.84}$$

Equation (6.83) can be derived as follows:

$$E(\lambda_b|\lambda_g) = \int_{\lambda_g}^{\infty} \lambda_b \; f(\lambda_b|\lambda_g) \; d\lambda_b,$$

or

$$E(\lambda_b|\lambda_g) = \int_{\lambda_g}^{\infty} \lambda_b \; \tau \; e^{-\tau \, (\lambda_b - \lambda_g)} \; d\lambda_b.$$

Let

$$x = \lambda_b - \lambda_g,$$

then

$$E(\lambda_b|\lambda_g) = \int_{0}^{\infty} (x + \lambda_g) \; \tau \; e^{-\tau \, x} \; dx,$$

$$= \lambda_g \int_{0}^{\infty} \tau \; e^{-\tau \, x} \; dx + \int_{0}^{\infty} x \; \tau \; e^{-\tau \, x} \; dx,$$

or

$$E(\lambda_b|\lambda_g) = \lambda_g + 1/\tau.$$

Equation (6.84) can be derived as follows: The marginal distribution of λ_b is

$$f(\lambda_b) = \int_0^{\lambda_b} f(\lambda_b, \lambda_g)\, d\lambda_g,$$

$$= \tau\, e^{-\tau\, \lambda_b} \int_0^{\lambda_b} \theta\, e^{-\lambda_g\, (\theta - \tau)}\, d\lambda_g,$$

$$= \tau\, e^{-\tau\, \lambda_b}\, \frac{\theta}{\theta - \tau} \left[1 - e^{-\lambda_b\, (\theta - \tau)} \right],$$

or

$$f(\lambda_b) = \frac{\theta\, \tau}{\theta - \tau} \left(e^{-\tau\, \lambda_b} - e^{-\theta\, \lambda_b} \right). \qquad (6.85)$$

Therefore,

$$E(\lambda_b) = \int_0^{\infty} \lambda_b\, f(\lambda_b)\, d\lambda_b,$$

$$= \int_0^{\infty} \lambda_b\, \frac{\theta\, \tau}{\theta - \tau} \left(e^{-\tau\, \lambda_b} - e^{-\theta\, \lambda_b} \right) d\lambda_b,$$

$$= \frac{\theta\, \tau}{\theta - \tau} \int_0^{\infty} \left[\frac{1}{\tau} \left(\lambda_b\, \tau\, e^{-\tau\, \lambda_b} \right) - \frac{1}{\theta} \left(\lambda_b\, \theta\, e^{-\theta\, \lambda_b} \right) \right] d\lambda_b,$$

$$= \frac{\theta\, \tau}{\theta - \tau} \left(\frac{1}{\tau^2} - \frac{1}{\theta^2} \right),$$

$$= \frac{\theta\, \tau}{\theta - \tau} \left(\frac{\theta^2 - \tau^2}{\theta^2 \tau^2} \right),$$

$$= \frac{\theta\, \tau}{\theta - \tau} \left[\frac{(\theta + \tau)(\theta - \tau)}{\theta^2 \tau^2} \right],$$

or

$$E(\lambda_b) = \frac{\theta + \tau}{\theta\, \tau}.$$

It may be seen that the marginal distribution of λ_b, $f(\lambda_b)$, given by Eq. (6.85), has a mode at

$$\lambda_b = [\log_e (\tau/\theta)]/(\tau - \theta)$$

for $\theta > \tau$.

In practical situations, where λ_g is much smaller than λ_b, the value of θ will be chosen much larger than the value of τ. The conditional exponential *pdf* of λ_b then becomes, if compared with the prior exponential *pdf* for λ_g, practically flat. The prior uncertainty for λ_g can be expressed through prior *pdf's* that make use of knowledge about production process standards, as contained in publications such as in Military Standards. On the other hand, the analyst is able to express his relatively much greater "ignorance" about λ_b, given λ_g, through an almost flat prior conditional *pdf* which is nevertheless proper. For values of θ much larger than the value of τ, the following relation shows how small the prior probability of having λ_b "close" to λ_g is – even if the mode of the marginal prior *pdf*, $f(\lambda_b, \lambda_g)$, is the origin; namely,

$$P(\lambda_b < M \ \lambda_g) = 1 - \frac{\theta}{\theta + \tau \ (M-1)}, \tag{6.86}$$

for every $M > 1$. This relationship can be derived as follows:

$$P(\lambda_b < M \ \lambda_g) = \int_0^\infty \left[\int_{\lambda_g}^{M\lambda_g} f(\lambda_b, \lambda_g) \, d\lambda_b \right] d\lambda_g,$$

$$= \int_0^\infty \left[\int_{\lambda_g}^{M\lambda_g} \tau \ e^{-\tau \ \lambda_b} \, d\lambda_b \right] \theta \ e^{-\lambda_g \ (\theta - \tau)} \, d\lambda_g,$$

$$= \int_0^\infty \left(e^{-\tau \ \lambda_g} - e^{-M \ \tau \ \lambda_g} \right) \theta \ e^{-\lambda_g \ (\theta - \tau)} \, d\lambda_g,$$

$$= \int_0^\infty \left\{ \theta \ e^{-\theta \ \lambda_g} - \theta \ e^{-[\theta + (M-1)\tau] \ \lambda_g} \right\} \, d\lambda_g,$$

or

$$P(\lambda_b < M\lambda_g) = 1 - \frac{\theta}{\theta + \tau \ (M-1)},$$

for every $M > 1$.

6.6.3 THE BAYESIAN COST MODEL AND THE OPTIMAL DURATION OF *ESS*

There is a proportion p of substandard parts in the assembly with N total parts. But the inspection of any part does not reveal whether it is substandard or not. This fact makes all parts look similar and entails a judgement of exchangeability of the parts with respect to quality and behavior under stress screening. In particular, for any part in the assembly, the probability that it is substandard is

$$E(p) = \alpha/(\alpha + \beta), \tag{6.87}$$

where E stands for expectation with respect to the beta prior for p.

The *conditional cost* per part of a screen of duration T_s at stress level $l = 1$ is, therefore, derived as follows:

$$C(T_s|p, \lambda_b, \lambda_g) = [(\text{cost of having a ``bad'' part escape the}$$
$$\text{screen}) \times P(\text{any part drawn at random}$$
$$\text{from the assembly is a ``bad'' part})$$
$$\times P(\text{a ``bad'' part escapes the screen})]$$
$$+ [(\text{cost of having a ``good'' part destroyed}$$
$$\text{by the screen}) \times P(\text{any part drawn at}$$
$$\text{random from the assembly is a ``good''}$$
$$\text{part}) \times P(\text{a ``good'' part fails during}$$
$$\text{the screen})],$$
$$= C_b \, p \, e^{-\lambda_b \, T_s} + C_g \, (1-p) \, [1 - e^{-\lambda_g \, T_s}],$$

or

$$C(T_s|p, \lambda_b, \lambda_g) = p \, C_b \, [e^{-\lambda_b \, T_s}] + (1-p) \, C_g \, [1 - e^{-\lambda_g \, T_s}]. \qquad (6.88)$$

The expected total cost with respect to the prior distribution for $(p, \lambda_b, \lambda_g)$ is

$$C(T_s) = \int_{p=0}^{1} \int_{\lambda_b=0}^{\infty} \int_{\lambda_g=0}^{\lambda_b} C(T_s|p, \lambda_b, \lambda_g) \, f(p, \lambda_b, \lambda_g) \, dp \, d\lambda_b \, d\lambda_g,$$

$$= \int_{p=0}^{1} \int_{\lambda_b=0}^{\infty} \int_{\lambda_g=0}^{\lambda_b} [p \, C_b \, e^{-\lambda_b \, T_s} + (1-p) \, C_g \, (1 - e^{-\lambda_g \, T_s})]$$
$$\cdot f(p) \, f(\lambda_b, \lambda_g) \, dp \, d\lambda_b \, d\lambda_g,$$

$$= C_g \int_{p=0}^{1} \int_{\lambda_b=0}^{\infty} \int_{\lambda_g=0}^{\lambda_b} (1-p) \, f(p) \, f(\lambda_b, \lambda_g) \, dp \, d\lambda_b \, d\lambda_g$$

$$+ C_b \int_{p=0}^{1} \int_{\lambda_b=0}^{\infty} \int_{\lambda_g=0}^{\lambda_b} p \, e^{-\lambda_b \, T_s} \, f(p) \, f(\lambda_b, \lambda_g) \, dp \, d\lambda_b \, d\lambda_g$$

$$- C_g \int_{p=0}^{1} \int_{\lambda_b=0}^{\infty} \int_{\lambda_g=0}^{\lambda_b} (1-p) \, e^{-\lambda_g \, T_s} \, f(p) \, f(\lambda_b, \lambda_g) \, dp \, d\lambda_b \, d\lambda_g,$$

or

$$C(T_s) = C_g \int_0^1 (1-p) \, f(p) \, dp \int_{\lambda_b=0}^{\infty} \int_{\lambda_g=0}^{\lambda_b} f(\lambda_b, \lambda_g) \, d\lambda_b \, d\lambda_g$$

$$+C_b \int_0^1 p \, f(p) \, dp \int_{\lambda_b=0}^{\infty} \int_{\lambda_g=0}^{\lambda_b} e^{-\lambda_b \, T_s} \, f(\lambda_b, \lambda_g) \, d\lambda_b \, d\lambda_g$$

$$-C_g \int_0^1 (1-p) \, f(p) \, dp \int_{\lambda_b=0}^{\infty} \int_{\lambda_g=0}^{\lambda_b} e^{-\lambda_g \, T_s} \, f(\lambda_b, \lambda_g) \, d\lambda_b \, d\lambda_g.$$

$$(6.89)$$

But the first term in Eq. (6.89) is

$$C_g \int_0^1 (1-p) \, f(p) \, dp \int_{\lambda_b=0}^{\infty} \int_{\lambda_g=0}^{\lambda_b} f(\lambda_b, \lambda_g) \, d\lambda_b \, d\lambda_g$$

$$= C_g \left[\int_0^1 f(p) \, dp - \int_0^1 p \, f(p) \, dp \right] \times 1,$$

$$= C_g \left(1 - \frac{\alpha}{\alpha + \beta} \right),$$

$$= C_g \left(\frac{\beta}{\alpha + \beta} \right).$$

The second term in Eq. (6.89) is

$$C_b \int_0^1 p \, f(p) \, dp \int_{\lambda_b=0}^{\infty} \int_{\lambda_g=0}^{\lambda_b} e^{-\lambda_b \, T_s} \, f(\lambda_b, \lambda_g) \, d\lambda_b \, d\lambda_g$$

$$= C_b \left(\frac{\alpha}{\alpha + \beta} \right) \int_{\lambda_b=0}^{\infty} \int_{\lambda_g=0}^{\lambda_b} e^{-\lambda_b \, T_s} \, \theta \, \tau \, e^{-\lambda_g(\theta-\tau)} e^{-\lambda_b \, \tau} \, d\lambda_b \, d\lambda_g,$$

$$= C_b \left(\frac{\alpha}{\alpha + \beta} \right) \int_{\lambda_b=0}^{\infty} e^{-\lambda_b \, T_s} \, \frac{\theta \, \tau}{\theta - \tau} \left[e^{-\tau \, \lambda_b} - e^{-\theta \, \lambda_b} \right] d\lambda_b,$$

$$= C_b \left(\frac{\alpha}{\alpha + \beta} \right) \int_{\lambda_b=0}^{\infty} \frac{\theta \, \tau}{\theta - \tau} \left[e^{-(\tau+T_s) \, \lambda_b} - e^{-(\theta+T_s) \, \lambda_b} \right] d\lambda_b,$$

$$= C_b \left(\frac{\alpha}{\alpha + \beta} \right) \frac{\theta \, \tau}{\theta - \tau} \left(\frac{1}{\tau + T_s} - \frac{1}{\theta + T_s} \right),$$

$$= C_b \left(\frac{\alpha}{\alpha + \beta} \right) \frac{\theta \, \tau}{(\tau + T_s)(\theta + T_s)}.$$

The third term in Eq. (6.89) is

$$C_g \int_0^1 (1 - p) \, f(p) \, dp \int_{\lambda_b=0}^{\infty} \int_{\lambda_g=0}^{\lambda_b} e^{-\lambda_g \, T_s} f(\lambda_b, \lambda_g) \, d\lambda_b \, d\lambda_g$$

$$= C_g \left(\frac{\beta}{\alpha + \beta} \right) \int_{\lambda_b=0}^{\infty} \int_{\lambda_g=0}^{\lambda_b} e^{-\lambda_g \, T_s} \, \theta \, \tau \, e^{-\lambda_g \, (\theta - \tau)} e^{-\lambda_b \, \tau} \, d\lambda_b \, d\lambda_g,$$

$$= C_g \left(\frac{\beta}{\alpha + \beta} \right) \int_{\lambda_b=0}^{\infty} \theta \, \tau \, e^{-\lambda_b \, \tau} \left[\int_{\lambda_g=0}^{\lambda_b} e^{-\lambda_g \, (T_s + \theta - \tau)} \, d\lambda_g \right] d\lambda_b,$$

$$= C_g \left(\frac{\beta}{\alpha + \beta} \right) \int_{\lambda_b=0}^{\infty} \left(\frac{\theta \, \tau}{T_s + \theta - \tau} \right) e^{-\lambda_b \, \tau} \left[1 - e^{-\lambda_b \, (T_s + \theta - \tau)} \right] d\lambda_b,$$

$$= C_g \left(\frac{\beta}{\alpha + \beta} \right) \left(\frac{\theta \, \tau}{T_s + \theta - \tau} \right) \int_{\lambda_b=0}^{\infty} \left[e^{-\lambda_b \, \tau} - e^{-\lambda_b \, (T_s + \theta)} \right] d\lambda_b,$$

$$= C_g \left(\frac{\beta}{\alpha + \beta} \right) \left(\frac{\theta \, \tau}{T_s + \theta - \tau} \right) \left(\frac{1}{\tau} - \frac{1}{T_s + \theta} \right),$$

$$= C_g \left(\frac{\beta}{\alpha + \beta} \right) \left(\frac{\theta}{T_s + \theta} \right).$$

Substituting the above into Eq. (6.89) yields

$$C(T_s) = C_g \left(\frac{\beta}{\alpha + \beta} \right) + C_b \left(\frac{\alpha}{\alpha + \beta} \right) \frac{\theta \, \tau}{(\tau + T_s)(\theta + T_s)}$$
$$- C_g \left(\frac{\beta}{\alpha + \beta} \right) \left(\frac{\theta}{T_s + \theta} \right).$$

To obtain the optimal T_s^* which minimizes the expected cost given by Eq. (6.90), we differentiate both sides of Eq. (6.90) with respect to T_s, set it equal to zero, solve it for $T_s = T_s^*$ and get

$$\frac{dC(T_s)}{dT_s} = C_b \left(\frac{\alpha}{\alpha+\beta}\right) (\theta\,\tau) \left[\frac{-(\tau + T_s + \theta + T_s)}{(\tau + T_s)^2(\theta + T_s)^2}\right]$$

$$-C_g \left(\frac{\beta}{\alpha+\beta}\right) \theta \left[\frac{-1}{(T_s + \theta)^2}\right],$$

$$= 0.$$

Simplifying yields

$$C_b\,\alpha\,\tau\,(\theta + \tau + 2\,T_s) = C_g\,\beta\,(T_s^2 + 2\,\tau\,T_s + \tau^2),$$

or

$$\theta + \tau + 2\,T_s = \left(\frac{C_g\,\beta}{C_b\,\alpha\,\tau}\right)(T_s^2 + 2\,\tau\,T_s + \tau^2). \qquad (6.90)$$

Let

$$K = \frac{1}{\tau}\,(b/a)\,(C_g/C_b). \qquad (6.91)$$

Then, Eq. (6.90) becomes

$$\theta + \tau + 2\,T_s = K\,(T_s^2 + 2\,\tau\,T_s + \tau^2),$$

or

$$K\,T_s^2 + 2\,(\tau\,K - 1)\,T_s + (K\,\tau^2 - \theta - \tau) = 0.$$

Solving this equation yields

$$T_s = \frac{-(K\,\tau - 1) \pm \sqrt{K\,\theta - K\,\tau + 1}}{K},$$

or

$$T_s^* = \frac{1}{K}\,[1 - K\,\tau + \sqrt{1 + K\,(\theta - \tau)}]. \qquad (6.92)$$

If the value of T_s^* in Eq. (6.92) is negative, then the optimal decision is, of course, not to screen the assembly. Note that T_s^* increases with θ for fixed τ and costs; i.e., the lower the failure rate for good parts, λ_g, the longer we should perform *ESS*.

Note that for the general case of $l \neq 1$, T_s^* would be written in the form $T_s^* = T_o^*/l$, with T_o^* being a constant independent of the acceleration of time; that is, T_o^* has the same value for all l. The minimum total expected cost of the screening process of optimal duration T_s^* for the whole assembly is $C(T_s^*, l)$ multiplied by N.

6.6.4   OTHER *ESS* MEASURES BASED ON BAYESIAN RESULTS

We now consider other measures of "goodness" relative to an *ESS* design. *The expected probability that a substandard part will escape the screen of duration T_s is*

$$E(e^{-\lambda_b T_s}) = \int_{\lambda_b=0}^{\infty} \int_{\lambda_g=0}^{\lambda_b} e^{-\lambda_b T_s} f(\lambda_b, \lambda_g) \, d\lambda_b \, d\lambda_g,$$

$$= \int_{\lambda_b=0}^{\infty} \int_{\lambda_g=0}^{\lambda_b} e^{-\lambda_b T_s} \, \theta \, \tau \, e^{-\lambda_g (\theta-\tau)} e^{-\lambda_b \tau} \, d\lambda_b \, d\lambda_g,$$

$$= \int_{\lambda_b=0}^{\infty} e^{-\lambda_b T_s} \frac{\theta \, \tau}{\theta-\tau} \left[e^{-\tau \, \lambda_b} - e^{-\theta \, \lambda_b} \right] d\lambda_b,$$

$$= \int_{\lambda_b=0}^{\infty} \frac{\theta \, \tau}{\theta-\tau} \left[e^{-(\tau+T_s) \, \lambda_b} - e^{-(\theta+T_s) \, \lambda_b} \right] d\lambda_b,$$

$$= \frac{\theta \, \tau}{\theta-\tau} \left(\frac{1}{\tau+T_s} - \frac{1}{\theta+T_s} \right),$$

or

$$E(e^{-\lambda_b T_s}) = \frac{\theta \, \tau}{(\tau+T_s)(\theta+T_s)}, \tag{6.93}$$

where E stands for expectation with respect to the prior joint *pdf* of λ_b and λ_g. Recall that $\lambda_b > \lambda_g$ so that Eq. (6.93) depends on both θ and τ. The expected number of substandard parts that will escape from *ESS*, also called the *Remaining Defect Density*, denoted by $D_R(T_s)$, can be obtained by substituting N for C_b in the second term of Eq. (6.90), as follows:

$$D_R(T_s) = N \left(\frac{\alpha}{\alpha+\beta} \right) \frac{\theta \, \tau}{(\tau+T_s)(\theta+T_s)}. \tag{6.94}$$

The probability that a substandard part will not escape from the screen of duration T_s is called the *Screening Strength*, denoted by $SS(T_s)$, and can be obtained from Eq. (6.93) as follows:

$$SS(T_s) = 1 - E(e^{-\lambda_b T_s}),$$

or

$$SS(T_s) = 1 - \frac{\theta \, \tau}{(\tau+T_s)(\theta+T_s)}. \tag{6.95}$$

On the other hand, the *probability that a good part will survive the screen* is

$$
E(e^{-\lambda_g \, T_s}) = \int\limits_{\lambda_b=0}^{\infty} \int\limits_{\lambda_g=0}^{\lambda_b} e^{-\lambda_g \, T_s} f(\lambda_b, \lambda_g) \, d\lambda_b \, d\lambda_g,
$$

$$
= \int\limits_{\lambda_b=0}^{\infty} \int\limits_{\lambda_g=0}^{\lambda_b} e^{-\lambda_g \, T_s} \, \theta \, \tau \, e^{-\lambda_g \, (\theta-\tau)} \, e^{-\lambda_b \, \tau} \, d\lambda_b \, d\lambda_g,
$$

$$
= \int\limits_{\lambda_b=0}^{\infty} \theta \, \tau \, e^{-\lambda_b \, \tau} \left[\int\limits_{\lambda_g=0}^{\lambda_b} e^{-\lambda_g \, (T_s+\theta-\tau)} \, d\lambda_g \right] d\lambda_b,
$$

$$
= \int\limits_{\lambda_b=0}^{\infty} \left(\frac{\theta \, \tau}{T_s+\theta-\tau} \right) e^{-\lambda_b \, \tau} \left[1 - e^{-\lambda_b \, (T_s+\theta-\tau)} \right] d\lambda_b,
$$

$$
= \left(\frac{\theta \, \tau}{T_s+\theta-\tau} \right) \int\limits_{\lambda_b=0}^{\infty} \left[e^{-\lambda_b \, \tau} - e^{-\lambda_b \, (T_s+\theta)} \right] d\lambda_b,
$$

$$
= \left(\frac{\theta \, \tau}{T_s+\theta-\tau} \right) \left(\frac{1}{\tau} - \frac{1}{T_s+\theta} \right),
$$

or

$$
E(e^{-\lambda_g \, T_s}) = \frac{\theta}{T_s+\theta}. \tag{6.96}
$$

The *expected number of good parts remaining in the assembly after the screen*, denoted by $G_R(T_s)$, can be obtained by substituting N for C_g in the third term of Eq. (6.90), as follows:

$$
G_R(T_s) = N \left(\frac{\beta}{\alpha+\beta} \right) \left(\frac{\theta}{T_s+\theta} \right).
$$

Another measure of interest in the Military Standards literature is the *Yield*, defined as the prior probability of having zero substandard parts remaining in the assembly after the screen. The *Yield*, denoted by $Y(T_s)$, is approximately

$$
Y(T_s) \cong exp \left\{ -\frac{\theta \, \tau \, [N\alpha/(\alpha+\beta)]}{[(\theta+T_s)(\tau+T_s)]} \right\} = e^{-D_R \, (T_s)}.
$$

The parameter of the approximating Poisson distribution is the *Remaining Defect Density* , $D_R(T_s)$, given by Eq. (6.94). Note that

$$
D_R(T_s) = [1 - SS(T_s)] \, D_{IN},
$$

where

$$D_{IN} = \text{expected number of substandard parts before the screen,}$$
$$= \textit{Incoming Defect Density},$$

or

$$D_{IN} = N\left(\frac{\alpha}{\alpha + \beta}\right),$$

and

$$1 - SS(T_s) = \frac{\theta \tau}{(\theta + T_s)(\tau + T_s)},$$

from Eq. (6.95).

EXAMPLE 6–4

Assume $l = 1$ and

$$E(\lambda_b) = 1/\tau + 1/\theta \cong 10^{-2} \text{ fr/hr},$$

while

$$E(\lambda_g) = 1/\theta = 10^{-5} \text{ fr/hr}.$$

Find the optimal ESS time, T_s^*, using the Bayesian approach.

SOLUTION TO EXAMPLE 6–4

From Eq. (6.92)

$$T_s^* = \frac{1}{K}\left[1 - K \tau + \sqrt{1 + K (\theta - \tau)}\right],$$

then,

$$(T_s^* + \tau) K - 1 = \sqrt{1 + K (\theta - \tau)}.$$

Squaring both sides yields

$$(T_s^* + \tau)^2 K^2 - 2(T_s^* + \tau)K + 1 = 1 + K(\theta - \tau),$$

or

$$(T_s^* + \tau)^2 K^2 = (2T_s^* + \theta + \tau)K.$$

Since $K \neq 0$, then

$$K = \frac{(2T_s^* + \theta + \tau)}{(T_s^* + \tau)^2}. \tag{6.97}$$

But from Eq. (6.91)

$$K = \frac{1}{\tau}(\beta/\alpha)(C_g/C_b),$$

or

$$K = \frac{1}{\tau}\left\{\frac{1}{[\alpha/(\alpha+\beta)]} - 1\right\}\frac{1}{C_b/C_g}. \tag{6.98}$$

Equating Eq. (6.98) to Eq. (6.97) yields

$$\frac{1}{\tau}\left\{\frac{1}{[\alpha/(\alpha+\beta)]} - 1\right\}\frac{1}{C_b/C_g} = \frac{(2T_s^* + \theta + \tau)}{(T_s^* + \tau)^2},$$

or

$$C_b/C_g = \frac{(T_s^* + \tau)^2}{\tau(2T_s^* + \theta + \tau)}\left\{\frac{1}{[\alpha/(\alpha+\beta)]} - 1\right\}. \tag{6.99}$$

From the given information

$$\theta = 10^5 \text{ hr/fr},$$

and

$$\tau \cong 10^2 \text{ hr/fr}.$$

Substituting the values of θ and τ into Eq. (6.99) yields

$$C_b/C_g = \frac{(T_s^* + 100)^2}{100(2T_s^* + 100,000 + 100)}\left\{\frac{1}{[\alpha/(\alpha+\beta)]} - 1\right\},$$

or

$$C_b/C_g = \frac{(T_s^* + 100)^2}{200(T_s^* + 50,050)}\left\{\frac{1}{[\alpha/(\alpha+\beta)]} - 1\right\}. \tag{6.100}$$

Plotting "C_b/C_g" versus "$\alpha/(\alpha + \beta)$" according to Eq. (6.100) with T_s^* as the varying parameter yields Figure 6.6. This figure gives iso-contours of optimal *ESS* durations, T_s^*, as a function of the expected fraction defective, $\alpha/(\alpha + \beta)$, on the x-axis and the ratio of the cost of a bad part escaping the screen to the cost of a good part failing during the screen, C_b/C_g, on the vertical axis. In Table 6.5, Column 3 lists the optimum *ESS* durations, T_s^*, computed from Eqs. (6.92) and (6.91), for various combinations of C_b/C_g and $\alpha/(\alpha + \beta)$.

The optimal duration for $l \neq 1$ is

$$T_s^{**} = T_s^*/l. \tag{6.101}$$

Values of T_s^{**} under stress levels $l = 2$ and $l = 4$ for various combinations of C_b/C_g and $\alpha/(\alpha + \beta)$ are listed in Columns 4 and 5, respectively, of Table 6.5.

TABLE 6.5– Optimum ESS durations, T_s^*, for $l = 1$ and T^{**} for $l \neq 1$, in hours, for various combinations of C_b/C_g and $\alpha/(\alpha+\beta)$, for Example 6–4.

1	2	3	4	5
C_b/C_g	$\alpha/(\alpha+\beta)$	T_s^* $(l=1)$	T_s^{**} $(l=2)$	T_s^{**} $(l=4)$
5	0.0005	58.32	29.16	14.58
5	0.001	124.11	62.06	31.03
5	0.002	217.39	108.70	54.35
5	0.003	289.19	144.60	72.30
5	0.004	349.90	174.95	87.48
10	0.0005	124.05	62.03	31.01
10	0.001	217.23	108.62	54.31
10	0.002	349.45	174.73	87.36
10	0.003	451.29	225.65	112.82
10	0.004	537.44	268.72	134.36
20	0.0005	217.15	108.58	54.29
20	0.001	349.22	174.61	87.31
20	0.002	536.79	268.40	134.20
20	0.003	681.41	340.71	170.35
20	0.004	803.84	401.92	200.96

Optimal stress levels, l^*, can also be determined for fixed durations, T_s, since the contours are the same for any $l\,\theta/l\,\tau = 10^5/10^2 = 10^3$. Therefore,

$$l^* = T_s^*/T_s. \qquad (6.102)$$

For example, the optimum ESS stress level for the specified ESS duration $T_s = 100$ hr, when $C_b/C_g = 10$ and $\alpha/(\alpha+\beta) = 0.002$, is

$$l^* = 349.45/100 = 3.4945 \cong 3.5,$$

which is actually the optimum acceleration factor.

6.7   COMMENTS

6.7.1   LIFE DISTRIBUTION ASSUMPTION

The times-to-failure distribution has been assumed to be the bimodal mixed-exponential in all cases of this chapter. From practical point of view, this assumption may be good enough for engineering planning purposes since this distribution has a decreasing failure rate (DFR) function which matches the

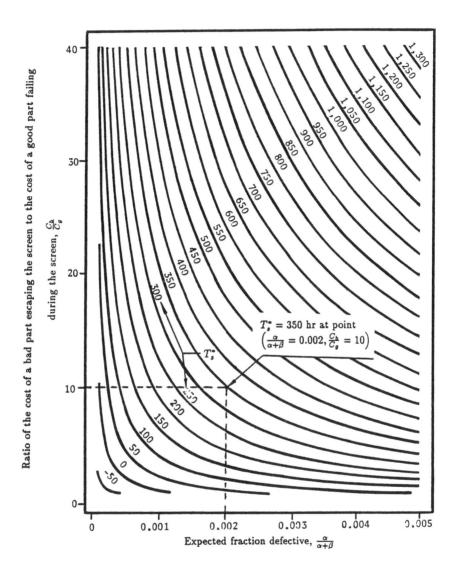

Fig. 6.6– Iso-contours of optimal *ESS* duration, T_s^*, when $\theta = 10^5$ hr/fr and $\tau = 10^2$ hr/fr for Example 6–4.

failure rate behavior during screening. But from failure mechanism point of view, this assumption may not be appropriate.

1. First, since thermal cycling and random vibration have been the most popular two stimuli applied in *ESS*, fatigue failures, either thermal, mechanical or thermal-mechanical, in addition to metalization and other failure modes, are among the most frequent occurences. The more general and appropriate distribution to describe the times to failure may be the mixed-Weibull with a *pdf* such as the following:

$$f(T_s) = p_g \left[\frac{\beta_g}{\eta_g} \left(\frac{T_s - \gamma_g}{\eta_g} \right)^{\beta_g - 1} e^{-\left(\frac{T_s - \gamma_g}{\eta_g} \right)^{\beta_g}} \right]$$
$$+ p_b \left[\frac{\beta_b}{\eta_b} \left(\frac{T_s - \gamma_b}{\eta_b} \right)^{\beta_b - 1} e^{-\left(\frac{T_s - \gamma_b}{\eta_b} \right)^{\beta_b}} \right].$$

(6.103)

where

β_g, β_b = Weibull shape parameters for the "good" and the "bad" subpopulations, respectively,

η_g, η_b = Weibull scale parameters for the "good" and the "bad" subpopulations, respectively,

and

γ_g, γ_b = Weibull location parameters for the "good" and the "bad" subpopulations, respectively.

Admittedly, the application of this general model brings some complexity to the quantification of the associated statistics, such as the renewal function, renewal rate and joint prior and marginal distributions in the Bayesian approach, etc., and to the optimization of the *ESS* duration. However, with the modern computation technology, this complexity should not be a big obstacle.

2. Second, the bimodal mixed distribution may not be able to adequately describe the actual failure modes during the screen, since during screen, in particular at system level, multiple failure modes may occur. To cope with this more general situation, a multi-modal mixed distribution with the following *pdf* may be a more physically rational alternative:

$$f(T_s) = \sum_{i=1}^{m} p_i \, f_i(T_s),$$

(6.104)

where

p_i = proportion for subpopulation i,

and

$f_i(\ ) = pdf$ for life of subpopulation i.

Note the form of $f_i(\ )$ can be either exponential, Weibull, or any other appropriate pdf.

6.7.2 MULTI-LEVEL SCREENING MODEL

The part-system, two-level screening model developed in [5] can be extended to the multi-level screening case, such as a part-module-unit-system four level screen. By quantifying the corresponding renewal function during screen and the associated screening cost at each assembly level of screen, and combining them into the cost model, the optimum screen duration at each level can be readily obtained by minimizing the total expected life-cycle cost using any multi-variate minimization routine; i.e.,

$$Min \quad C(T_{s1}, T_{s2}, \cdots, T_{sn}; T_w), \qquad (6.105)$$

where

$C(\ ) = $ total expected life-cycle cost which is a function

of both screening duration at each level, T_{si},

for $i = 1, 2, \cdots, n$, and warranty period, T_w,

and

$n = $ total number of assembly levels of screen.

In the mean time one can also determine the optimum warranty period, T_w^*, if it is not specified in advance, such that the total life-cycle cost is minimized.

6.7.3 INTERPRETATION OF THE OPTIMUM *ESS* DURATION

In all the literature where the statistical quantification models, as presented in this chapter, are developed, no interpretation is made as to how the so-called *ESS* duration is related with the actual thermal cycles and/or random vibration time. One argument may be the following:

The optimum ESS time corresponds to the time of screening with those stresses and their corresponding levels under which the failure time data were originally collected and the subpopulation proportions and failure rates estimated.

However, this interpretation implies a uniform-stress environment, such as a single-stress or a simultaneously-combined-stress environment, where the same stress profile is applied throughout the whole screening process, as discussed in Section 6.2 of this chapter. The sequential stress screening and other non-stress-uniform environment situations should be excluded from this interpretation. Otherwise, the optimum ESS duration has no physical meaning at all since there is no way to tell whether this single value of time represents the thermal cycling time, or random vibration time or the screening time for a combined environment.

In other words, the statistical quantification and optimization models presented in this chapter are only valid for stress-uniform environments, such as single-stress or simultaneously-combined-stress screening.

6.7.4 DETERMINATION OF LIFE DISTRIBUTION PARAMETERS

Since the mixed life distribution has been widely accepted as the appropriate model for the times-to-failure distribution during ESS, even though the bimodal mixed-exponential may not be the best choice, the determination of the distribution parameters becomes a very important matter. If there are no historical data available, MIL-HDBK-217 can be used as a starting point for the approximate estimation, if it is assumed that the times-to-failure distribution is the exponential requiring the knowledge of the failure rate only, or λ. If historical data are available, the distribution parameters can be determined either graphically or analytically. An extensive coverage of the parameter estimation methods for the mixed life distribution is provided in [7; 8; 9] where two graphical methods and two analytical methods are presented and illustrated with numerical examples.

REFERENCES

1. Saari, A. E., et al, *Environmental Stress Screening, RADC TR-86-149*, September 1986.

2. Perlstein, H. J., Littlefield, J. W. and Bazovsky, I., "The Quantification of Environmental Stress Screening," *Proceedings of the Institute of Environmental Sciences*, San Jose, California, pp. 202-208, 1987.

3. Perlstein, H. J., Littlefield, J. W. and Bazovsky, I., "*ESS* Quantification for Complex Systems," *Proceedings of the Institute of Environmental Sciences*, King of Prussia, Pennsylvania, pp. 50-57, 1988.

4. Cox, D. R., *Renewal Theory*, Methuen Press, London, 142 pp., 1962.

5. Reddy, R. K. and Dietrich, D. L., "A Two-level *ESS* Model: A Mixed Distribution Approach," *IEEE Transactions on Reliability*, Vol. 43, No. 1, pp. 85-90, March 1994.

6. Barlow, R. E., Bazovsky, I. and Wechsler, S., "Classical and Bayes Approaches to Environmental Stress Screening (*ESS*): A Comparison," *Proceedings of the Annual Reliability and Maintainability Symposium*, Los Angeles, California, pp. 81-84, 1990.

7. Kececioglu, Dimitri and Sun, Feng-Bin, *Burn-in – Its Quantification and Optimization*, to be published by Prentice Hall, Englewood Cliffs, New Jersey, 300 pp., 1995.

8. Kececioglu, Dimitri and Sun, Feng-Bin, "Mixed-Weibull Parameter Estimation for Burn-in Data Using the Bayesian Approach," *Microelectronics and Reliability*, Vol. 34, No. 10, pp. 1657-1679, 1994.

9. Kececioglu, Dimitri, *Reliability Engineering Handbook*, Prentice Hall, Englewood Cliffs, New Jersey, Vol. 1, fourth printing, 726 pp., 1995.

APPENDIX 6A
COMPUTER PROGRAM FOR EXAMPLE 6–2

```
C
C $$$$$$$$$$$$$$$$$$$$$$$$$$$$$$$$$$$$$$$$$$$$$$$$$$$$$$$$$$$$$$$$$$
C $                                                              $
C $  This program is developed for ESS duration quantification   $
C $  for a specified quality factor, zeta, which is based upon   $
C $  the conventioal renewal theory.     Elementary iteration    $
C $  technique is utilized in this program for the screen        $
C $  duration optimization interactively.                        $
C $                                                              $
C $  Input data information:                                     $
C $          1. Number of element types and their label names.   $
C $          2. Quantity of items for each element type.         $
C $          3. Initial proportion of standard elements.         $
C $          4. Failure rates under the ESS environment          $
C $                of standard and substandard elements.         $
C $                                                              $
C $$$$$$$$$$$$$$$$$$$$$$$$$$$$$$$$$$$$$$$$$$$$$$$$$$$$$$$$$$$$$$$$$$
C
C
C==============================================================
C DIMENSION DEFINITION FOR:
C  1. CATEGORY LABELS: CAT*50
C  2. CATEGORY QUANTITY: QTY(50)
C  3. FAILURE RATE FOR GOOD COMPONENTS: LAMBG(50)
C  4. FAILURE RATE FOR WEAK COMPONENTS: LAMBE(50)
C  5. PROPORTION OF GOOD COMPONENTS: PG(50)
C  6. PROPORTION OF WEAK COMPONENTS: PE(50)
C  7. INITIAL RENEWAL RATE: RNI(50)
C  8. FINAL RENEWAL RATE: RNF(50)
C  9. EXPONENTIAL DECAY FACTOR: A(50)
C 10. COMPONENT OF ZETA: EPS(50)
C 11. TITLE/HEADER SWITCH: TSW=0 OR 1
C==============================================================
C
        DIMENSION CAT(50),QTY(50), LAMBG(50), LAMBE(50)
        DIMENSION PG(50), PE(50), RNI(50), RNF(50), A(50)
        REAL LAMBG, LAMBE
        INTEGER TSW
        CHARACTER CAT*20
        CHARACTER OPTION*1
```

```
          COMMON ESYS,TSW,NC,RR,RNF,RNI,A,TM,QTY,X,
        1 RNFSYS,DIFXE,ZR,OPTION
          OPEN(6,FILE='ESS-DIX-6A.OUT',STATUS='NEW')
        1 WRITE(*,*)
          WRITE(*,*)' OPTIMUM ESS DURATION DETERMINATION'
          WRITE(*,*)'FOR A SPECIFIED SCREEN RESIDUE, ZETA'
          WRITE(*,*)'==================================='
          WRITE(*,*)
          WRITE(*,*)'LAMBG=FAILURE RATE OF GOOD ITEMS IN fr/hr'
          WRITE(*,*)'LAMBE=FAILURE RATE OF WEAK ITEMS IN fr/hr'
          WRITE(*,*)'PG=INITIAL PROPORTION OF GOOD ITEMS'
          WRITE(*,*)'PE=INITIAL PROPORTION OF WEAK ITEMS'
          WRITE(*,*)
          WRITE(6,*)
          WRITE(6,*)' OPTIMUM ESS DURATION DETERMINATION'
          WRITE(6,*)'FOR A SPECIFIED SCREEN RESIDUE, ZETA'
          WRITE(6,*)'==================================='
          WRITE(6,*)
          WRITE(6,*)'LAMBG=FAILURE RATE OF GOOD ITEMS IN fr/hr'
          WRITE(6,*)'LAMBE=FAILURE RATE OF WEAK ITEMS IN fr/hr'
          WRITE(6,*)'PG=INITIAL PROPORTION OF GOOD ITEMS'
          WRITE(6,*)'PE=INITIAL PROPORTION OF WEAK ITEMS'
          WRITE(6,*)
          TSW=0
        C
        C=========================================================
        C     INPUT QUANTITIES OF ITEMS IN EACH CATEGORY
        C=========================================================
        C
          WRITE(*,*) 'PLEASE ENTER NUMBER OF DISTINCT CATEGORIES
        1 (MAX=50)!'
          WRITE(6,*) 'PLEASE ENTER NUMBER OF DISTINCT CATEGORIES
        1 (MAX=50)!'
          READ(*,*) NC
          WRITE(*,*)
          WRITE(*,*)'THE INPUT DATA:'
          WRITE(*,*)'==============='
          WRITE(6,*)
          WRITE(6,*)'THE INPUT DATA:'
          WRITE(6,*)'==============='
          RNFSYS=0
          DO 10 IC=1,NC
          WRITE(*,*) 'PLEASE ENTER THE CATEGORY CODE
        1 FOR CATEGORY',IC,':'
```

```
          WRITE(6,*) 'PLEASE ENTER THE CATEGORY CODE
     1    FOR CATEGORY',IC,':'
          READ(*,'(A)') CAT(IC)
          WRITE(*,*) 'CAT(',IC,')=',CAT(IC)
          WRITE(*,*) 'PLEASE ENTER THE QUANTITY OF CATEGORY',IC,':'
          WRITE(6,*) 'CAT(',IC,')=',CAT(IC)
          WRITE(6,*) 'PLEASE ENTER THE QUANTITY OF CATEGORY',IC,':'
          READ(*,*) QTY(IC)
          WRITE(*,*) 'QTY(',IC,')=',QTY(IC)
          WRITE(*,*) 'PLEASE ENTER LAMBDA_GOOD FOR CATEGORY',IC,':'
          WRITE(6,*) 'QTY(',IC,')=',QTY(IC)
          WRITE(6,*) 'PLEASE ENTER LAMBDA_GOOD FOR CATEGORY',IC,':'
          READ(*,*) LAMBG(IC)
          WRITE(*,*) 'LAMBG(',IC,')=',LAMBG(IC)
          WRITE(*,*) 'PLEASE ENTER LAMBDA_WEAK FOR CATEGORY',IC,':'
          WRITE(6,*) 'LAMBG(',IC,')=',LAMBG(IC)
          WRITE(6,*) 'PLEASE ENTER LAMBDA_WEAK FOR CATEGORY',IC,':'
          READ(*,*) LAMBE(IC)
          WRITE(*,*) 'LAMBE(',IC,')=',LAMBE(IC)
          WRITE(*,*) 'PLEASE ENTER PROPORTION OF GOOD PARTS FOR
     1    CATEGORY',IC
          WRITE(*,*) 'IN DECIMAL:'
          WRITE(6,*) 'LAMBE(',IC,')=',LAMBE(IC)
          WRITE(6,*) 'PLEASE ENTER PROPORTION OF GOOD PARTS FOR
     1    CATEGORY',IC
          WRITE(6,*) 'IN DECIMAL:'
          READ(*,*) PG(IC)
          WRITE(*,*) 'PG(',IC,')=',PG(IC)
          WRITE(6,*) 'PG(',IC,')=',PG(IC)
          PE(IC)=1-PG(IC)
          WRITE(*,*) 'PE(',IC,')=',PE(IC)
          WRITE(6,*) 'PE(',IC,')=',PE(IC)
          A(IC)=LAMBE(IC)*PG(IC) + LAMBG(IC)*PE(IC)
          RNI(IC)=LAMBE(IC)*PE(IC) + LAMBG(IC)*PG(IC)
          RNF(IC)=LAMBE(IC)*LAMBG(IC)/A(IC)
          RNFSYS=RNFSYS + RNF(IC)*QTY(IC)
    10    CONTINUE
          WRITE(*,*)
          WRITE(6,*)
    C
    C==========================================
    C     CALCULATION OF THE MAIN FACTORS
    C==========================================
    C
```

```
15        WRITE(*,*) 'PLEASE ENTER SPECIFIED VALUE FOR SCREEN
     1    RESIDUE ZETA, ZR!'
          WRITE(6,*) 'PLEASE ENTER SPECIFIED VALUE FOR SCREEN
     1    RESIDUE ZETA, ZR!'
          READ(*,*) ZR
          WRITE(*,*)
          WRITE(*,*) 'SPECIFIED SCREEN RESIDUE IS',ZR
          WRITE(*,*)
          WRITE(*,*) '          PRELIMINARY CALCULATIONS:'
          WRITE(*,*) '============================================='
          WRITE(*,*) 'SYSTEM FINAL RENEWAL RATE=',RNFSYS
          WRITE(*,*)
          WRITE(*,*) '    TIME       SYSTEM RENEWAL        ZETA
     1    ZETA'
          WRITE(*,*) '    (HRS)          RATE         (CALCULATED)
     1    DIFFERENCE'
          WRITE(*,*)
          WRITE(6,*)
          WRITE(6,*) 'SPECIFIED SCREEN RESIDUE IS',ZR
          WRITE(6,*)
          WRITE(6,*) '          PRELIMINARY CALCULATIONS:'
          WRITE(6,*) '============================================='
          WRITE(6,*) 'SYSTEM FINAL RENEWAL RATE=',RNFSYS
          WRITE(6,*)
          WRITE(6,*) '    TIME       SYSTEM RENEWAL        ZETA
     1    ZETA'
          WRITE(6,*) '    (HRS)          RATE         (CALCULATED)
     1    DIFFERENCE'
          WRITE(6,*)
          DO 30 ITM=0,300,50
          ESYS=0.0
          DO 20 IC=1,NC
          RR=RNF(IC)+(RNI(IC)-RNF(IC))*EXP(-A(IC)*ITM)
          ESYS=ESYS+RR*QTY(IC)
20        CONTINUE
          X=(ESYS-RNFSYS)/RNFSYS
          DIFXE=X-ZR
          WRITE(*,25) ITM, ESYS, X, DIFXE
          WRITE(6,25) ITM, ESYS, X, DIFXE
25        FORMAT(1X, I9, 6X, F12.6, 6X, F9.4, 6X, F9.4)
          WRITE(*,*)
          WRITE(6,*)
30        CONTINUE
35        WRITE(*,*)
```

```
      WRITE(*,*) 'PLEASE SELECT ANY ONE OF THE FOLLOWING OPTIONS
   1  TO CONTINUE:'
      WRITE(*,*)
      WRITE(*,*) 'B -- GO TO BEGINNING'
      WRITE(*,*) 'T -- SPECIFY NEW TIME'
      WRITE(*,*) 'Z -- SPECIFY NEW ZETA'
      WRITE(*,*) 'O -- OPTIMIZE ESS TIME'
      WRITE(*,*) 'X -- EXIT PROGRAM'
      WRITE(6,*)
      WRITE(6,*) 'PLEASE SELECT ANY ONE OF THE FOLLOWING OPTIONS
   1  TO CONTINUE:'
      WRITE(6,*)
      WRITE(6,*) 'B -- GO TO BEGINNING'
      WRITE(6,*) 'T -- SPECIFY NEW TIME'
      WRITE(6,*) 'Z -- SPECIFY NEW ZETA'
      WRITE(6,*) 'O -- OPTIMIZE ESS TIME'
      WRITE(6,*) 'X -- EXIT PROGRAM'
      READ(*,'(A)') OPTION
      WRITE(*,*)
      WRITE(*,*) 'YOUR SELECTED OPTION IS  ', OPTION
      WRITE(*,*)
      WRITE(6,*)
      WRITE(6,*) 'YOUR SELECTED OPTION IS  ', OPTION
      WRITE(6,*)
      IF((OPTION.EQ.'B').OR.(OPTION.EQ.'b')) GOTO 1
      IF((OPTION.EQ.'T').OR.(OPTION.EQ.'t')) GOTO 800
      IF((OPTION.EQ.'Z').OR.(OPTION.EQ.'z')) GOTO 15
      IF((OPTION.EQ.'O').OR.(OPTION.EQ.'o')) GOTO 40
      IF((OPTION.EQ.'X').OR.(OPTION.EQ.'x')) GOTO 1000
      GOTO 35
C
C==================================================================
C ESS DURATION OPTIMIZATION FOR A SPECIFIED SCREEN RESIDUE, ZETA
C==================================================================
C
40    NPS=0
      NNS=0
      TM=100
      WRITE(*,*)
      WRITE(*,*) '        OPTIMIZATION CALCULATION:'
      WRITE(*,*) '========================================'
      WRITE(6,*)
      WRITE(6,*) '        OPTIMIZATION CALCULATION:'
      WRITE(6,*) '========================================'
```

```
        CALL TIT1
        DTEMP=DIFXE
C
C====================================================
C          IDENTIFY THE FIRST PLUS/MINUS
C====================================================
C
        DO 180 I=1,100
        IF(DTEMP.GT.0)GOTO 80
        IF(DTEMP.LT.0)GOTO 170
        IF(DTEMP.EQ.0)GOTO 900
        GOTO 1000
C
C===================================================
C              PLUS IDENTIFIED
C===================================================
C
80      NPS=NPS+1
        T1=TM
        DIF1=DTEMP
        IF(NNS.GT.0)GOTO 190
        TM=TM*2.0
        CALL TIT1
        GOTO 175
C
C=========================================
C              MINUS IDENTIFIED
C=========================================
C
170     NNS=NNS+1
        T2=TM
        DIF2=DTEMP
        IF(NPS.GT.0)GOTO 190
        TM=TM/2.0
        CALL TIT1
175     DTEMP=DIFXE
180     CONTINUE
        GOTO 1000
C
C=========================================
C    FIND ZERO BETWEEN PLUS/MINUS POINTS
C=========================================
C
190     TM=T1+(DIF1/(DIF1-DIF2))*(T2-T1)
```

```
         CALL TIT1
         DTEMP=DIFXE
         DO 250 I=1,100
         IF(DTEMP.GT.0)GOTO 200
         IF(DTEMP.LT.0)GOTO 210
         IF(DTEMP.EQ.0)GOTO 900
         GOTO 1000
C
C============================================
C        NEW PLUS IDENTIFIED
C============================================
C
200      T1=TM
         DIF1=DTEMP
         GOTO 220
C
C============================================
C        NEW MINUS IDENTIFIED
C============================================
C
210      T2=TM
         DIF2=DTEMP
         IF(ABS(DIFXE).LE.0.000005)GOTO 900
220      IF(ABS(T2-T1).LE.0.05)GOTO 900
         TM=T1+(DIF1/(DIF1-DIF2))*(T2-T1)
         CALL TIT1
         DTEMP=DIFXE
250      CONTINUE
         WRITE(*,*)'OVER 100 ITERATIONS !!!'
         WRITE(6,*)'OVER 100 ITERATIONS !!!'
         GOTO 35
C
C============================================
C        TRIAL CALCULATIONS
C============================================
C
800      IF(TSW.EQ.1)GOTO 850
         WRITE(*,*)
         WRITE(*,*)'TRIAL CALCULATIONS:'
         WRITE(*,*)'=================='
         WRITE(6,*)
         WRITE(6,*)'TRIAL CALCULATIONS:'
         WRITE(6,*)'=================='
850      CALL TIT2
```

```
        GOTO 35
C
C===========================================
C     PRINT OPTIMIZED ESS TIME VALUE
C===========================================
C
900     WRITE(*,925) TM, ESYS, X, DIFXE
        WRITE(6,925) TM, ESYS, X, DIFXE
925     FORMAT(1X, F9.3, 4X, F12.6, 7X, F9.4, 6X, F12.6)
        TSW=0
        GOTO 35
1000    STOP
        END




C
C $$$$$$$$$$$$$$$$$$$$$$$$$$$$$$$$$$$$$$$$$$$$$$$$$$$$$$$
C $   SUBPROGRAM OF CALCULATING QUALITY FACTORS     $
C $$$$$$$$$$$$$$$$$$$$$$$$$$$$$$$$$$$$$$$$$$$$$$$$$$$$$$$
C
        SUBROUTINE TIT1
        DIMENSION QTY(50), RNI(50), RNF(50), A(50)
        INTEGER TSW
        CHARACTER OPTION*1
        COMMON ESYS,TSW,NC,RR,RNF,RNI,A,TM,QTY,X,
     1  RNFSYS,DIFXE,ZR,OPTION
        ESYS=0
        IF(TSW.NE.0) GOTO 145
        WRITE(*,*)
        WRITE(*,*) '    TIME       SYSTEM RENEWAL        ZETA
     1  ZETA'
        WRITE(*,*) '    (HRS)          RATE           (CALCULATED)
     1  DIFFERENCE'
        WRITE(*,*)
        WRITE(6,*)
        WRITE(6,*) '    TIME       SYSTEM RENEWAL        ZETA
     1  ZETA'
        WRITE(6,*) '    (HRS)          RATE           (CALCULATED)
     1  DIFFERENCE'
        WRITE(6,*)
145     TSW=1
        DO 150 IC=1,NC
        RR=RNF(IC) + (RNI(IC)-RNF(IC))*EXP(-A(IC)*TM)
```

```
         ESYS=ESYS + RR*QTY(IC)
150      CONTINUE
         X=(ESYS-RNFSYS)/RNFSYS
         DIFXE=X-ZR
         IF((OPTION.EQ.'T').OR.(OPTION.EQ.'t')) GOTO 160
         GOTO 180
160      WRITE(*,170) TM, ESYS, X, DIFXE
         WRITE(6,170) TM, ESYS, X, DIFXE
170      FORMAT(1X, F9.3, 4X, F12.6, 7X, F9.4, 6X, F12.6)
180      RETURN
         END




C
C $$$$$$$$$$$$$$$$$$$$$$$$$$$$$$$$$$$$$$$$$$$$$$$$$$$$$$$$$$$$$$$$$$$$$$$$
C $   SUBPROGRAM OF CALCULATING QUALITY FACTORS FOR SPECIFIC ESS TIME
C $$$$$$$$$$$$$$$$$$$$$$$$$$$$$$$$$$$$$$$$$$$$$$$$$$$$$$$$$$$$$$$$$$$$$$$$
C
         SUBROUTINE TIT2
         DIMENSION QTY(50), RNI(50), RNF(50), A(50)
         INTEGER TSW
         CHARACTER OPTION*1
         COMMON ESYS,TSW,NC,RR,RNF,RNI,A,TM,QTY,X,RNFSYS,DIFXE,ZR,OPT
         WRITE(*,*)'PLEASE ENTER SPECIFIC ESS TIME OF INTEREST!'
         WRITE(6,*)'PLEASE ENTER SPECIFIC ESS TIME OF INTEREST!'
         READ(*,*)TM
         ESYS=0
         IF(TSW.NE.0) GOTO 145
         WRITE(*,*)
         WRITE(*,*) '      TIME      SYSTEM RENEWAL        ZETA
     1 ZETA'
         WRITE(*,*) '      (HRS)         RATE          (CALCULATED)
     1 DIFFERENCE'
         WRITE(*,*)
         WRITE(6,*)
         WRITE(6,*) '      TIME      SYSTEM RENEWAL        ZETA
     1 ZETA'
         WRITE(6,*) '      (HRS)         RATE          (CALCULATED)
     1 DIFFERENCE'
         WRITE(6,*)
145      TSW=1
         DO 150 IC=1,NC
         RR=RNF(IC) + (RNI(IC)-RNF(IC))*EXP(-A(IC)*TM)
         ESYS=ESYS + RR*QTY(IC)
```

```
150     CONTINUE
        X=(ESYS-RNFSYS)/RNFSYS
        DIFXE=X-ZR
        IF((OPTION.EQ.'T').OR.(OPTION.EQ.'t')) GOTO 160
        GOTO 180
160     WRITE(*,170) TM, ESYS, X, DIFXE
        WRITE(6,170) TM, ESYS, X, DIFXE
170     FORMAT(1X, F9.3, 4X, F12.6, 7X, F9.4, 6X, F12.6)
180     RETURN
        END
```

APPENDIX 6B
SAMPLE OUTPUT FOR APPENDIX 6A

```
OPTIMUM ESS DURATION DETERMINATION
FOR A SPECIFIED SCREEN RESIDUE, ZETA
======================================


LAMBG=FAILURE RATE OF GOOD ITEMS IN fr/hr
LAMBE=FAILURE RATE OF WEAK ITEMS IN fr/hr
PG=INITIAL PROPORTION OF GOOD ITEMS
PE=INITIAL PROPORTION OF WEAK ITEMS


PLEASE ENTER NUMBER OF DISTINCT CATEGORIES (MAX=50)!

THE INPUT DATA:
===============
PLEASE ENTER THE CATEGORY CODE FOR CATEGORY 1:
CAT(          1)=PARTS
PLEASE ENTER THE QUANTITY OF CATEGORY          1:
QTY(          1)=   3750.000
PLEASE ENTER LAMBDA_GOOD FOR CATEGORY          1:
LAMBG(         1)=   1.1000000E-06
PLEASE ENTER LAMBDA_WEAK FOR CATEGORY          1:
LAMBE(         1)=   1.5000000E-02
PLEASE ENTER PROPORTION OF GOOD PARTS FOR CATEGORY 1
IN DECIMAL:
PG(           1)=   0.9998250
PE(           1)=   1.7499924E-04
PLEASE ENTER THE CATEGORY CODE FOR CATEGORY 2:
CAT(          2)=CONNECTIONS
PLEASE ENTER THE QUANTITY OF CATEGORY          2:
QTY(          2)=   22000.00
PLEASE ENTER LAMBDA_GOOD FOR CATEGORY          2:
LAMBG(         2)=   2.2499999E-07
PLEASE ENTER LAMBDA_WEAK FOR CATEGORY          2:
LAMBE(         2)=   4.5000002E-02
PLEASE ENTER PROPORTION OF GOOD PARTS FOR CATEGORY 2
IN DECIMAL:
PG(           2)=   0.9999770
PE(           2)=   2.3007393E-05


PLEASE ENTER SPECIFIED VALUE FOR SCREEN RESIDUE ZETA, ZR!
```

SPECIFIED SCREEN RESIDUE IS 0.2000000

PRELIMINARY CALCULATIONS:
===
SYSTEM FINAL RENEWAL RATE= 9.0758363E-03

TIME (HRS)	SYSTEM RENEWAL RATE	ZETA (CALCULATED)	ZETA DIFFERENCE
0	0.041695	3.5941	3.3941
50	0.016126	0.7769	0.5769
100	0.011526	0.2699	0.0699
150	0.010140	0.1173	-0.0827
200	0.009569	0.0543	-0.1457
250	0.009308	0.0256	-0.1744
300	0.009185	0.0121	-0.1879

PLEASE SELECT ANY ONE OF THE FOLLOWING OPTIONS TO CONTINUE:

B -- GO TO BEGINNING
T -- SPECIFY NEW TIME
Z -- SPECIFY NEW ZETA
O -- OPTIMIZE ESS TIME
X -- EXIT PROGRAM

YOUR SELECTED OPTION IS O

OPTIMIZATION CALCULATION:
===

TIME (HRS)	SYSTEM RENEWAL RATE	ZETA (CALCULATED)	ZETA DIFFERENCE
117.160	0.010891	0.2000	-0.000001

PLEASE SELECT ANY ONE OF THE FOLLOWING OPTIONS TO CONTINUE:

```
B -- GO TO BEGINNING
T -- SPECIFY NEW TIME
Z -- SPECIFY NEW ZETA
O -- OPTIMIZE ESS TIME
X -- EXIT PROGRAM
```

YOUR SELECTED OPTION IS Z

PLEASE ENTER SPECIFIED VALUE FOR SCREEN RESIDUE ZETA, ZR!

SPECIFIED SCREEN RESIDUE IS 0.3000000

PRELIMINARY CALCULATIONS:
===
SYSTEM FINAL RENEWAL RATE= 9.0758363E-03

TIME (HRS)	SYSTEM RENEWAL RATE	ZETA (CALCULATED)	ZETA DIFFERENCE
0	0.041695	3.5941	3.2941
50	0.016126	0.7769	0.4769
100	0.011526	0.2699	-0.0301
150	0.010140	0.1173	-0.1827
200	0.009569	0.0543	-0.2457
250	0.009308	0.0256	-0.2744
300	0.009185	0.0121	-0.2879

PLEASE SELECT ANY ONE OF THE FOLLOWING OPTIONS TO CONTINUE:

```
B -- GO TO BEGINNING
T -- SPECIFY NEW TIME
Z -- SPECIFY NEW ZETA
O -- OPTIMIZE ESS TIME
X -- EXIT PROGRAM
```

YOUR SELECTED OPTION IS O

```
        OPTIMIZATION CALCULATION:
========================================

TIME        SYSTEM RENEWAL         ZETA              ZETA
(HRS)           RATE           (CALCULATED)        DIFFERENCE

94.242       0.011799            0.3000            -0.000002
```

PLEASE SELECT ANY ONE OF THE FOLLOWING OPTIONS TO CONTINUE:

```
B -- GO TO BEGINNING
T -- SPECIFY NEW TIME
Z -- SPECIFY NEW ZETA
O -- OPTIMIZE ESS TIME
X -- EXIT PROGRAM
```

YOUR SELECTED OPTION IS Z

PLEASE ENTER SPECIFIED VALUE FOR SCREEN RESIDUE ZETA, ZR!

SPECIFIED SCREEN RESIDUE IS 0.1500000

```
             PRELIMINARY CALCULATIONS:
==================================================
SYSTEM FINAL RENEWAL RATE=  9.0758363E-03
```

TIME (HRS)	SYSTEM RENEWAL RATE	ZETA (CALCULATED)	ZETA DIFFERENCE
0	0.041695	3.5941	3.4441
50	0.016126	0.7769	0.6269
100	0.011526	0.2699	0.1199
150	0.010140	0.1173	-0.0327
200	0.009569	0.0543	-0.0957
250	0.009308	0.0256	-0.1244
300	0.009185	0.0121	-0.1379

PLEASE SELECT ANY ONE OF THE FOLLOWING OPTIONS TO CONTINUE:

B -- GO TO BEGINNING
T -- SPECIFY NEW TIME
Z -- SPECIFY NEW ZETA
O -- OPTIMIZE ESS TIME
X -- EXIT PROGRAM

YOUR SELECTED OPTION IS O

          OPTIMIZATION CALCULATION:
==

| TIME | SYSTEM RENEWAL | ZETA | ZETA |
(HRS)	RATE	(CALCULATED)	DIFFERENCE
134.572	0.010437	0.1500	-0.000004

PLEASE SELECT ANY ONE OF THE FOLLOWING OPTIONS TO CONTINUE:

B -- GO TO BEGINNING
T -- SPECIFY NEW TIME
Z -- SPECIFY NEW ZETA
O -- OPTIMIZE ESS TIME
X -- EXIT PROGRAM

YOUR SELECTED OPTION IS X

FORTRAN STOPPED

Chapter 7

PHYSICAL QUANTIFICATION OF *ESS* BY THERMAL CYCLING

7.1 THE ARRHENIUS MODEL AND ITS PITFALLS

7.1.1 THE ARRHENIUS MODEL AND THE CORRESPONDING ACCELERATION FACTOR

Elevated temperature is the most commonly used environmental stress for both accelerated life testing and stress screening of electronic devices. As discussed in Chapter 4 thermal cycling and random vibration are the most effective screening environments in *ESS*. The effect of elevated temperature upon electronic devices is generally modelled using the Arrhenius reaction rate equation. This equation is

$$V = Ae^{-\frac{E_A}{KT^*}}, \tag{7.1}$$

where

V = reaction rate measured in moles/(volume×time),

A = proportionality constant, or the so-called 'frequency factor',

E_A = activation energy measured in electron-volts, eV,

T^* = absolute temperature in °K,

and

$$K = \text{Boltzmann's constant} = 8.623 \times 10^5 \ eV/^\circ K.$$

Equation (7.1) expresses the time rate of device degradation as a function of operating temperature. The degradation is assumed to correspond to chemical or physical reactions among device constituents and contaminants. In those cases where the failure mechanism follows the Arrhenius relationship and the life distribution is exponential, the constant failure rate, $\lambda(T^*)$, is directly proportional to the reaction rate; thus, from Eq. (7.1)

$$\lambda(T^*) = D \ e^{-\frac{E_A}{KT^*}}, \tag{7.2}$$

where D is a constant. Under the exponential life distribution assumption, the mean time to failure or mean life, $L(T^*)$, is given by

$$L(T^*) = \frac{1}{\lambda(T^*)} = C \ e^{\frac{E_A}{KT^*}},$$

where

$$C = \frac{1}{D} = \text{a constant.}$$

Correspondingly, the failure rates and mean lives at operational temperature T_o^* and at accelerated temperature T_a^* are given by

$$\begin{cases} \lambda(T_o^*) = D \ e^{-\frac{E_A}{KT_o^*}}, \\ \lambda(T_a^*) = D \ e^{-\frac{E_A}{KT_a^*}}, \end{cases}$$

and

$$\begin{cases} L(T_o^*) = C \ e^{\frac{E_A}{KT_o^*}}, \\ L(T_a^*) = C \ e^{\frac{E_A}{KT_a^*}}, \end{cases}$$

respectively. The acceleration factor, A_F, is defined as the ratio of the mean life at the operating temperature to that at the accelerated temperature, or

$$A_F = \frac{L(T_o^*)}{L(T_a^*)} = e^{\frac{E_A}{K}\left(\frac{1}{T_o^*} - \frac{1}{T_a^*}\right)},$$

which is identical to the ratio of the failure rate at the accelerated temperature to that at the operating temperature; i.e.,

$$A_F = \frac{L(T_o^*)}{L(T_a^*)} = \frac{\lambda(T_a^*)}{\lambda(T_o^*)} = e^{\frac{E_A}{K}\left(\frac{1}{T_o^*} - \frac{1}{T_a^*}\right)}. \tag{7.3}$$

Note that Eq. (7.3) is valid only when the activation energy, E_A, is independent of the applied temperature. This is not the case in reality as may be seen later on.

Given the failure rate or mean life at the accelerated temperature, $\lambda(T_a^*)$ or $L(T_a^*)$, the failure rate or mean life at the operating temperature is given by

$$\lambda(T_o^*) = \frac{\lambda(T_a^*)}{A_F} = \lambda(T_a^*) \, e^{-\frac{E_A}{K}\left(\frac{1}{T_o^*} - \frac{1}{T_a^*}\right)},$$

or

$$L(T_o^*) = L(T_a^*) \, A_F = L(T_a^*) \, e^{\frac{E_A}{K}\left(\frac{1}{T_o^*} - \frac{1}{T_a^*}\right)}.$$

It should be noted that the values of E_A, D and C can be assumed to be constant *provided that the temperature variation during the test is not significant* [1]. However, this last innocent sounding statement has unfortunately been overlooked by many reliability practitioners. The Arrhenius equation has been used to convert failure rates, or lifetimes, from one temperature to another using the same value for the activation energy, E_A. Admittedly, it is very difficult to state over *how large* a temperature variation the activation energy may be assumed constant. Now let us look at the real meaning of the activation energy and show the contradiction between the Arrhenius model and the age acceleration equation derived from the Arrhenius model, which is caused by the assumption that E_A is constant over any temperature range.

7.1.2 THE PHYSICAL MEANING OF ACTIVATION ENERGY

The activation energy is what the term says it is; i.e., the energy input required to cause a molecule of a constituent to participate in the reaction. By the analogy with chemical reactions, Figure 7.1 illustrates its meaning [2].

In Figure 7.1, the abscissa is the failure (reaction) progress axis and the ordinate is the energy axis. The values of energy constitute internal energy of defects at ambient temperature. These values may vary from defect to defect, depending on their nature, size, shape, location within the structure of the component, electric potentials across them, and the like, as well as ambient temperature. Values of the internal energy are always finite and greater than zero. In our case, the value of the activation energy, E_i, is

$$E(A) \leq E_i \leq E(B),$$

where

$$E(B) = \text{energy level at point } B,$$

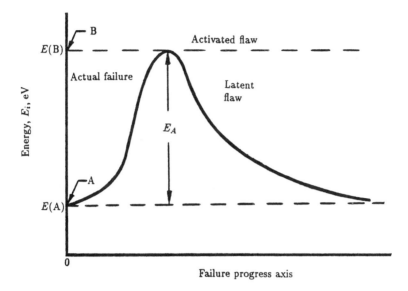

Fig. 7.1– Activation energy and the failure process.

and

$$E(A) = \text{energy level at point } A.$$

Let's take the case of $E_i = E(A)$, as an example. At this energy level, we received a product which was tested "good" but which we know has some flaws. We failed to detect them only because our means were limited. Since we cannot detect these flaws, we call them latent (inactive) flaws. To make a flaw detectable, we must raise its energy level to, at least, the level of $E_i = E(B)$, which means we would have to activate it by providing additional energy in some form, for example, heat. By maintaining the flaw at or above the level of activation, we intensify its effect to the point of failure. Thus, a flaw which was originally latent, according to our test, now becomes active and detectable.

The above considerations suggest that *activation energy, E_A, is an amount of energy necessary to surmount an energy barrier which separates latent flaws from actual failures.* This means that the condition $E_i \geq E(B)$ must be met in order to cause a failure. The values of the activation energy E_A can range from $E_A = 0$ to $E_A = \infty$. Now let us fix the temperature, T^*, at some arbitrary value and vary the value of E_A in Eq. (7.2) to compare the outcomes of both the Arrhenius model and the acceleration equation derived from it.

7.1.3 THE PITFALLS OF THE ARRHENIUS MODEL

According to the Arrhenius model given by Eq. (7.1), if $E_A \longrightarrow 0$, or the energy to activate a particular flaw equals zero, then all the components with these flaws are certain to fail spontaneously without any action on our part. As the required activation energy increases, the rate of failure precipitation decreases. Particularly if $E_A \longrightarrow \infty$, the failure rate drops to zero; thus, no component is going to fail no matter how much action (temperature) we apply. The above brief analysis leads to the following conclusion:

Given the temperature level, the lower the value of the activation energy, the higher the probability of failure, and, consequently, the higher the failure rate. On the other hand, the higher the activation energy, the lower the probability of failure; therefore, the lower the failure rate.

Now, let us look at the aging acceleration obtained by the use of elevated temperature, as compared to that at the nominal operating temperature. If E_A is constant over any temperature range, then the aging acceleration factor is given by Eq. (7.3), or

$$A_F = \frac{L(T_o^*)}{L(T_a^*)} = \frac{\lambda(T_a^*)}{\lambda(T_o^*)} = e^{\frac{E_A}{K} \left(\frac{1}{T_o^*} - \frac{1}{T_a^*} \right)}. \tag{7.4}$$

It follows from Eq. (7.4) that

$$L(T_a^*) = \frac{L(T_o^*)}{A_F}. \tag{7.5}$$

Given T_o^* and T_a^*, if $E_A \longrightarrow 0$, then from Eq. (7.4) $A_F \longrightarrow 1$, which means no acceleration. From Eq. (7.5) the time of stress screening approaches the actual life time of the product in the field, regardless of the applied temperature, provided it is less than infinity. As the activation energy increases, the acceleration factor increases correspondingly. Then, the life at elevated temperature decreases; i.e., the probability of failure increases. Particularly, if $E_A \longrightarrow \infty$, then $A_F \longrightarrow \infty$. From Eq. (7.5) the life at elevated temperature will go to zero, which means that the desired level of screening should be accomplished instantly without any action on our part, regardless of the temperature, provided it is greater than zero. This analysis leads to the following conclusion:

Given the nominal and accelerated temperature levels, the lower the value of the activation energy, the lower the failure rate. On the other hand, the higher the activation energy, the higher the failure rate. This conclusion is obviously contradictory to that drawn from the basic Arrhenius law as stated earlier.

Therefore, the present use of the acceleration formula of Eq. (7.4) is inconsistent with the Arrhenius Law. According to the Arrhenius model, increasing the activation energy implies an increase both in time and in action to precipitate a predetermined fraction of defects. In contrast, according to the acceleration equation, increasing the activation energy implies a decrease both in time and in action to screen a predetermined fraction of defects. To avoid this inconsistency, the activation energy should be considered as the energy provided by the test condition. In other words, the *provided activation energy* is strictly temperature dependent, while the *required activation energy* is a characteristic constant for a given failure mechanism which is independent of the screening temperature.

7.2 MODIFICATION AND PARAMETER ESTIMATION OF THE ACCELERATION FACTOR EQUATION

Equation (7.4) can be modified by substituting E_A by $E_A(T^*)$ for $T^* = T_o^*$ and $T^* = T_a^*$ in Eq. (7.2), respectively, or

$$A_F = e^{\frac{1}{K}\left[\frac{E_A(T_o^*)}{T_o^*} - \frac{E_A(T_a^*)}{T_a^*}\right]}. \tag{7.6}$$

Another modification on Eq. (7.4) was made by Pugacz-Muraszkiewicz [2] as follows:

$$A_F = e^{\frac{[E_P(T_a^*)]^2}{E_R\,K}\left(\frac{1}{T_o^*} - \frac{1}{T_a^*}\right)}, \tag{7.7}$$

where

$E_P(T_a^*)$ = energy provided by the test in eV, which is a function of the applied temperature, T_a^*,

and

E_R = energy required by the failure mechanism in eV, which is independent of the applied temperature, T_a^*.

It may be seen that, when $E_P(T^*) = E_R$, Eq. (7.7) reduces to Eq. (7.4).
Assume the provided activation energy, $E_P(T_a^*)$, can be expressed by

$$E_P(T_a^*) = a\,(T_a^*)^b. \tag{7.8}$$

Then, Eq. (7.7) can be rewritten as

$$A_F = e^{\frac{[E_P(T_a^*)]^2}{E_R\,K}\left(\frac{1}{T_o^*} - \frac{1}{T_a^*}\right)},$$

$$= e^{\frac{[a\,(T_a^*)^b]^2}{E_R\,K}\left(\frac{1}{T_o^*} - \frac{1}{T_a^*}\right)},$$

$$= e^{\left(\frac{a^2}{E_R\,K}\right)(T_a^*)^{2b}\left(\frac{1}{T_o^*} - \frac{1}{T_a^*}\right)},$$

or

$$A_F = e^{C\,(T_a^*)^B \left(\frac{1}{T_o^*} - \frac{1}{T_a^*}\right)}, \tag{7.9}$$

where

$C =$ a constant independent of the applied temperature,

or

$$C = \frac{a^2}{E_R\,K}, \tag{7.10}$$

and

$$B = 2\,b. \tag{7.11}$$

Taking the logarithm twice of both sides of Eq. (7.9) yields

$$\log_e \log_e A_F = \log_e \left[C\,(T_a^*)^B \left(\frac{1}{T_o^*} - \frac{1}{T_a^*}\right) \right],$$

$$= \log_e C + B\,\log_e T_a^* + \log_e \left(\frac{1}{T_o^*} - \frac{1}{T_a^*}\right),$$

or

$$\log_e \log_e A_F - \log_e \left(\frac{1}{T_o^*} - \frac{1}{T_a^*}\right) = \log_e C + B\,\log_e T_a^*. \tag{7.12}$$

Let

$$Y = \log_e \log_e A_F - \log_e \left(\frac{1}{T_o^*} - \frac{1}{T_a^*}\right), \tag{7.13}$$

$$A = \log_e C, \tag{7.14}$$

and

$$X = \log_e T_a^*. \tag{7.15}$$

Then, Eq. (7.12) becomes

$$Y = A + BX. \tag{7.16}$$

For a given random sample $\{(T_{a1}^*, A_{F1}), (T_{a2}^*, A_{F2}), \cdots, (T_{an}^*, A_{Fn})\}$, the least-squares estimates for $C = \frac{a^2}{E_R\,K}$ and B, or $\widehat{C}$ and $\widehat{B}$, are

$$\widehat{C} = e^{\widehat{A}}, \tag{7.17}$$

and

$$\widehat{B} = \frac{L_{XY}}{L_{XX}}, \tag{7.18}$$

with the correlation coefficient of

$$R_{XY} = \frac{L_{XY}}{\sqrt{L_{XX}\,L_{YY}}},$$ (7.19)

where

$$\hat{A} = \overline{Y} - \hat{B}\overline{X},$$

$$L_{XY} = \sum_{i=1}^{n} (X_i - \overline{X})(Y_i - \overline{Y}),$$

$$L_{XX} = \sum_{i=1}^{n} (X_i - \overline{X})^2,$$

$$L_{YY} = \sum_{i=1}^{n} (Y_i - \overline{Y})^2,$$

$$\overline{X} = \frac{1}{n}\sum_{i=1}^{n} X_i,$$

$$\overline{Y} = \frac{1}{n}\sum_{i=1}^{n} Y_i,$$

$$X_i = \log_e T_{ai}^*,$$

and

$$Y_i = \log_e \log_e A_{Fi} - \log_e \left(\frac{1}{T_o^*} - \frac{1}{T_{ai}^*}\right).$$

Therefore, the modified acceleration factor Eq. (7.7), using the least-squares parameter estimates, becomes

$$A_F = e^{\hat{C}\,(T_a^*)^{\hat{B}}\left(\frac{1}{T_o^*} - \frac{1}{T_a^*}\right)},$$ (7.20)

The following is an example illustrating the application of this least squares estimation method.

EXAMPLE 7–1

To obtain the equivalent acceleration factor due to thermal cycling in ESS, it is often necessary to quantify the aging acceleration due to the reaction rate stress–the elevated temperature. To accomplish this, accelerated life tests should be so conducted that the acceleration factor equation's parameters can be estimated. Given in Table 7.1 are two random samples from the accelerated life tests on two different electronic components. The nominal operating temperature is 25°C, or 298°K. The nominal lives of these two electronic components under this temperature are $L_{o1} = 1.5 \times 10^5$ hr and

TABLE 7.1– Results of two accelerated life tests on two different electronic components, for Example 7–1.

i	Component 1			Component 2		
	T_{ai}^*, °K	L_{ai}, 10^5 hr	A_{Fi}[†]	T_{ai}^*, °K	L_{ai}, 10^6 hr	A_{Fi}[†]
1	323	0.33333	4.5	323	0.80000	2.5
2	348	0.09146	16.4	348	0.37037	5.4
3	373	0.02988	50.2	373	0.13048	10.5
4	398	0.01128	133.0	398	0.10338	18.8
5	423	0.00476	315.0	423	0.06329	31.6

† $A_{Fi} = L_o/L_{ai}$.

$L_{o2} = 2.0 \times 10^6$ hr, respectively. Determine the acceleration factor equations by the least-squares method for each component using Eq. (7.7).

SOLUTION TO EXAMPLE 7–1

Substituting the given data into Eqs. (7.17), (7.18) and (7.19) yields the following results:

1. Sample 1:

$$\widehat{C}_1 = 1.196859 \times 10^{-8},$$
$$\widehat{B}_1 = 4.358275,$$

and

$$R_{XY} = 0.967715.$$

Substituting these parameters into Eq. (7.9) yields the modified acceleration factor equation for Component 1; i.e.,

$$A_{F1} = e^{\widehat{C}_1 \, (T_a^*)^{\widehat{B}_1} \left(\frac{1}{T_o^*} - \frac{1}{T_a^*}\right)},$$

or

$$A_{F1} = e^{1.196859 \times 10^{-8} \, (T_a^*)^{4.358275} \left(\frac{1}{298} - \frac{1}{T_a^*}\right)}.$$

2. Sample 2:

$$\widehat{C}_2 = 9.824996 \times 10^{-9},$$
$$\widehat{B}_2 = 4.306011,$$

and

$$R_{XY} = 0.968630.$$

Substituting these parameters into Eq. (7.9) yields the modified acceleration factor equation for Component 2; i.e.,

$$A_{F2} = e^{\widehat{C}_2 \, (T_a^*)^{\widehat{B}_2} \left(\frac{1}{T_o^*} - \frac{1}{T_a^*} \right)},$$

or

$$A_{F2} = e^{9.824996 \times 10^{-9} \, (T_a^*)^{4.306011} \left(\frac{1}{298} - \frac{1}{T_a^*} \right)}.$$

From this analysis, the following conclusion may be drawn:

Although the Arrhenius equation adequately describes the effect of elevated temperature, it does not provide a complete model of thermal stress under thermal cycling, because the activation energy is temperature dependent. In addition, the use of elevated temperature implies a heating and cooling interval not usually considered in evaluating aging acceleration. Often, a significant fraction of device failures occurs during heating and cooling. The basic Arrhenius equation does not fully represent this effect of thermal cycling!

In the following section, a general model for computing the age acceleration due to thermal cycling is presented. This model accounts for Arrhenius-type reaction effects during heating and cooling, as well as during the elevated temperature soak, and includes a further term to reflect ramp rate effects. The model for ramp rate is general and permits any monotonic heating or cooling profile. In addition, the model allows for nonconstant activation energies.

7.3   TEMPERATURE PROFILE AND THE MODEL FOR ACCELERATION FACTOR

Consider a typical thermal cycle, as shown in Fig. 7.2. On the time scale, heating occurs in the intervals $(0, t_1)$ and (t_5, t_6) and cooling occurs during (t_2, t_4). The interval (t_1, t_2) is the elevated temperature dwell (or soak) time. The interval (t_4, t_5) is the lower temperature dwell time. T_o^* represents the nominal operating temperature and T_u^* and T_l^* are the upper and the lower temperature extremes, respectively. If the ramps displayed in Fig. 7.2 are linear, then the heating and cooling processes have uniform rates as shown in Fig. 7.3. In reality, temperature ramps are rarely uniform.

In general, ramps are monotonic but may have various forms. To represent a general temperature profile over time, let a typical temperature cycle

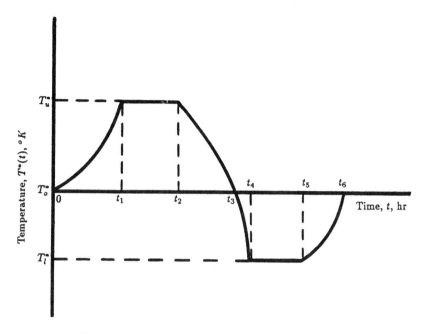

Fig. 7.2– A typical temperature cycle in *ESS*.

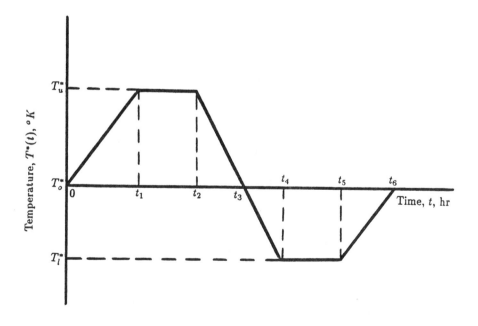

Fig. 7.3– A temperature cycle in *ESS* with uniform heating and cooling.

be represented by the following:

$$T^*(t) = \begin{cases} T_o^* + (T_u^* - T_o^*)(t/t_1)^{\beta_h}, & 0 \le t < t_1, \\ T_u^*, & t_1 \le t < t_2, \\ T_u^* - (T_u^* - T_l^*)[(t - t_2)/(t_4 - t_2)]^{\beta_c}, & t_2 \le t < t_4, \\ T_l^*, & t_4 \le t < t_5, \\ T_l^* + (T_o^* - T_l^*)[(t - t_5)/(t_6 - t_5)]^{\beta_h}, & t_5 \le t \le t_6, \end{cases}$$

$$(7.21)$$

where

T_o^* = nominal operating temperature,

T_u^* = upper temperature extreme,

T_l^* = lower temperature extreme,

β_h = shape parameter of the heating profile,

and

β_c = shape parameter of the cooling profile.

Equation (7.21) represents many monotonic heating and cooling profiles. Different choices of the shape parameters, β_h and β_c, yield different forms for the heating and cooling ramps. It is suggested that the shape provided by $\beta_h = 2$ with various choices of β_c occur frequently [3]. Note that $\beta_h = \beta_c = 1$ yields uniform heating and cooling.

It is plausible that different rates of expansion and contraction in the constituent materials of electronic devices result in mechanical degradation of the devices during heating and cooling. This may be responsible for the frequently observed failures during temperature changes. To represent this behavior, heating and cooling rates are viewed as additional stresses. They are treated as independent of the reaction rate stress, as suggested by Nachlas, Binney and Gruber [4]. Therefore, the aging acceleration of the thermal cycling is the combined effect of two independent stresses; i.e., reaction rate stress governed by the Arrhenius Law, and the temperature change rate stress during heating and cooling. *The equivalent acceleration factor in one typical temperature cycle is the average value of the products of the acceleration factors from these two independent stresses*; i.e.,

$$A_F^* = \frac{1}{t_6} \int_0^{t_6} A_{F1}[T^*(t)] \cdot A_{F2}\left[\frac{dT^*(t)}{dt}\right] dt, \qquad (7.22)$$

where

A_F^* = equivalent acceleration factor in one typical temperature cycle,

$A_{F1}[T^*(t)]$ = acceleration factor due to reaction rate stress, which is a function of $T^*(t)$,

and

$$A_{F2}\left[\frac{dT^*(t)}{dt}\right] = \text{acceleration factor due to temperature}$$

change rate stress, which is a function

of $\frac{dT^*(t)}{dt}$.

The acceleration factor due to reaction rate stress, $A_{F1}[T^*(t)]$, can be obtained readily by replacing T_a^* by $T^*(t)$ in Eq. (7.6), or in Eq. (7.7); i.e.,

$$A_{F1}[T^*(t)] = e^{\frac{1}{K}\left[\frac{E_A(T_o^*)}{T_o^*} - \frac{E_A[T^*(t)]}{T^*(t)}\right]}, \tag{7.23}$$

or

$$A_{F1}[T^*(t)] = e^{\frac{\{E_P[T^*(t)]\}^2}{E_R K}\left(\frac{1}{T_o^*} - \frac{1}{T^*(t)}\right)}. \tag{7.24}$$

The acceleration factor due to temperature change rate stress, A_{F2}, can be evaluated from [4]

$$A_{F2}\left[\frac{dT^*(t)}{dt}\right] = e^{\eta\left|\frac{dT^*(t)}{dt}\right|}, \tag{7.25}$$

where

$\eta = $ a constant.

Substituting Eqs. (7.23) and (7.25) into Eq. (7.22) yields the equivalent acceleration factor using Eqs. (7.23), the first modified form for the acceleration factor due to reaction rate stress only; i.e.,

$$A_F^* = \frac{1}{t_6}\int_0^{t_6} e^{\frac{1}{K}\left\{\frac{E_A(T_o^*)}{T_o^*} - \frac{E_A[T^*(t)]}{T^*(t)}\right\}} e^{\eta\left|\frac{dT^*(t)}{dt}\right|} dt,$$

or

$$A_F^* = \frac{1}{t_6}e^{\frac{E_A(T_o^*)}{KT_o^*}}\int_0^{t_6} e^{-\left\{\frac{E_A[T^*(t)]}{KT^*(t)} - \eta\left|\frac{dT^*(t)}{dt}\right|\right\}} dt. \tag{7.26}$$

Substituting Eqs. (7.24) and (7.25) into Eq. (7.22) yields the equivalent acceleration factor using Eqs. (7.24), the second modified form for the acceleration factor due to reaction rate stress only; i.e.,

$$A_F^* = \frac{1}{t_6}\int_0^{t_6} e^{\frac{\{E_P[T^*(t)]\}^2}{E_R K}\left[\frac{1}{T_o^*} - \frac{1}{T^*(t)}\right]} e^{\eta\left|\frac{dT^*(t)}{dt}\right|} dt,$$

or

$$A_F^* = \frac{1}{t_6} \int_0^{t_6} e^{\frac{\{E_P[T^*(t)]\}^2}{E_R K}\left[\frac{1}{T_o^*} - \frac{1}{T^*(t)}\right] + \eta \left|\frac{dT^*(t)}{dt}\right|} dt. \qquad (7.27)$$

Note that since $dT^*(t)/dt = 0$ in the intervals (t_1, t_2) and (t_4, t_5), the integrals implied by Eqs. (7.26) and (7.27) in these two intervals correspond to the modified Arrhenius models Eqs. (7.6) and (7.7) evaluated at $T_a^* = T_u^*$ and $T_a^* = T_l^*$, respectively.

7.4 EQUIVALENT ACCELERATION FACTOR EVALUATION USING EQ. (7.26)

If Eq. (7.26) is used and the activation energy is assumed to be in the form of Eq. (7.8), or

$$E_A(T^*) = a(T^*)^b;$$

then, Eq. (7.26) becomes

$$A_F^* = \frac{1}{t_6} e^{\frac{E_A(T_o^*)}{KT_o^*}} \int_0^{t_6} e^{-\left\{\frac{E_A[T^*(t)]}{KT^*(t)} - \eta \left|\frac{dT^*(t)}{dt}\right|\right\}} dt,$$

or

$$A_F^* = \frac{1}{t_6} e^{\frac{a(T_o^*)^{b-1}}{K}} \int_0^{t_6} e^{-\left\{\frac{a[T^*(t)]^{b-1}}{K} - \eta \left|\frac{dT^*(t)}{dt}\right|\right\}} dt.$$

Now,

$$\left|\frac{dT^*(t)}{dt}\right| = \begin{cases} \beta_h(T_u^* - T_o^*)(t/t_1)^{\beta_h - 1}\left(\frac{1}{t_1}\right), & 0 \le t < t_1, \\ 0, & t_1 \le t < t_2, \\ \beta_c(T_u^* - T_l^*)[(t - t_2)/(t_4 - t_2)]^{\beta_c - 1}\left(\frac{1}{t_4 - t_2}\right), & t_2 \le t < t_4, \\ 0, & t_4 \le t < t_5, \\ \beta_h(T_o^* - T_l^*)[(t - t_5)/(t_6 - t_5)]^{\beta_h - 1}\left(\frac{1}{t_6 - t_5}\right), & t_5 \le t \le t_6. \end{cases}$$

$$(7.28)$$

Therefore,

$$A_F^* = \frac{1}{t_6} e^{\frac{a(T_o^*)^{b-1}}{K}} \left\{ \int_0^{t_1} exp\left\{ -\left\{ \frac{a}{K}\left[T_o^* + (T_u^* - T_o^*)\left(\frac{t}{t_1}\right)^{\beta_h} \right]^{b-1} \right. \right. \right.$$

$$-\frac{\eta\beta_h(T_u^* - T_o^*)}{t_1}\left(\frac{t}{t_1}\right)^{\beta_h-1}\Bigg\}\Bigg\}\, dt$$

$$+\int_{t_1}^{t_2} e^{-\frac{a}{K}(T_u^*)^{b-1}}\, dt$$

$$+\int_{t_2}^{t_4} exp\left\{-\left\{\frac{a}{K}\left[T_u^* - (T_u^* - T_l^*)\left(\frac{t-t_2}{t_4-t_2}\right)^{\beta_c}\right]^{b-1}\right.\right.$$

$$\left.\left.-\frac{\eta\beta_c(T_u^* - T_l^*)}{t_4 - t_2}\left(\frac{t-t_2}{t_4-t_2}\right)^{\beta_c-1}\right\}\right\}\, dt$$

$$+\int_{t_4}^{t_5} e^{-\frac{a}{K}(T_l^*)^{b-1}}\, dt$$

$$+\int_{t_5}^{t_6} exp\left\{-\left\{\frac{a}{K}\left[T_l^* + (T_o^* - T_l^*)\left(\frac{t-t_5}{t_6-t_5}\right)^{\beta_h}\right]^{b-1}\right.\right.$$

$$\left.\left.-\frac{\eta\beta_h(T_o^* - T_l^*)}{t_6 - t_5}\left(\frac{t-t_5}{t_6-t_5}\right)^{\beta_h-1}\right\}\right\}\, dt\Bigg\},$$

or

$$A_F^* = \frac{1}{t_6}e^{\frac{a(T_o^*)^{b-1}}{K}}\left\{\int_0^{t_1} exp\left\{-\left\{\frac{a}{K}\left[T_o^* + (T_u^* - T_o^*)\left(\frac{t}{t_1}\right)^{\beta_h}\right]^{b-1}\right.\right.\right.$$

$$\left.\left.-\frac{\eta\beta_h(T_u^* - T_o^*)}{t_1}\left(\frac{t}{t_1}\right)^{\beta_h-1}\right\}\right\}\, dt$$

$$+(t_2 - t_1)e^{-\frac{a}{K}(T_u^*)^{b-1}}$$

$$+\int_{t_2}^{t_4} exp\left\{-\left\{\frac{a}{K}\left[T_u^* - (T_u^* - T_l^*)\left(\frac{t-t_2}{t_4-t_2}\right)^{\beta_c}\right]^{b-1}\right.\right.$$

$$\left.\left.-\frac{\eta\beta_c(T_u^* - T_l^*)}{t_4 - t_2}\left(\frac{t-t_2}{t_4-t_2}\right)^{\beta_c-1}\right\}\right\}\, dt$$

$$+(t_5 - t_4)e^{-\frac{a}{K}(T_l^*)^{b-1}}$$

$$+\int_{t_5}^{t_6} exp\left\{-\left\{\frac{a}{K}\left[T_l^* + (T_o^* - T_l^*)\left(\frac{t-t_5}{t_6-t_5}\right)^{\beta_h}\right]^{b-1}\right.\right.$$

$$\left.\left.-\frac{\eta\beta_h(T_o^* - T_l^*)}{t_6 - t_5}\left(\frac{t-t_5}{t_6-t_5}\right)^{\beta_h-1}\right\}\right\}\, dt\Bigg\}. \qquad (7.29)$$

If $\beta_h = \beta_c = 1$, or the heating and cooling processes are uniform, then Eq. (7.29) becomes

$$A_F^* = \frac{1}{t_6} e^{\frac{a(T_o^*)^{b-1}}{K}} \left\{ \int_0^{t_1} e^{-\left\{ \frac{a}{K} \left[T_o^* + (T_u^* - T_o^*) \left(\frac{t}{t_1} \right) \right]^{b-1} - \frac{\eta \beta_h (T_u^* - T_o^*)}{t_1} \right\}} dt \right.$$

$$+ (t_2 - t_1) e^{-\frac{a}{K}(T_u^*)^{b-1}}$$

$$+ \int_{t_2}^{t_4} e^{-\left\{ \frac{a}{K} \left[T_u^* - (T_u^* - T_l^*) \left(\frac{t - t_2}{t_4 - t_2} \right) \right]^{b-1} - \frac{\eta \beta_c (T_u^* - T_l^*)}{t_4 - t_2} \right\}} dt$$

$$+ (t_5 - t_4) e^{-\frac{a}{K}(T_l^*)^{b-1}}$$

$$\left. + \int_{t_5}^{t_6} e^{-\left\{ \frac{a}{K} \left[T_l^* + (T_o^* - T_l^*) \left(\frac{t - t_5}{t_6 - t_5} \right) \right]^{b-1} - \frac{\eta \beta_h (T_o^* - T_l^*)}{t_6 - t_5} \right\}} dt \right\},$$

or

$$A_F^* = \frac{1}{t_6} e^{\frac{a(T_o^*)^{b-1}}{K}} \left\{ e^{\frac{\eta \beta_h (T_u^* - T_o^*)}{t_1}} \int_0^{t_1} e^{-\frac{a}{K} \left[T_o^* + \left(\frac{T_u^* - T_o^*}{t_1} \right) t \right]^{b-1}} dt \right.$$

$$+ (t_2 - t_1) e^{-\frac{a}{K}(T_u^*)^{b-1}}$$

$$+ e^{\frac{\eta \beta_c (T_u^* - T_l^*)}{t_4 - t_2}} \int_{t_2}^{t_4} e^{-\frac{a}{K} \left[T_u^* - \left(\frac{T_u^* - T_l^*}{t_4 - t_2} \right)(t - t_2) \right]^{b-1}} dt$$

$$+ (t_5 - t_4) e^{-\frac{a}{K}(T_l^*)^{b-1}}$$

$$\left. + e^{\frac{\eta \beta_h (T_o^* - T_l^*)}{t_6 - t_5}} \int_{t_5}^{t_6} e^{-\frac{a}{K} \left[T_l^* + (T_o^* - T_l^*) \left(\frac{t - t_5}{t_6 - t_5} \right) \right]^{b-1}} dt \right\}.$$

$$(7.30)$$

Admittedly, both Eqs. (7.29) and (7.30) are two complicated expressions. In fact, there is no closed form equivalent for the general model. However, Eqs. (7.29) and (7.30) are well behaved and can be integrated numerically with little effort.

As a special case, if $b = 1$ in Eq. (7.30), or the activation energy is a linear function of the applied temperature, then

$$V = Ae^{-\frac{aT}{KT}} = Ae^{-\frac{a}{K}} \equiv \text{constant},$$

and

$$A_{F1}[T^*(t)] = e^{\frac{a}{K} \left[\frac{T_o^*}{T_o^*} - \frac{T^*(t)}{T^*(t)} \right]} \equiv 1,$$

or the reaction rate stress has no effect on the ageing acceleration no matter how high the applied temperature is. In this case, a closed form solution can be obtained as follows:

$$
A_F^* = \frac{1}{t_6} e^{\frac{a}{K}} \left\{ e^{\frac{\eta \beta_h (T_u^* - T_o^*)}{t_1}} \int_0^{t_1} e^{-\frac{a}{K}} dt \right.
$$

$$
+ (t_2 - t_1) e^{-\frac{a}{K}}
$$

$$
+ e^{\frac{\eta \beta_c (T_u^* - T_l^*)}{t_4 - t_2}} \int_{t_2}^{t_4} e^{-\frac{a}{K}} dt
$$

$$
+ (t_5 - t_4) e^{-\frac{a}{K}}
$$

$$
\left. + e^{\frac{\eta \beta_h (T_o^* - T_l^*)}{t_6 - t_5}} \int_{t_5}^{t_6} e^{-\frac{a}{K}} dt \right\},
$$

$$
= \frac{1}{t_6} e^{\frac{a}{K}} \left\{ e^{\frac{\eta \beta_h (T_u^* - T_o^*)}{t_1}} \left[t_1 e^{-\frac{a}{K}} \right] \right.
$$

$$
+ (t_2 - t_1) e^{-\frac{a}{K}}
$$

$$
+ e^{\frac{\eta \beta_c (T_u^* - T_l^*)}{t_4 - t_2}} \left[(t_4 - t_2) e^{-\frac{a}{K}} \right]
$$

$$
+ (t_5 - t_4) e^{-\frac{a}{K}}
$$

$$
\left. + e^{\frac{\eta \beta_h (T_o^* - T_l^*)}{t_6 - t_5}} \left[(t_6 - t_5) e^{-\frac{a}{K}} \right] \right\},
$$

or

$$
A_F^* = \frac{1}{t_6} \left[t_1 e^{\frac{\eta \beta_h (T_u^* - T_o^*)}{t_1}} + (t_2 - t_1) \right.
$$

$$
+ (t_4 - t_2) e^{\frac{\eta \beta_c (T_u^* - T_l^*)}{t_4 - t_2}} + (t_5 - t_4)
$$

$$
\left. + (t_6 - t_5) e^{\frac{\eta \beta_h (T_o^* - T_l^*)}{t_6 - t_5}} \right].
$$

(7.31)

   In summary, Eqs. (7.29), (7.30) and (7.31) are equivalent acceleration factor equations using the first modified form for the reaction rate stress acceleration factor, or Eq. (7.26). Equation (7.29) is the general model. Equation (7.30) is a special case of Eq. (7.29) when $\beta_h = \beta_c = 1$, or the heating and cooling processes are uniform. Equation (7.31) is another special case of Eq. (7.29) when $\beta_h = \beta_c = 1$ and $b = 1$, or the heating and cooling processes are uniform and the activation energy is a linear function of the applied temperature.

7.5 EQUIVALENT ACCELERATION FACTOR EVALUATION USING EQ. (7.27)

If Eq. (7.27) is used and the activation energy is assumed to be in the form of Eq. (7.8), or

$$E_A(T^*) = a(T^*)^b,$$

then the acceleration factor term, due to reaction rate stress only, in Eq. (7.27), should be substituted by Eq. (7.9) or

$$A_{F1}[T^*(t)] = e^{C\,[T^*(t)]^B\left[\frac{1}{T_o^*} - \frac{1}{T^*(t)}\right]},$$

and Eq. (7.27) becomes

$$A_F^* = \frac{1}{t_6} \int_0^{t_6} e^{\frac{\{E_P[T^*(t)]\}^2}{E_R K}\left[\frac{1}{T_o^*} - \frac{1}{T^*(t)}\right] + \eta\left|\frac{dT^*(t)}{dt}\right|} \, dt,$$

$$= \frac{1}{t_6} \int_0^{t_6} e^{C[T^*(t)]^B\left[\frac{1}{T_o^*} - \frac{1}{T^*(t)}\right] + \eta\left|\frac{dT^*(t)}{dt}\right|} \, dt,$$

$$= \frac{1}{t_6} \left\{ \int_0^{t_1} exp\left\{ C\left\{ \left[T_o^* + (T_u^* - T_o^*)\left(\frac{t}{t_1}\right)^{\beta_h}\right]^B \left(\frac{1}{T_o^*}\right) \right. \right. \right.$$

$$\left. - \left[T_o^* + (T_u^* - T_o^*)\left(\frac{t}{t_1}\right)^{\beta_h}\right]^{B-1} \right\}$$

$$\left. + \frac{\eta \beta_h(T_u^* - T_o^*)}{t_1}\left(\frac{t}{t_1}\right)^{\beta_h - 1} \right\} dt$$

$$+ \int_{t_1}^{t_2} e^{C(T_u^*)^B\left(\frac{1}{T_o^*} - \frac{1}{T_u^*}\right)} \, dt$$

$$+ \int_{t_2}^{t_4} exp\left\{ C\left\{ \left[T_u^* - (T_u^* - T_l^*)\left(\frac{t - t_2}{t_4 - t_2}\right)^{\beta_c}\right]^B \left(\frac{1}{T_o^*}\right) \right. \right.$$

$$\left. - \left[T_u^* - (T_u^* - T_l^*)\left(\frac{t - t_2}{t_4 - t_2}\right)^{\beta_c}\right]^{B-1} \right\}$$

$$\left. + \frac{\eta \beta_c(T_u^* - T_l^*)}{t_4 - t_2}\left(\frac{t - t_2}{t_4 - t_2}\right)^{\beta_c - 1} \right\} dt$$

$$+ \int_{t_4}^{t_5} e^{C(T_l^*)^B \left(\frac{1}{T_o^*} - \frac{1}{T_l^*} \right)} dt$$

$$+ \int_{t_5}^{t_6} exp \left\{ C \left\{ \left[T_l^* + (T_o^* - T_l^*) \left(\frac{t - t_5}{t_6 - t_5} \right)^{\beta_h} \right]^B \left(\frac{1}{T_o^*} \right) \right. \right.$$

$$- \left[T_l^* + (T_o^* - T_l^*) \left(\frac{t - t_5}{t_6 - t_5} \right)^{\beta_h} \right]^{B-1} \right\}$$

$$+ \frac{\eta \beta_h (T_o^* - T_l^*)}{t_6 - t_5} \left(\frac{t - t_5}{t_6 - t_5} \right)^{\beta_h - 1} \left. \right\} dt \left. \right\},$$

or

$$A_F^* = \frac{1}{t_6} \left\{ \int_0^{t_1} exp \left\{ C \left\{ \left[T_o^* + (T_u^* - T_o^*) \left(\frac{t}{t_1} \right)^{\beta_h} \right]^B \left(\frac{1}{T_o^*} \right) \right. \right. \right.$$

$$- \left[T_o^* + (T_u^* - T_o^*) \left(\frac{t}{t_1} \right)^{\beta_h} \right]^{B-1} \right\}$$

$$+ \frac{\eta \beta_h (T_u^* - T_o^*)}{t_1} \left(\frac{t}{t_1} \right)^{\beta_h - 1} \right\} dt$$

$$+ (t_2 - t_1) e^{C(T_u^*)^B \left(\frac{1}{T_o^*} - \frac{1}{T_u^*} \right)}$$

$$+ \int_{t_2}^{t_4} exp \left\{ C \left\{ \left[T_u^* - (T_u^* - T_l^*) \left(\frac{t - t_2}{t_4 - t_2} \right)^{\beta_c} \right]^B \left(\frac{1}{T_o^*} \right) \right. \right.$$

$$- \left[T_u^* - (T_u^* - T_l^*) \left(\frac{t - t_2}{t_4 - t_2} \right)^{\beta_c} \right]^{B-1} \right\}$$

$$+ \frac{\eta \beta_c (T_u^* - T_l^*)}{t_4 - t_2} \left(\frac{t - t_2}{t_4 - t_2} \right)^{\beta_c - 1} \right\} dt$$

$$+ (t_5 - t_4) e^{-C(T_l^*)^B \left(\frac{1}{T_l^*} - \frac{1}{T_o^*} \right)}$$

$$+ \int_{t_5}^{t_6} exp \left\{ C \left\{ \left[T_l^* + (T_o^* - T_l^*) \left(\frac{t - t_5}{t_6 - t_5} \right)^{\beta_h} \right]^B \left(\frac{1}{T_o^*} \right) \right. \right.$$

$$- \left[T_l^* + (T_o^* - T_l^*) \left(\frac{t - t_5}{t_6 - t_5} \right)^{\beta_h} \right]^{B-1} \right\}$$

$$+\frac{\eta\beta_h(T_o^* - T_l^*)}{t_6 - t_5}\left(\frac{t - t_5}{t_6 - t_5}\right)^{\beta_h-1}\Bigg\}\, dt\Bigg\}.$$

$$(7.32)$$

If $\beta_h = \beta_c = 1$, or the heating and cooling processes are uniform, then Eq. (7.32) becomes

$$A_F^* = \frac{1}{t_6}\Bigg\{\int_0^{t_1} exp\Bigg\{C\Bigg\{\left[T_o^* + (T_u^* - T_o^*)\left(\frac{t}{t_1}\right)\right]^B\left(\frac{1}{T_o^*}\right)$$

$$-\left[T_o^* + (T_u^* - T_o^*)\left(\frac{t}{t_1}\right)\right]^{B-1}\Bigg\} + \frac{\eta(T_u^* - T_o^*)}{t_1}\Bigg\}\, dt$$

$$+(t_2 - t_1)e^{C(T_u^*)^B\left(\frac{1}{T_o^*} \cdot \frac{1}{T_u^*}\right)}$$

$$+\int_{t_2}^{t_4} exp\Bigg\{C\Bigg\{\left[T_u^* - (T_u^* - T_l^*)\left(\frac{t - t_2}{t_4 - t_2}\right)\right]^B\left(\frac{1}{T_o^*}\right)$$

$$-\left[T_u^* - (T_u^* - T_l^*)\left(\frac{t - t_2}{t_4 - t_2}\right)\right]^{B-1}\Bigg\} + \frac{\eta(T_u^* - T_l^*)}{t_4 - t_2}\Bigg\}\, dt$$

$$+(t_5 - t_4)e^{-C(T_l^*)^B\left(\frac{1}{T_l^*} - \frac{1}{T_o^*}\right)}$$

$$+\int_{t_5}^{t_6} exp\Bigg\{C\Bigg\{\left[T_l^* + (T_o^* - T_l^*)\left(\frac{t - t_5}{t_6 - t_5}\right)\right]^B\left(\frac{1}{T_o^*}\right)$$

$$-\left[T_l^* + (T_o^* - T_l^*)\left(\frac{t - t_5}{t_6 - t_5}\right)\right]^{B-1}\Bigg\} + \frac{\eta(T_o^* - T_l^*)}{t_6 - t_5}\Bigg\}\, dt\Bigg\},$$

or

$$A_F^* = \frac{1}{t_6}\Bigg\{e^{\frac{\eta(T_u^* - T_o^*)}{t_1}}\int_0^{t_1} exp\Bigg\{C\Bigg\{\left[T_o^* + (T_u^* - T_o^*)\left(\frac{t}{t_1}\right)\right]^B\left(\frac{1}{T_o^*}\right)$$

$$-\left[T_o^* + (T_u^* - T_o^*)\left(\frac{t}{t_1}\right)\right]^{B-1}\Bigg\}\Bigg\}\, dt$$

$$+(t_2 - t_1)e^{C(T_u^*)^B\left(\frac{1}{T_o^*} - \frac{1}{T_u^*}\right)}$$

$$+e^{\frac{\eta(T_u^* - T_l^*)}{t_4 - t_2}}\int_{t_2}^{t_4} exp\Bigg\{C\Bigg\{\left[T_u^* - (T_u^* - T_l^*)\left(\frac{t - t_2}{t_4 - t_2}\right)\right]^B\left(\frac{1}{T_o^*}\right)$$

$$-\left[T_u^* - (T_u^* - T_l^*)\left(\frac{t-t_2}{t_4-t_2}\right)\right]^{B-1}\Bigg\}\Bigg\} dt$$

$$+(t_5 - t_4)e^{-C(T_l^*)^B\left(\frac{1}{T_l^*}-\frac{1}{T_o^*}\right)}$$

$$+e^{\frac{\eta(T_o^* - T_l^*)}{t_6-t_5}}\int_{t_5}^{t_6} exp\Bigg\{ C\Bigg\{\left[T_l^* + (T_o^* - T_l^*)\left(\frac{t-t_5}{t_6-t_5}\right)\right]^B\left(\frac{1}{T_o^*}\right)$$

$$-\left[T_l^* + (T_o^* - T_l^*)\left(\frac{t-t_5}{t_6-t_5}\right)\right]^{B-1}\Bigg\}\Bigg\} dt\Bigg\}.$$

(7.33)

Admittedly, both Eqs. (7.32) and (7.33) are also two complicated expressions. In fact, there is no closed form equivalent for the general model. However, Eqs. (7.32) and (7.33) are also well behaved and can be integrated numerically with little effort.

As a special case, if $b = 0.5$ or $B = 2b = 1$ in Eq. (7.33), then

$$A_{F1}[T^*(t)] = e^{C\,[T^*(t)]^1\left[\frac{1}{T_o^*}-\frac{1}{T^*(t)}\right]},$$

or

$$A_{F1}[T^*(t)] = e^{C\left[\frac{T^*(t)}{T_o^*}-1\right]}.$$

In this case, a closed form solution can be obtained as follows:

$$A_F^* = \frac{1}{t_6}\Bigg\{ e^{\frac{\eta(T_u^* - T_o^*)}{t_1}}\int_0^{t_1} e^{C\left\{\left[T_o^* + (T_u^* - T_o^*)\left(\frac{t}{t_1}\right)\right]\left(\frac{1}{T_o^*}\right)-1\right\}} dt$$

$$+(t_2 - t_1)e^{CT_u^*\left(\frac{1}{T_o^*}-\frac{1}{T_u^*}\right)}$$

$$+e^{\frac{\eta(T_u^* - T_l^*)}{t_4-t_2}}\int_{t_2}^{t_4} e^{C\left\{\left[T_u^* - (T_u^* - T_l^*)\left(\frac{t-t_2}{t_4-t_2}\right)\right]\left(\frac{1}{T_o^*}\right)-1\right\}} dt$$

$$+(t_5 - t_4)e^{-CT_l^*\left(\frac{1}{T_l^*}-\frac{1}{T_o^*}\right)}$$

$$+e^{\frac{\eta(T_o^* - T_l^*)}{t_6-t_5}}\int_{t_5}^{t_6} e^{C\left\{\left[T_l^* + (T_o^* - T_l^*)\left(\frac{t-t_5}{t_6-t_5}\right)\right]\left(\frac{1}{T_o^*}\right)-1\right\}} dt\Bigg\},$$

$$= \frac{1}{t_6}\Bigg\{ e^{\left[\frac{\eta(T_u^* - T_o^*)}{t_1}-C\right]}\int_0^{t_1} e^{C\left[1+\frac{T_u^* - T_o^*}{T_o^* t_1}t\right]} dt$$

$$+(t_2 - t_1)e^{CT_u^*\left(\frac{1}{T_o^*}-\frac{1}{T_u^*}\right)}$$

$$+e^{\left[\frac{\eta(T_u^*-T_l^*)}{t_4-t_2}-C\right]}\int_{t_2}^{t_4}e^{C\left[\frac{T_u^*}{T_o^*}-\frac{(T_u^*-T_l^*)}{T_o^*(t_4-t_2)}(t-t_2)\right]}dt$$

$$+(t_5 - t_4)e^{-CT_l^*\left(\frac{1}{T_l^*}-\frac{1}{T_o^*}\right)}$$

$$+\left. e^{\left[\frac{\eta(T_o^*-T_l^*)}{t_6-t_5}-C\right]}\int_{t_5}^{t_6}e^{C\left[\frac{T_l^*}{T_o^*}+\frac{(T_o^*-T_l^*)}{T_o^*(t_6-t_5)}(t-t_5)\right]}dt\right\},$$

$$=\frac{1}{t_6}\left\{e^{\frac{\eta(T_u^*-T_o^*)}{t_1}}\int_0^{t_1}e^{\frac{C(T_u^*-T_o^*)}{T_o^*t_1}t}dt\right.$$

$$+(t_2 - t_1)e^{CT_u^*\left(\frac{1}{T_o^*}-\frac{1}{T_u^*}\right)}$$

$$+e^{\left[\frac{\eta(T_u^*-T_l^*)}{t_4-t_2}-C+\frac{CT_u^*}{T_o^*}\right]}\int_{t_2}^{t_4}e^{-\frac{C(T_u^*-T_l^*)}{T_o^*(t_4-t_2)}(t-t_2)}dt$$

$$+(t_5 - t_4)e^{-CT_l^*\left(\frac{1}{T_l^*}-\frac{1}{T_o^*}\right)}$$

$$+\left. e^{\left[\frac{\eta(T_o^*-T_l^*)}{t_6-t_5}-C+\frac{CT_l^*}{T_o^*}\right]}\int_{t_5}^{t_6}e^{\frac{C(T_o^*-T_l^*)}{T_o^*(t_6-t_5)}(t-t_5)}dt\right\},$$

or

$$A_F^* = \frac{1}{t_6}\left\{\frac{T_o^* t_1}{C(T_u^*-T_o^*)}e^{\frac{\eta(T_u^*-T_o^*)}{t_1}}\left[e^{\frac{C(T_u^*-T_o^*)}{T_o^*}}-1\right]\right.$$

$$+(t_2 - t_1)e^{CT_u^*\left(\frac{1}{T_o^*}-\frac{1}{T_u^*}\right)}$$

$$+\frac{T_o^*(t_4-t_2)}{C(T_u^*-T_l^*)}e^{\left[\frac{\eta(T_u^*-T_l^*)}{t_4-t_2}-C\right]}\left[e^{\frac{CT_u^*}{T_o^*}}-e^{\frac{CT_l^*}{T_o^*}}\right]$$

$$+(t_5 - t_4)e^{-CT_l^*\left(\frac{1}{T_l^*}-\frac{1}{T_o^*}\right)}$$

$$+\left.\frac{T_o^*(t_6-t_5)}{C(T_o^*-T_l^*)}e^{\left[\frac{\eta(T_o^*-T_l^*)}{t_6-t_5}-C\right]}\left[e^C-e^{\frac{CT_l^*}{T_o^*}}\right]\right\}.$$

(7.34)

In summary, Eqs. (7.32), (7.33) and (7.34) are equivalent acceleration factor equations using the second modified form for the reaction rate stress acceleration factor, or Eq. (7.27). Equation (7.32) is the general model. Equation (7.33) is a special case of Eq. (7.29) when $\beta_h = \beta_c = 1$, or when the heating and cooling processes are uniform. Equation (7.34) is another special case of Eq. (7.32) when $\beta_h = \beta_c = 1$ and $b = 0.5$.

7.6   APPLICATION OF THE AGING ACCELERATION MODELS

The general model for age acceleration under thermal cycling has several possible uses. The first of these is to compute the acceleration attained in each cycle for any specific device. Given a particular component and corresponding values of the model parameters the acceleration may be computed. In addition, the temperature dependence of the activation energy, and/or the ramp rate effects may be assumed not to exist for a particular device and the model may be solved for the resulting simplified case to again compute the age acceleration attained in each cycle. The following is an example illustrating this application.

EXAMPLE 7–2

Given are the following parameters of a temperature cycle profile:

$T_o^* = 298°K$, $T_u^* = 338°K$, $T_l^* = 233°K$, $t_1 = \frac{1}{3}$ hr,
$t_2 = \frac{4}{3}$ hr, $\quad t_4 = 2$ hr, $\quad\quad t_5 = \frac{7}{3}$ hr, $\quad t_6 = \frac{8}{3}$ hr,
$\beta_h = 2$, $\quad\quad \beta_c = 2$.

The constant, η, for the acceleration factor, due to the temperature change stress, is given by

$\eta = 0.01$.

The parameters C and B, for the acceleration factor due to the reaction rate stress, have been estimated in Example 7–1 using the accelerated life test data of the second sample; i.e.,

$C = 9.824996 \times 10^{-9}$,

and

$B = 4.306011$.

Do the following:

1. Evaluate the equivalent acceleration factor in one temperature cycle using Eq. (7.29) assuming $a = 10^{-10}, 10^{-9}, 10^{-8}, 10^{-7}$, and 10^{-6}.

2. Evaluate the equivalent acceleration factor in one temperature cycle using Eq. (7.32).

TABLE 7.2– Equivalent acceleration factors using Eq. (7.29) for different values of a for Example 7–2.

No.	a	A_{F1}^*
1	1.0E-10	4.346206
2	1.0E-09	4.347811
3	1.0E-08	4.364098
4	1.0E-07	4.551372
5	1.0E-06	10.338300

3. Compare the results obtained in Cases 1 and 2, and decide which value is appropriate for a.

4. Redo Cases 1 and 2, assuming $b = 0$ (constant activation energy).

5. Redo Cases 1 and 2, assuming $\eta = 0$ (ignore the ramp effect).

SOLUTION TO EXAMPLE 7–2

1. According to Eq. (7.11),

$$b = \frac{B}{2} = \frac{4.306011}{2} = 2.1530055.$$

Then, the provided activation energy, $E_P(T_a^*)$, becomes

$$E_P(T_a^*) = a(T_a^*)^b = a(T_a^*)^{2.1530055}.$$

Note that Boltzmann's constant, K, is

$$K = 8.623 \times 10^{-5} \text{ eV/}^\circ\text{K}.$$

The computer program, given in Appendix 7A, has been developed for the evaluation of the equivalent acceleration factors in one temperature cycle for Eqs. (7.29) and (7.32). The resultant equivalent acceleration factors using Eq. (7.29) for different values of a are listed in Table 7.2. It may be seen that the higher the value of a the higher the equivalent acceleration factor provided that the other parameters are fixed.

2. Applying the same computer program also yields the equivalent acceleration factor using Eq. (7.32), which is

$$A_{F2}^* = 4.370327.$$

3. By comparing the results in Cases 1 and 2, it may be seen that the equivalent acceleration factors using Eq. (7.29) for different values of a are different from that using Eq. (7.32). But, theoretically both Eqs.

(7.29) and (7.32) should yield close results, if not exactly the same, since a is an unknown but fixed equipment-inherent parameter. That is, only one value is appropriate for a. However, it may be observed from Table 7.2 that a value of a between 10^{-8} and 10^{-7} corresponds to the equivalent acceleration factor of $A_F^* = 4.370327$.

Running the computer program again by trial-and-error, with various values of a between 10^{-8} and 10^{-7}, yields

$$A_{F1}^*(a = 1.3381 \times 10^{-8}) = 4.370327.$$

Therefore,

$$a = 1.3381 \times 10^{-8}$$

is the appropriate value for a because it yields the same value for A_{F1}^* as for A_{F2}^*.

As a by-product, the energy required by the failure mechanism, E_R, can be estimated using Eq. (7.10); i.e.,

$$E_R = \frac{a^2}{CK},$$

$$= \frac{(1.3381 \times 10^{-8})^2}{(9.824996 \times 10^{-9})(8.623 \times 10^{-5})},$$

or

$$E_R = 2.1131 \times 10^{-4} \text{ eV}.$$

4. If the activation energy is assumed constant, or

$$b = 0,$$

then

$$B = 2b = 0.$$

Running the same computer program with $b = B = 0$, and the other parameters unchanged, yields

$$A_{F1}^* = A_{F2}^* = 4.346028,$$

which is less than that when the activation energy is not constant but dependent on the applied temperature profile.

5. If the ramp effect is ignored, or

$$\eta = 0;$$

then, all of the acceleration is contributed solely by the reaction rate stress which is governed by the modified Arrhenius Law with temperature dependent activation energy.

Running the same computer program with $\eta = 0$, with the other parameters unchanged, yields

$$A^*_{F1} = 0.9980928,$$

and

$$A^*_{F2} = 1.138102.$$

It may be seen that A^*_{F1} and A^*_{F2} are different under the assumption of $\eta = 0$. The reason for this is that η, a and b (or B) are the aging characteristic parameters of the screened equipment and they are correlated. The change of the value of one parameter will require the corresponding change of the values of the other parameters. But, in this case, η is assumed to be zero (though it is not so actually) without the corresponding changes on a and b (or B). This is why we can not make an arbitrary assumption without being aware of its consequences. However since A^*_{F1} and A^*_{F2} are close enough in this case, we may take the average of them as the approximate result; i.e.,

$$A^*_F \cong \frac{A^*_{F1} + A^*_{F2}}{2} = \frac{0.9980928 + 1.138102}{2} = 1.0680974,$$

which is much less than those of Cases 1, 2 and 3 when the ramp effect is considered. Therefore, the temperature change rate stress (ramp effect) plays a key role in the aging acceleration process of temperature cycling. The equivalent acceleration factor is largerly determined by this stress.

Other uses of the model can be defined in terms of the model parameters. These parameters may be grouped into information sets. For example, given the values of the other parameters, T^*_u and t_6 may be treated as decision variables. The model may then be used to select the elevated temperature and the cycle length required to achieve a specific acceleration. In this manner, the model will support a test design effort.

On the other hand, if the value of η is not known, or if the values of a and b (or B) are in question, the model can be applied to experimental data to provide estimates for these parameters.

Perhaps one of the most valuable uses of the model is in equipment selection. Assuming that the available thermal chambers can be characterized in terms of their achievable temperatures, T^*_u, and their heating and cooling specifications, β_h and β_c, the model may be used to determine which equipment will provide a desired acceleration. In addition, given appropriate cost information, the model can be used to support equipment purchase decisions.

In summary, taking the model parameters in information sets such as $\{T^*_u, t_6\}$, $\{\eta\}$, $\{a, b\}$ and $\{T^*_u, \beta_h, \beta_c\}$ allows the use of the model to analyze questions affected by the values of the parameters in any set. In each case, the same general approach is used and the analysis can be based upon any assumed type of equipment behavior.

7.7   OPTIMUM NUMBER OF THERMAL CYCLES FOR A SPECIFIED FIELD $MTBF$ GOAL

In the preceding sections, the following have been discussed:

1. The Arrhenius model and the acceleration factor.

2. The physical meaning of activation energy.

3. The pitfalls of the Arrhenius model.

4. Modification and parameter estimation of the acceleration factor equation.

5. Thermal cycling profile and the models for equivalent acceleration factors using different forms for the activation energy.

The next task is to determine the optimum number of thermal cycles, N_{therm}, for a given temperature profile and for a specified field $MTBF$ goal.

Assume that an electronic equipment under a given thermal cycling profile has a bimodal mixed life distribution with the following *pdf* and reliability function:

$$f(T) = p_b \; f_b(T) + p_g \; f_g(T),$$

and

$$R(T) = p_b \; R_b(T) + p_g \; R_g(T).$$

Let T_s be the screening time, then the post-screen reliability function for a mission time of t is given by

$$R(T_s, t) = \frac{R(T_s + t)}{R(T_s)} = \frac{p_b \; R_b(T_s + t) + p_g \; R_g(T_s + t)}{p_b \; R_b(T_s) + p_g \; R_g(T_s)}.$$

Then, the post-screen $MTBF$ under the given thermal cycling profile environment, $MTBF_s$, is given by

$$MTBF_s = \int_0^\infty R(T_s, t) \; dt,$$

$$= \frac{p_b \left[\int_0^\infty R_b(T_s + t) \; dt \right] + p_g \left[\int_0^\infty R_g(T_s + t) \; dt \right]}{p_b \; R_b(T_s) + p_g \; R_g(T_s)}.$$

Let $\tau = T_s + t$, then $d\tau = dt$ and

$$MTBF_s = \frac{p_b \left[\int_{T_s}^\infty R_b(\tau) \; d\tau \right] + p_g \left[\int_{T_s}^\infty R_g(\tau) \; d\tau \right]}{p_b \; R_b(T_s) + p_g \; R_g(T_s)}.$$

Therefore, the equivalent $MTBF$ in the field, $MTBF_{field}$, will be given by

$$MTBF_{field} = A_F \; MTBF_s,$$

or

$$MTBF_{field} = A_F \left\{ \frac{p_b \left[\int_{T_s}^{\infty} R_b(\tau) \, d\tau \right] + p_g \left[\int_{T_s}^{\infty} R_g(\tau) \, d\tau \right]}{p_b \, R_b(T_s) + p_g \, R_g(T_s)} \right\} . \quad (7.35)$$

If the bimodal mixed-exponential life distribution governs, then

$$R_b(T) = e^{-\lambda_b \, T},$$

and

$$R_g(T) = e^{-\lambda_g \, T}.$$

Correspondingly,

$$\int_{T_s}^{\infty} R_b(\tau) \, d\tau = \int_{T_s}^{\infty} e^{-\lambda_b \, \tau} \, d\tau = \frac{e^{-\lambda_b \, T_s}}{\lambda_b},$$

and

$$\int_{T_s}^{\infty} R_g(\tau) \, d\tau = \int_{T_s}^{\infty} e^{-\lambda_g \, \tau} \, d\tau = \frac{e^{-\lambda_g \, T_s}}{\lambda_g}.$$

Consequently,

$$MTBF_{field} = A_F \left[\frac{p_b \left(\frac{e^{-\lambda_b \, T_s}}{\lambda_b} \right) + p_g \left(\frac{e^{-\lambda_g \, T_s}}{\lambda_g} \right)}{p_b \, e^{-\lambda_b \, T_s} + p_g \, e^{-\lambda_g \, T_s}} \right],$$

or

$$MTBF_{field} = A_F \left[\frac{\left(\frac{p_b \, e^{-\lambda_b \, T_s}}{\lambda_b} \right) + \left(\frac{p_g \, e^{-\lambda_g \, T_s}}{\lambda_g} \right)}{p_b \, e^{-\lambda_b \, T_s} + p_g \, e^{-\lambda_g \, T_s}} \right].$$

Note that both A_F and $MTBF_{field}$ are functions of screening time T_s. The optimum screen time, T_s^*, for a specified field $MTBF$ goal, $MTBF_{field}^*$, can be obtained by solving the following equation for T_s^*:

$$MTBF_{field}^* = A_F(T_s^*) \left[\frac{\left(\frac{p_b \, e^{-\lambda_b \, T_s^*}}{\lambda_b} \right) + \left(\frac{p_g \, e^{-\lambda_g \, T_s^*}}{\lambda_g} \right)}{p_b \, e^{-\lambda_b \, T_s^*} + p_g \, e^{-\lambda_g \, T_s^*}} \right]. \quad (7.36)$$

Assume that each complete thermal cycle produces an identical life acceleration effect; i.e., provides the same acceleration factor value. Then,

$$A_F(T_s^*) = N_{therm}^* \, A_F, \quad (7.37)$$

where

A_F = equivalent acceleration factor produced by one typical thermal cycle, which is given by Eq. (7.29) if the first modification form of the Arrhenius model is used, and is given by Eq. (7.32) if the second modification form of the Arrhenius model is used,

N^*_{therm} = number of thermal cycles during $(0, T^*_s)$,

$$N^*_{therm} = \frac{T^*_s}{t_6}, \tag{7.38}$$

and

t_6 = duration of one typical thermal cycle as shown in Fig. 7.3.

Substituting Eqs. (7.37) and (7.38) into Eq. (7.36) yields

$$MTBF^*_{field} = \left(\frac{A_F}{t_6}\right) T^*_s \left[\frac{\left(\frac{p_b\, e^{-\lambda_b\, T^*_A}}{\lambda_b}\right) + \left(\frac{p_g\, e^{-\lambda_g\, T^*_A}}{\lambda_g}\right)}{p_b\, e^{-\lambda_b\, T^*_A} + p_g\, e^{-\lambda_g\, T^*_A}} \right]. \tag{7.39}$$

Since this is a complicated expression and the evaluation of A_F relies on the computer program given in Appendix 7A, no closed form solution for T^*_s is available. However, an iteration relation for T^*_s can be readily derived from Eq. (7.39) as follows:

$$T^*_s = \left(\frac{MTBF^*_{field}\, t_6}{A_F}\right) \left[\frac{p_b\, e^{-\lambda_b\, T^*_A} + p_g\, e^{-\lambda_g\, T^*_A}}{\left(\frac{p_b\, e^{-\lambda_b\, T^*_A}}{\lambda_b}\right) + \left(\frac{p_g\, e^{-\lambda_g\, T^*_A}}{\lambda_g}\right)} \right]. \tag{7.40}$$

A computer program, which is based on Eq. (7.40) with an initial value of $T^*_s = t_6$, and which requires the value of A_F obtained by running the computer program given in Appendix 7A, is developed and is given in Appendix 7B. The application of this Appendix is illustrated by an example as given next.

EXAMPLE 7–3

Given is a temperature cycle profile with the same parameters as given in Example 7–2; i.e.,

$T^*_o = 298°K$, $T^*_u = 338°K$, $T^*_l = 233°K$, $t_1 = \frac{1}{3}$ hr,
$t_2 = \frac{4}{3}$ hr, $t_4 = 2$ hr, $t_5 = \frac{7}{3}$ hr, $t_6 = \frac{8}{3}$ hr,
$\beta_h = 2$, $\beta_c = 2$.

The constant, η, for the acceleration factor, due to the temperature change stress, is given by

$\eta = 0.01$.

The parameters C and B, for the acceleration factor due to the reaction rate stress, have been estimated in Example 7-1 using the accelerated life test data of the second sample; i.e.,

$C = 9.824996 \times 10^{-9}$,

and

$B = 4.306011$.

The data analysis on the historical times to failure, which were observed during thermal cycling using the same temperature profile as given, yields the mixed-exponential life distribution with the following parameters:

$p_b = 0.01, \quad \lambda_b = 0.15 \text{ fr/hr}$,

$p_g = 0.99, \quad \lambda_b = 0.0005 \text{ fr/hr}$.

Determine the optimum thermal cycling time, T_s^*, or equivalently, the optimum number of thermal cycles, N_{therm}^*, for a specified field $MTBF$ of 50,000 hr. The desired convergence accuracy for T_s^* is 0.000001.

SOLUTION TO EXAMPLE 7-3

According to Example 7-2, the equivalent acceleration factor produced by a typical temperature cycle of length $t_6 = 8/3$ hr is

$A_F = 4.37$.

Running the computer program given in Appendix 7B with the given input information yields the following results:

Iteration step	T_s^*, hr
0	$t_6 = 8/3$
1	15.37776
2	15.29003
3	15.29024
4	15.29024

Therefore, the optimum thermal cycling time is

$T_s^* = 15.29$ hr.

Equivalently, the corresponding optimum number of thermal cycles is

$N_{therm}^* = \dfrac{T_s^*}{t_6} = 15.29 \, (3/8) = 5.73 \text{ or } 6 \text{ cycles.}$

7.8 A SUMMARY OF SOME USEFUL THERMAL FATIGUE LIFE PREDICTION MODELS FOR ELECTRONIC EQUIPMENT

It is well known that the standard Eyring and Arrhenius models can be used to describe and quantify the effect of temperature on chemical reaction rates provided that the dominant component failure mechanisms depend on the steady-state temperature. However, it is often difficult to confirm the correlation between steady-state temperature and the failure mechanisms [5, p. 61]. Usually an electronic equipment contains various components of different thermal masses and different thermal characteristics, and various interconnections, which may result in all kinds of failures even under the same stress environment. Multiple failure mechanisms of an electronic equipment under dynamic temperature environment may require specific physical models to describe the failure behavior and predict the corresponding time to failure for each failure mechanism.

During thermal cycling, mechanical failures occur from differential thermal expansion between bonded materials, rate and amplitude of time-dependent temperature changes, and large spatial temperature gradients, all of which can cause tensile, compressive, bending, fatigue and fracture failures. The time to failure for each failure mechanism may be governed by multiple factors, such as geometry, physical properties, environmental stress levels, etc. Various physical models have been derived in the literature for the associated failure mechanisms to predict the mean time to failure, based on the stress-strain analysis, cumulative damage evaluation and/or fracture mechanics analysis. The following is a summary of some useful life prediction models for mechanical failures under thermal cycling environment.

7.8.1 WIRE FATIGUE LIFE PREDICTION

Wires in an electronic equipment are usually used to connect die bond pads to leads, or die bond pads to die bond pads for hybrid packages. The cyclic temperature changes during temperature and power cycling cause repeated flexing of the wire due to the differential coefficient of thermal expansion between the wire and the package, which may lead to wire failures. It has been confirmed that the main failure position is at the heel of the wire [5; 6; 7; 8].

The average number of cycles to failure, N_f, for a wire undergoing temperature cycling is dependent on the temperature change magnitude, the geometry of the wire and the wire material properties and is given by [5; 9]

$$N_f = A \, (\epsilon_f)^n, \tag{7.41}$$

where

$$\epsilon_f = \frac{r}{\rho_0} \left[\frac{\cos^{-1}\{\cos \lambda_0 [1 - (\alpha_w - \alpha_s) \, \Delta T^*]\}}{\theta_0} - 1 \right],$$

r = wire radius, m (meter),

θ_0 = angle of the wire with the substrate, radians,

α_w = coefficient of thermal expansion of the wire, (m/m)/°C,

α_s = coefficient of thermal expansion of the substrate, (m/m)/°C,

ΔT^* = average magnitude of temperature change in °C during temperature cycling; i.e., the absolute difference between the maximum and the minimum temperature extremes,

ρ_0 = initial radius of curvature of the wire, m (meter),

A = material constant,

and

n = material constant.

The typical values for the average number of cycles to failure are the following [10]:

1. 18,000 cycles for 0.008 inch diameter 99.99% pure aluminum bonds on a 2N4863 power transistor subjected to temperatures from 25 °C to 125 °C.

2. Over 200,000 power cycles for 0.002 inch diameter aluminum, 1% silicon, ultrasonic bonds with loop heights greater than 25% of the bond to bond spacing, for temperatures ranging from 38°C to 170 °C.

7.8.2   WIRE BOND FATIGUE LIFE PREDICTION

In an electronic equipment, the wire bonds connect the bond pad and its associated die metalization to a wire which is connected, in turn, to either a lead, or to other bond pads in the case of a hybrid package. When subjected to thermal cycling a wire bond will experience cyclic shear stresses between the wire bond itself, and the wire and the substrate due to differential thermal expansion resulting in cumulative fatigue damage until a failure occurs [11; 12]. The fatigue damage induced by thermal cycling results from the initiation and growth of the cracks to fracture. Cracks may be initiated from defects or microcracks inherent in the wire or bond pad. Some cracks in the bond pad or substrate may be introduced during the bonding process, or from ultrasonic vibrations or thermal energy applied at the wire-pad interface during manufacture and assembly. Usually these cracks go undetected during

THERMAL FATIGUE LIFE MODELS

inspection and will propagate under the thermal cycling environment. It has been found that a significant number of modules fail due to wire bond failure during the thermal cycling process [12].

1. **Fatigue Life of the Bond Pad due to Cyclic Shear Between the Bond Pad and the Substrate and Between the Bond Pad and the Wire**

   The average number of cycles to failure of the bond pad, $N_{f,p}$, due to cyclic shear between the bond pad and the substrate, and between the bond pad and the wire during thermal cycling is given by [10]:

   $$N_{f,p} = C_p \, (\tau_{p,max})^{-m_p}, \tag{7.42}$$

   where

   $\quad C_p, m_p$ = shear fatigue properties for the bond pad materials,

   $\quad \tau_{p,max}$ = maximum shear stress in the bond pad due to shear between the bond pad and the substrate, and between the bond pad and the wire,

   $\quad \tau_{p,max} = Q \, \Delta T^*$,

   $\quad \Delta T^*$ = average magnitude of temperature changes during thermal cycling, °C,

   $$Q = \left(\frac{G_p}{b_p \, Z}\right) \left\{ (\alpha_w - \alpha_s) - \frac{(\alpha_s - \alpha_p)}{\left[1 + \frac{E_s \, A_s}{E_p \, A_p \, (1-\nu_s)}\right]} \right\},$$

   $\quad G_p$ = shear modulus of the bond pad materials, N/m^2,

   $\quad b_p$ = bond pad thickness, m,

   $\quad b_s$ = substrate thickness, m,

   $\quad \alpha_s$ = coefficient of thermal expansion for the substrate, (m/m)/°C,

   $\quad \alpha_p$ = coefficient of thermal expansion for the bond pad, (m/m)/°C,

   $\quad \alpha_w$ = coefficient of thermal expansion for the wire, (m/m)/°C,

   $\quad E_p$ = modulus of elasticity of the pad material, N/m^2,

   $\quad E_s$ = modulus of elasticity of the substrate material, N/m^2,

   $\quad \nu_s$ = Poisson's ratio for the substrate material, which is dimensionless,

   $\quad A_p$ = cross-sectional area of the pad, m^2,

   $\quad A_s$ = effective cross-sectional area of the substrate, m^2,

   $\quad A_s = \dfrac{b_s \, (W_p + W_s)}{2}$,

W_p = width of the bond pad, m,

and

W_s = width of the substrate, m.

2. **Fatigue Life of the Wire due to Cyclic Shear Between the Wire and the Bond Pad**

The average number of cycles to failure of the wire, $N_{f,w}$, due to cyclic shear between the wire and the bond pad during thermal cycling is given by [12]:

$$N_{f,w} = C_w \left(\tau_{w,max} \right)^{-m_w}, \tag{7.43}$$

where

C_w, m_w = shear fatigue properties of the wire material,

$\tau_{w,max}$ = maximum shear stress between wire and bond pad,

$$\tau_{w,max} = \left\{ \frac{r^2}{4\,Z^2\,A_w^2} \left[\frac{\cosh(Z\,x_w)}{\cosh(Z\,l_w)} - 1 \right]^2 \right.$$
$$\left. + \frac{\sinh^2(Z\,x_w)}{\cosh^2(Z\,l_w)\,Q^2} \right\}^{\frac{1}{2}} \Delta T^*,$$

$$Z = \left\{ \left(\frac{G_p}{b_p} \right) \left[\frac{r}{E_w\,A_w} + \frac{(1 - \nu_s)\,W_p}{E_s\,A_s} \right] \right\}^{\frac{1}{2}},$$

r = wire radius, m,

A_w = cross-sectional area of the wire,

E_w = modulus of elasticity of the wire,

x_w = location of the maximum shear stress in wire,

$$x_w = \pm \arctanh \left(\frac{A_w}{r} \right),$$

and

l_w = bonded length of the wire, m.

The rest of the nomenclature is as defined in the case of bond pad fatigue life prediction.

3. **Fatigue Life of the Substrate due to Cyclic Shear Between the Substrate and the Bond Pad**

The average number of cycles to failure of the substrate, $N_{f,s}$, due to cyclic shear between the substrate and the bond pad during thermal cycling is given by [12]:

$$N_{f,s} = C_s \left(\tau_{s,max} \right)^{-m_s}, \tag{7.44}$$

where

$$C_s, m_s = \text{shear fatigue properties of the substrate material,}$$

$$\tau_{s,max} = \text{maximum shear stress between substrate and bond pad,}$$

$$\tau_{s,max} = \left\{ \left[\frac{W_p \, Q}{2 \, Z \, A_s} \left(1 - \frac{\cosh(Z \, x_s)}{\cosh(Z \, l_s)} \right) \right. \right.$$
$$\left. + \frac{(\alpha_s - \alpha_p)}{(1 - \nu_s)/E_s + A_s/(E_p \, A_p)} \right]^2$$
$$\left. + Q^2 \, \frac{\sinh^2(Z \, x_s)}{\cosh^2(Z \, l_s)} \right\}^{\frac{1}{2}} \Delta T^*,$$

$$x_s = \text{location of the maximum shear stress in the substrate,}$$

$$x_s = \mp \operatorname{arctanh}\left(\frac{A_s}{W_p} \right),$$

and

$$l_s = \text{bonding length of the substrate, m.}$$

The rest of the nomenclature is as defined in the previous two cases of fatigue life prediction for bond pad and wire, respectively.

7.8.3 DIE FRACTURE AND FATIGUE LIFE PREDICTION

Generally, die and substrate or lead frame have different thermal expansion coefficients because they are made of different materials. For example, dies are usually made of silicon, gallium arsenide or indium phosphide; but the substrate is typically alumina, berylia or copper. Tensile stress in the central portion of the die and shear stress at the edges of the die are developed during thermal and power cycling, which may lead to vertical and horizontal cracking of the die, respectively. As the surface cracks at the center or at the edge of the die propagate and finally reach their critical sizes, a sudden ultimate fracture of the brittle die may occur without any plastic deformation.

The average number of cycles to fracture failure, N_f, of the die is given by [5]:

$$N_f = \frac{2}{(2 - n) \, A \, (\Delta\sigma_{app})^n \, \pi^{n/2}} \left(a_c^{1 - \frac{n}{2}} - a_i^{1 - \frac{n}{2}} \right), \qquad (7.45)$$

where

$$A, n = \text{die material dependent coefficient and exponent,}$$
$$\text{respectively, which governs the die crack propagation}$$
$$\text{behavior as represented by Paris' Power Law [12]; i.e.,}$$

$$\frac{da}{dN} = A \, (\Delta K)^n$$

N = number of temperature cycles applied,

a = crack size, m,

a_i = initial crack size, m,

a_c = critical crack size, m,

K = stress intensity factor at or near the die crack tip,

$\Delta K = K_{max} - K_{min}$,

$\quad\;\; = \sigma_{app} (\pi a)^{\frac{1}{2}}$, for an infinite solid,

σ_{app} = applied stress amplitude, N/m^2,

$\sigma_{app} = 10^{-6} k |\alpha_s - \alpha_d| \Delta T^* \sqrt{\dfrac{E_s E_a L}{X}}$,

k = dimensionless geometric constant,

α_s = coefficient of thermal expansion of substrate, (m/m)/°C,

α_d = coefficient of thermal expansion of die, (m/m)/°C,

E_s = tensile modulus of substrate, N/m^2,

E_a = tensile modulus of adhesive, N/m^2,

L = diagonal length of die, m,

ΔT^* = temperature change magnitude during thermal cycling, °C,

and

X = adhesive bond thickness, m.

7.8.4 DIE AND SUBSTRATE ADHESION FATIGUE LIFE PREDICTION

During a temperature or power cycling, temperature differences and temperature gradient differences will occur between the die, the die attach and the die substrate. The bond between the die and the substrate will experience fatigue failure due to the differential thermal expansion.

Voids are the most common die attach defects. These voids in the die attach, in particular when they are relatively large in size, will induce high longitudinal stresses during thermal or power cycling since the heat must flow around the void creating a large temperature gradient in the silicon. They act as microcracks which propagate and lead to debonding of the die from the substrate or the substrate from the case.

1. Fatigue Life for the Die Attach

The average number of cycles to failure of the die attach, N_f, as a result of plastic strain cycling is given by the Coffin-Manson relation

[5, p. 80] as follows:

$$N_f = 0.5 \left(\frac{\gamma_a}{\gamma_f}\right)^{\frac{1}{c}},$$ (7.46)

where

$c =$ Coffin-Manson coefficient,

$\gamma_f =$ fatigue ductility coefficient of the die attach which is defined as the shear strain required to cause failure in one load reversal,

$\gamma_a =$ actual plastic strain amplitude,

$\gamma_a = \dfrac{L \, |\alpha_s - \alpha_d| \, \Delta T^*}{X}$,

$L =$ diagonal die length, m,

$\alpha_s =$ coefficient of thermal expansion of the substrate, $(\text{m/m})/°\text{C}$,

$\alpha_d =$ coefficient of thermal expansion of the die, $(\text{m/m})/°\text{C}$,

$\Delta T^* =$ temperature change amplitude during thermal cycling, $°\text{C}$,

and

$X =$ height of die attach, m.

2. Fatigue Life for the Substrate Attach

The average number of cycles to fatigue failure of the substrate attach is also given by the Coffin-Manson relation [5, p. 81; 13; 14] as follows:

$$N_f = 0.5 \left(\frac{\gamma_a}{\gamma_f}\right)^{\frac{1}{c}},$$ (7.47)

where

$c =$ Coffin-Manson coefficient,

$\gamma_f =$ fatigue ductility coefficient of the substrate which is defined as the shear strain required to cause failure in one load reversal,

$\gamma_a =$ actual plastic strain amplitude,

$\gamma_a = \dfrac{L_s \, |\alpha_c - \alpha_s| \, \Delta T^*}{h_{sa}}$,

$L_s =$ diagonal length of the substrate, m,

$\alpha_s =$ coefficient of thermal expansion of the substrate, (m/m)/°C,

α_c = coefficient of thermal expansion of the case,
(m/m)/°C,

ΔT = temperature change amplitude during thermal
cycling, °C,

and

h_{sa} = thickness of the substrate attach, m.

7.8.5 SOLDER JOINT FATIGUE LIFE PREDICTION

It was indicated in [15] that most of the electronic components used for surface-mounted technology (SMT) applications rely upon the structural integrity of the solder to support these components in various thermal, vibration, and shock environments. The SMT solder joint mechanically attaches the components to the printed wiring board (PWB) and provides electrical and thermal continuity. The major cause of failure is the low-cycle thermal fatigue of the solder joints [16], though vibration-induced high cycle fatigue is believed to make a secondary contribution to fatigue damage [16; 17; 18]. In low-cycle thermal fatigue, the thermal strains are caused by different coefficients of thermal expansion for materials used in PWB assembly and also by the temperature differentials due to internal heat generation or the external temperature variations caused by component, system load fluctuations, and/or on-off cycles, and by environmental temperature changes. Experience has shown [5, p. 263] that solder can creep at a rate of about 0.002 to 0.003 inches per year when subjected to compression loads.

The most often used solder has been the eutectic solder which is composed of 63% tin and 37% lead, the so-called Sn-Pb solder. The melting temperature of this Sn-Pb solder is only about 184°C, which is low enough to protect the sensitive electronic components from thermal damage when they are attached to the printed circuit board (PCB). When the stress levels are above approximately 800 psi at room temperature conditions, this solder tends to creep extensively. These creep effects can cause premature failures in the solder unless the stress levels are kept under about 400 psi in thermal cycling and vibration environments and they are temperature sensitive. When the temperature is above about 85°C, the creep increases very significantly. However, when the temperature is below about −20°C, the creep reduces sharply.

A widely accepted thermal fatigue life prediction model for near eutectic Sn-Pb solders under thermal cycling conditions was proposed by Engelmaier [15], which includes an empirical description of the effect of cyclic frequency and mean temperature value. This model can be expressed as

$$N_f = \frac{1}{2}\left(\frac{\Delta\gamma}{2\,\varepsilon_f}\right)^{\frac{1}{c}},$$ (7.48)

where

N_f = mean number of thermal cycles to failure,

$\Delta\gamma$ = total shear strain range in one cycle of loading,

ε_f = fatigue ductility coefficient = 0.325,

c = fatigue ductility exponent,

$c = -0.442 - 6 \times 10^{-4}\,T^*_{mean} + 1.74 \times 10^{-2}\,\log_e(1+f)$,

T^*_{mean} = mean cyclic temperature in the solder joint, °C,

$T^*_{mean} = \frac{1}{2}(T^*_{max} + T^*_{min})$,

T^*_{max} = maximum cyclic temperature, °C,

T^*_{min} = minimum cyclic temperature, °C,

and

f = cyclic frequency, $1 < f < 1,000$ cycles/day.

7.8.6 COMMENTS

The physical models for the thermal fatigue life prediction are summarized for various failure mechanisms which may occur during thermal cycling. Each of them gives the average number of cycles to failure for a particular failure mechanism in terms of the associated physical characteristic parameters, geometries, applied temperature change amplitude during thermal cycling and temperature-induced stress. However, the effect of the rate of temperature change on the thermal fatigue life is not reflected in these models. Furthermore, the randomness in the physical characteristics, geometries and applied temperature profiles will cause the randomness both in the temperature-induced stress and the fatigue life; i.e., the thermal fatigue life should be statistically distributed. Multiple failure mechanisms each being governed by a life distribution in an electronic equipment during the thermal cycling may lead to a mixed life distribution for failures of the whole equipment. More extensive investigation is needed in the future in these areas.

7.9 CONCLUSIONS

Conventional practice in evaluating the effect of elevated temperature in accelerated life testing and in stress screening is based upon the standard Arrhenius equation and thereby ignores several aspects of the stress. A general model is provided here to include these additional aspects of the stress. This model combines the basic reaction rate stress during the cooling and heating intervals and the temperature change stress (ramp effect), which yields an equivalent acceleration factor in one temperature cycle. The classical Arrhenius model is modified by incorporating a temperature-dependent and therefore cycling-time-dependent activation energy in place of the traditionally assumed temperature independent constant activation energy. The resultant models with two modified versions of the Arrhenius model are rather complicated expressions which must be evaluated numerically, but provide a more realistic portrayal of the effect of thermal cycling upon electronic devices. This expanded but more general and realistic model presented in this chapter provides greater visibility to the parameters of a thermal cycle.

A model is derived to determine the optimum number of thermal cycles for a given temperature profile and a specified field $MTBF$ goal. This model is based on the result of equivalent acceleration factor of a typical temperature cycle and assumes the knowledge of the equipment's life distribution at the screening stress level.

Finally, several useful physical models for thermal fatigue life prediction are presented. These models serve the purpose of predicting the average time to failure for various specific failure mechanisms in an electronic equipment during thermal cycling. The application of these models requires the knowledge of the associated physical properties and geometries.

REFERENCES

1. Jensen, F., "Activation Energies and the Arrhenius Equation," *Quality and Reliability International*, Vol. 1, pp. 13-17, 1985.

2. Pugacz-Muraszkiewicz, I., "Arrhenius Law in Its Application to the Stress Screening of Electronic Hardware – A Model with a Pitfall," *Proceedings of the Institute of Environmental Sciences*, pp. 779-783, 1990.

3. Nachlas, J. A., "A General Model for Age Acceleration During Thermal Cycling," *Quality and Reliability International*, Vol. 2, pp. 3-6, 1986.

4. Nachlas, J. A., Binney, B. A. and Gruber, S. S., "Aging Acceleration Under Multiple Stresses," *Proceedings of the Annual Reliability and Maintainability Symposium*, Philadelphia, Pennsylvania, pp. 438-440, 1985.

5. Bar-Cohen, A. and Kraus, A. D., *Advances in Thermal Modeling of Electronic Components and Systems*, Vol. 3, ASME Press, New York, IEEE Press, New York, 402 pp., 1993.

6. Gaffeny, J., "Internal Lead Fatigue Through Thermal Expansion in Semiconductor Devices," *IEEE Transactions on Electronic Devices*, ED-15, p. 617, 1968.

7. Ravi, K. V. and Philosky, E. M., "Reliability Improvement of Wire Bonds Subjected to Fatigue Stresses," *Proceedings of the 10th IEEE Annual Reliability Physics Symposium*, Las Vegas, Nevada, pp. 143-149, 1972.

8. Phillips, W. E., "Microelectronic Ultrasonic Bonding," edited by Harman, G. G., National Bureau of Standards (US), Special Publication 400-2, pp. 80-86, 1974.

9. Pecht, M., Lall, P. and Dasgupta, A., "A Failure Prediction Model for Wire Bonds," *Proceedings of 1989 International Symposium on Hybrid Microelectronics*, pp. 607-613, 1989.

10. Harman, G. G., "Metallurgical Failure Modes of Wire Bonds," *Proceedings of 12th Annual International Reliability Physics Symposium*, pp. 131-141, 1974.

11. Hu, J., Pecht, M. and Dasgupta, A., "A Probabilistic Approach for Predicting Thermal Fatigue Life of Wirebonding in Microelectronics," *Journal of Electronic Packaging*, Vol. 113, pp. 275-285, September 1991.

12. Paris, P. C., Gomez, M. P. and Anderson, W. E., "A Rational Analytical Theory of Fatigue," *The Trend in Engineering*, Vol. 13, University of Washington, Seattle, Washington, pp. 9-24, 1961.

13. Bolger, J. C., "Polyimide Adhesive to Reduce Thermal Stress in LSI Ceramic Packages," *Proceedings of 14th National SAMPE Technical Conference*, pp. 12-14, October, 1982.

14. Bolger, J. C. and Mooney, C. T., "Die Attach in Hi-rel DIPS: Polyimides or Low Chloride Epoxies," *IEEE Transactions on Composite Hybrid Manufacturing Techniques*, CHMT-7 (4), December, 1984.

15. Engelmaier, W., "Fatigue Life of Leadless Chip Carrier Solder Joints During Power Cycling," *IEEE Transactions on Components, Hybrids, and Manufacturing Technology*, Vol. CHMT-6, No. 3, pp. 232-237, 1985.

16. Barker, D. B., Sharif, I., Dasgupta, A. and Pecht, M. G., "Effect of SMC Lead Dimensional Variabilities on Lead Compliance and Solder Joint Fatigue Life," *Journal of Electronic Packaging*, Vol. 114, pp. 177-184, 1992.

17. Blanks, H. S., "Accelerated Vibration Fatigue Life Testing of Leads and Solder Joint," *Microelectronics and Reliability*, Vol. 15, pp. 213-219, 1976.

18. Steinberg, D. S., *Vibration Analysis for Electronic Equipment*, John Wiley, New York, 443 pp., 1988.

APPENDIX 7A
COMPUTER PROGRAM FOR EXAMPLE 7-2

```
C
C $$$$$$$$$$$$$$$$$$$$$$$$$$$$$$$$$$$$$$$$$$$$$$$$$$$$$$$$$$$$$$$$$$$$$$$
C $ THIS PROGRAM IS DEVELOPED FOR EVALUATING THE EQUIVALENT $
C $ ACCELERATION FACTORS WHEN THE REACTION RATE STRESS AND  $
C $ THE TEMPERATURE CHANGE ARE BOTH PRESENT.                $
C $                                                         $
C $ TWO MODIFICATION FORMS FOR THE ARRHENIUS EQUATION ARE   $
C $ USED WHICH SHOULD YIELD CLOSE RESULTS.                  $
C $$$$$$$$$$$$$$$$$$$$$$$$$$$$$$$$$$$$$$$$$$$$$$$$$$$$$$$$$$$$$$$$$$$$$$$
C
        EXTERNAL FUNC1,FUNC2,FUNC3,FUNC4,FUNC5,FUNC6
C
C       'AS' == SMALL A, 'a'
C       'BS' == SMALL B, 'b'
C       'CB' == BIG C, 'C'
C       'BB' == BIG B, 'B'
C
        COMMON /PAR1/AS,BS,CB,BB,ETA
        COMMON /PAR2/T1,T2,T4,T5,T6,T0,TL,TU,BETAH,BETAC
        OPEN(5,FILE='ESSEXP4-6.IN',STATUS='UNKNOWN')
        READ(5,*)AS,BS,CB,BB,ETA,T1,T2,T4,T5,T6,T0,TL,
      1 TU,BETAH,BETAC
        WRITE(*,*)'                      THE INPUT PARAMETERS'
        WRITE(*,*)('*',I=1,78)
        WRITE(*,*)'AS=',AS,'BS=',BS,'CB=',CB,'BB=',BB
        WRITE(*,*)'ETA=',ETA,'T1=',T1,'T2=',T2,'T4=',T4
        WRITE(*,*)'T5=',T5,'T6=',T6,'T0=',T0,'TL=',TL
        WRITE(*,*)'TU=',TU,'BETAH=',BETAH,'BETAC=',BETAC
        WRITE(*,*)('*',I=1,78)
        CALL ACC1(AF1)
        CALL ACC2(AF2)
        WRITE(*,*)
        WRITE(*,*)'                              RESULTS'
        WRITE(*,*)('*',I=1,78)
        WRITE(*,*)'EQUIVALENT ACCELERATION FACTOR USING
      1 FIRST MODIFICATION:'
        WRITE(*,*)'                 AF1==',AF1
        WRITE(*,*)'EQUIVALENT ACCELERATION FACTOR USING
      1 SECOND MODIFICATION:'
        WRITE(*,*)'                 AF2==',AF2
```

```
      STOP
      END
C
C     SUBPROGRAM EVALUATING THE EQUIVALENT ACCELERATION
C         FACTOR USING THE FIRST VERSION OF MODIFIED
C                    ARRHENIUS EQUATION
C
      SUBROUTINE ACC1(AF1)
      EXTERNAL FUNC1,FUNC2,FUNC3,FUNC4,FUNC5,FUNC6
      REAL K
      COMMON /PAR1/AS,BS,CB,BB,ETA
      COMMON /PAR2/T1,T2,T4,T5,T6,TO,TL,TU,BETAH,BETAC
      K=8.623E-5
      CALL QGAUS(FUNC1,0.0,T1,S1)
      E1=(T2-T1)*EXP(-AS/K*TU**(BS-1.0))
      CALL QGAUS(FUNC2,T2,T4,S2)
      E2=(T5-T4)*EXP(-AS/K*TL**(BS-1.0))
      CALL QGAUS(FUNC3,T5,T6,S3)
      E0=1.0/T6*EXP(AS/K*TO**(BS-1.0))
      AF1=E0*(S1+E1+S2+E2+S3)
      RETURN
      END
C
C     SUBPROGRAM EVALUATING THE EQUIVALENT ACCELERATION
C         FACTOR USING THE SECOND VERSION OF MODIFIED
C                    ARRHENIUS EQUATION
C
      SUBROUTINE ACC2(AF2)
      EXTERNAL FUNC1,FUNC2,FUNC3,FUNC4,FUNC5,FUNC6
      REAL K
      COMMON /PAR1/AS,BS,CB,BB,ETA
      COMMON /PAR2/T1,T2,T4,T5,T6,TO,TL,TU,BETAH,BETAC
      K=8.623E-5
      CALL QGAUS(FUNC4,0.0,T1,S1)
      E1=(T2-T1)*EXP(CB*TU**BB*(1.0/TO-1.0/TU))
      CALL QGAUS(FUNC5,T2,T4,S2)
      E2=(T5-T4)*EXP(-CB*TL**BB*(1.0/TL-1.0/TO))
      CALL QGAUS(FUNC6,T5,T6,S3)
      E0=1.0/T6
      AF2=E0*(S1+E1+S2+E2+S3)
      RETURN
      END
C
C     SUBPROGRAM EVALUATING THE INTEGRATION
```

```
C               USING GAUSS METHOD
C
      SUBROUTINE QGAUS(FUNC,A,B,SS)
      DIMENSION X(5),W(5)
      DATA X/0.1488743389,0.4333953941,0.6794095682,
     1      0.8650633666,0.9739065285/
      DATA W/0.2955242247,0.2692667193,0.2190863625,
     1      0.1494513491,0.0666713443/
      XM=0.5*(B+A)
      XR=0.5*(B-A)
      SS=0.0
      DO 11 J=1,5
      DX=XR*X(J)
      SS=SS+W(J)*(FUNC(XM+DX)+FUNC(XM-DX))
11    CONTINUE
      SS=XR*SS
      RETURN
      END
C
C     FUNC1 THROUGH FUNC3 ARE THE INTEGRANDS FOR
C                  SUBPROGRAM ACC1(AF1)
C
      FUNCTION FUNC1(Z)
      REAL K
      COMMON /PAR1/AS,BS,CB,BB,ETA
      COMMON /PAR2/T1,T2,T4,T5,T6,T0,TL,TU,BETAH,BETAC
      K=8.623E-5
      P1=AS/K*(T0+(TU-T0)*(Z/T1)**BETAH)**(BS-1.0)
      P2=ETA*BETAH*(TU-T0)/T1*(Z/T1)**(BETAH-1.0)
      FUNC1=EXP(-(P1-P2))
      RETURN
      END
C
      FUNCTION FUNC2(Z)
      REAL K
      COMMON /PAR1/AS,BS,CB,BB,ETA
      COMMON /PAR2/T1,T2,T4,T5,T6,T0,TL,TU,BETAH,BETAC
      K=8.623E-5
      P1=AS/K*(TU-(TU-TL)*((Z-T2)/(T4-T2))**BETAC)**(BS-1.0)
      P2=ETA*BETAC*(TU-TL)/(T4-T2)*((Z-T2)/(T4-T2))**(BETAC-1.0)
      FUNC2=EXP(-(P1-P2))
      RETURN
      END
C
```

```
      FUNCTION FUNC3(Z)
      REAL K
      COMMON /PAR1/AS,BS,CB,BB,ETA
      COMMON /PAR2/T1,T2,T4,T5,T6,T0,TL,TU,BETAH,BETAC
      K=8.623E-5
      P1=AS/K*(TL+(T0-TL)*((Z-T5)/(T6-T5))**BETAH)**(BS-1.0)
      P2=ETA*BETAH*(T0-TL)/(T6-T5)*((Z-T5)/(T6-T5))**(BETAH-1.0)
      FUNC3=EXP(-(P1-P2))
      RETURN
      END
C
C     FUNC4 THROUGH FUNC6 ARE THE INTEGRANDS
C           FOR SUBPROGRAM ACC2(AF2)
C
      FUNCTION FUNC4(Z)
      REAL K
      COMMON /PAR1/AS,BS,CB,BB,ETA
      COMMON /PAR2/T1,T2,T4,T5,T6,T0,TL,TU,BETAH,BETAC
      K=8.623E-5
      P1=(1.0/T0)*(T0+(TU-T0)*(Z/T1)**BETAH)**BB
      P2=(T0+(TU-T0)*(Z/T1)**BETAH)**(BB-1.0)
      P3=ETA*BETAH*(TU-T0)/T1*(Z/T1)**(BETAH-1.0)
      FUNC4=EXP(CB*(P1-P2)+P3)
      RETURN
      END
C
      FUNCTION FUNC5(Z)
      REAL K
      COMMON /PAR1/AS,BS,CB,BB,ETA
      COMMON /PAR2/T1,T2,T4,T5,T6,T0,TL,TU,BETAH,BETAC
      K=8.623E-5
      P1=(1.0/T0)*(TU-(TU-TL)*((Z-T2)/(T4-T2))**BETAC)**BB
      P2=(TU-(TU-TL)*((Z-T2)/(T4-T2))**BETAC)**(BB-1.0)
      P3=ETA*BETAC*(TU-TL)/(T4-T2)*((Z-T2)/(T4-T2))**(BETAC-1.0)
      FUNC5=EXP(CB*(P1-P2)+P3)
      RETURN
      END
C
      FUNCTION FUNC6(Z)
      REAL K
      COMMON /PAR1/AS,BS,CB,BB,ETA
      COMMON /PAR2/T1,T2,T4,T5,T6,T0,TL,TU,BETAH,BETAC
      K=8.623E-5
      P1=(1.0/T0)*(TL+(T0-TL)*((Z-T5)/(T6-T5))**BETAH)**BB
```

```
P2=(TL+(TO-TL)*((Z-T5)/(T6-T5))**BETAH)**(BB-1.0)
P3=ETA*BETAH*(TO-TL)/(T6-T5)*((Z-T5)/(T6-T5))**(BETAH-1.0)
FUNC6=EXP(CB*(P1-P2)+P3)
RETURN
END
```

APPENDIX 7B
COMPUTER PROGRAM FOR EXAMPLE 7–3

```
C
C $$$$$$$$$$$$$$$$$$$$$$$$$$$$$$$$$$$$$$$$$$$$$$$$$$$$$$$$$$$$$$$$$$$$$$$$$$
C $ THIS PROGRAM IS DEVELOPED FOR THE DETERMINATION OF THE OPTIMUM $
C $ THERMAL CYCLING TIME, OR EQUIVALENTLY, THE OPTIMUM NUMBER OF   $
C $ THERMAL CYCLES TO ACHIEVE A SPECIFIED FIELD MTBF GOAL. THE     $
C $ ITERATION TECHNIQUE IS USED IN DEVELOPING THIS PROGRAM.        $
C $$$$$$$$$$$$$$$$$$$$$$$$$$$$$$$$$$$$$$$$$$$$$$$$$$$$$$$$$$$$$$$$$$$$$$$$$$
C
      REAL MFIELD, LAMBDAB,LAMBDAG,NCYC
      WRITE(*,*)'PLEASE ENTER THE SPECIFIED FIELD MTBF GOAL (hr) !'
      READ(*,*)MFIELD
      WRITE(*,*)'PLEASE ENTER THE TYPICAL THERMAL CYCLE LENGTH (hr) !'
      READ(*,*)T6
      WRITE(*,*)'PLEASE ENTER THE EQUIVALENT ACCELERATION FACTOR'
      WRITE(*,*)'PRODUCED IN ONE TYPICAL THERMAL CYCLE!'
      READ(*,*)AF
      WRITE(*,*)'PLEASE ENETER THE MIXED LIFE DISTRIBUTION'
      WRITE(*,*)'PARAMETERS, Pb, LAMBDAb (fr/hr), Pg, LAMBDAg'
      WRITE(*,*)'(fr/hr), AS OBTAINED AT THE SCREENING'
      WRITE(*,*)'STRESS LEVEL!'
      READ(*,*)PB,LAMBDAB,PG,LAMBDAG
      WRITE(*,*)'PLEASE ENTER THE DESIRED CONVERGENCE ACCURACY!'
      READ(*,*)EPS
      TS0=T6
      WRITE(*,*)'The initial value for TS is TS0=',TS0
10    R=PB*EXP(-LAMBDAB*TS0) + PG*EXP(-LAMBDAG*TS0)
      RR=PB*EXP(-LAMBDAB*TS0)/LAMBDAB
   1 + PG*EXP(-LAMBDAG*TS0)/LAMBDAG
      TS1=(MFIELD*T6/AF)*(R/RR)
      WRITE(*,*)'TS1=',TS1
      DELTA=ABS(TS0-TS1)
      ERROR=DELTA/TS0
      IF(ERROR.LE.EPS) GOTO 100
      TS0=TS1
      GOTO 10
100   TS=0.5*(TS0+TS1)
      WRITE(*,*)'THE OPTIMUM THERMAL CYCLING TIME IS'
      WRITE(*,*)'TS==',TS
      NCYC=TS/T6
      WRITE(*,*)'THE OPTIMUM NUMBER OF THERMAL CYCLES IS'
```

```
WRITE(*,*)'NCYC==',NCYC
STOP
END
```

Chapter 8

PHYSICAL QUANTIFICATION AND OPTIMIZATION OF *ESS* BY RANDOM VIBRATION

8.1 INTRODUCTION

As stated before, random vibration is one of the most effective screening stresses used in *ESS*. It is considered particularly effective in exposing mechanical defects, such as loose solder, improper bonds, printed circuit board shorts, and others. Among three principal types of vibration, namely, sine wave with fixed frequency, sine wave with swept frequency, and random vibration, random vibration is the most effective. Random vibration involves the excitation of a product with a predetermined profile over a wide frequency range, usually from 20 to 2,000 Hz. Product stress is created through simultaneous excitation of all resonant frequencies within the profile range.

Random vibration is defined as such vibration where either the excitation or the time variation of the system parameters, or both, are random. The random excitation may be classified into three categories: narrow band, wide band, and white noise. Essentially the quantitative analysis of random vibration is a combination of probability theory, stochastic calculus of random processes and applied dynamics. It is assumed that the readers have the knowledge of probability theory and mechanical vibration. Some fundamental concepts and equations associated with a random process, which are useful in random vibration, are briefly reviewed next.

8.2 FUNDAMENTAL THEORY OF A RANDOM PROCESS

8.2.1 DEFINITION OF A RANDOM PROCESS

The random response of dynamic systems is usually given in the form of a time history record. Each record is called a *realization* or a *sample function*. The collection of all possible records (sample functions) is called the *ensemble* or the *random process* [1; 2]. Figure 8.1 portrays a typical example of the realizations of a random process, $X(t)$, where t is the parameter, say, time. The superscript $j = 1, 2, \cdots$ in $x^j(t)$ represents the jth realization of the random process $X(t)$. At fixed times $t = t_1$ and t_2, $X(t_1)$ and $X(t_2)$ are random variables, and $X^j(t_1)$ and $X^j(t_2)$ are their jth realizations.

A random process, $X(\omega, t)$, which in the literature is also called a *random function*, a *stochastic process*, or a *time series* if the index parameter t is time, is defined as a parameted family of random variables with two parameters (arguments) $\omega \in \Omega$ denoting that ω belongs to the parent set Ω, and $t \in T$ denoting that t belongs to the parent set T, where Ω is the sample space of the family of random variables $X(., t)$ in which "." means the whole set of real numbers and T is the indexing set of the parameter t. When t is time, T represents the entire time domain of the random process.

8.2.2 CLASSIFICATION OF A RANDOM PROCESS

A random process $X(\omega, t)$ may be described in two different ways. It may be treated as a family of random variables $\{X(., t) : t \in T\}$ or as a set of functions $\{X(\omega, .) : \omega \in \Omega\}$ on T where "." means the whole set of real numbers. If the indexing set T is finite, a random process is a *random vector*; it is a *random sequence* if T is countably infinite. A random process is said to be discrete or continuous depending on whether it is a family of discrete or of continuous random variables. Thus a random process may be classified into one of the following four categories:

1. Continuously parametered continuous random process.

2. Continuously parametered discrete random process.

3. Discretely parametered continuous random process.

4. Discretely parametered discrete random process.

Usually the argument ω is dropped and the random process $X(\omega, t)$ is simply denoted by $X(t)$. Many random processes are characterized by more than one indexing parameter. Such processes are called *multiparametered random processes* or *random fields*. For example, $X(t, s)$, $t \in T$, $s \in S$, may describe the random process model of the pressure field due to jet noise that varies randomly with time t and space coordinate s over the domain T and S, respectively.

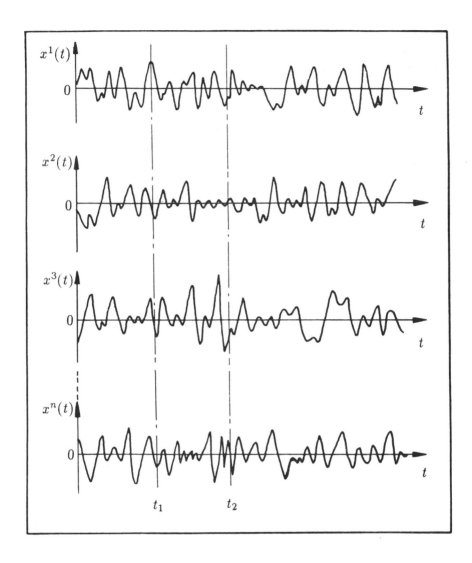

Fig. 8.1– Ensemble of a random process.

8.2.3   PROBABILITY DESCRIPTION OF A RANDOM PROCESS

In general, a single sample record may not be adequate to represent the actual behavior of the system except for ergodic processes where the ensemble statistical properties are equal to those of each sample record. For ergodic processes the statistical properties are determined along the sample provided that the sample exhibits all statistical characteristics of the system. For non-ergodic processes the statistical properties are determined across the ensemble.

Since a random process $X(t)$ reduces to a set of random variables at fixed instants of time $t = t_1, t_2, \cdots, t_n$, its probability structure may be defined by a hierarchy of joint probability density functions:

$$f(x_1, t_1),$$
$$f(x_1, t_1; x_2, t_2),$$
$$\vdots$$
$$f(x_1, t_1; x_2, t_2; ...; x_n, t_n), \tag{8.1}$$

where

$$x_i = x(t_i), \quad i = 1, 2, ..., n.$$

In Eq. (8.1), the definition of the probability density function of a random variable is extended to include the effect of the parameter t, so that

$$f(x_1, t_1)dx_1 = \text{probability that } X(t) \text{ lies in the}$$
$$\text{interval } (x_1, x_1 + dx_1) \text{ at time } t_1,$$

and

$$f(x_1, t_1; x_2, t_2)dx_1 dx_2 = \text{joint probability that } X(t) \text{ lies in}$$
$$\text{the interval } (x_1, x_1 + dx_1) \text{ at}$$
$$\text{time } t_1, \text{ and in the interval}$$
$$(x_2, x_2 + dx_2) \text{ at time } t_2,$$

.
.
.

8.2.4   ENSEMBLE AVERAGES

Consider a fixed time $t = t_1$, and the ensemble of values $x(t_1)$, or simply x. Let $g(x)$ be a known function of x. The average of $g(x)$ taken across the ensemble is called the *ensemble average* which can be obtained either experimentally or theoretically.

First, let us look at the experimental case where n sample values $x^j(t_1)$, for $j = 1, 2, \cdots, n$, are available. Under the assumption that those n samples adequately represent the process, the average of g across the ensemble would be simply the sum of the g values divided by the number of samples; i.e.,

$$\overline{g(x)} = \frac{1}{n} \sum_{j=1}^{n} g[x^j(t_1)]. \tag{8.2}$$

Second, let us consider the probability density function of x, $f(x)$. The fraction of samples for which the x values lie between x and $x + dx$ is $f(x)dx$. Then, the continuous analog of the discrete average given by Eq. (8.2) is

$$E[g(x)] = \int_{-\infty}^{\infty} g(x) \, f(x)dx. \tag{8.3}$$

When $g(x)$ is simply x itself, the ensemble average given by Eq. (8.3) becomes

$$E[x] = \int_{-\infty}^{\infty} x \, f(x) \, dx, \tag{8.4}$$

which defines the *ensemble mean* of x or the *ensemble expected value* of x. When $g(x)$ is the function x^2, Eq. (8.3) becomes

$$E[x^2] = \int_{-\infty}^{\infty} x^2 \, f(x) \, dx, \tag{8.5}$$

which defines the *ensemble mean square value* of x. The square root of Eq. (8.5) is called the *ensemble root mean square value* or the *ensemble rms value*.

Note that $E[x^2]$ represents the *average energy* of a random process, because in many physical problems, $X(t)$ may represent an energylike quantity. For example, if $X(t)$ is the displacement of a linear spring, $[X(t)]^2$ is proportional to the strain energy stored in the spring. If $X(t)$ represents the velocity of a rigid body, $[X(t)]^2$ is proportional to the kinetic energy of the body.

The ensemble average of the square of the deviation of x from its mean $E[x]$ is the *ensemble variance* of x, σ^2; i.e.,

$$\sigma^2 = E\left[(x - E[x])^2\right],$$
$$= \int_{-\infty}^{\infty} (x - E[x])^2 \, f(x)dx,$$

or

$$\sigma^2 = E[x^2] - (E[x])^2.$$

For a zero-mean process, or when $E[x] = 0$,

$$\sigma^2 = E[x^2] = \text{average energy}.$$

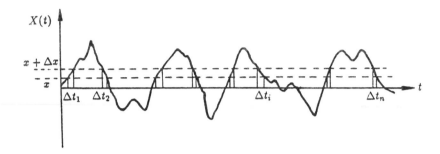

Fig. 8.2– A sample record of a random process.

8.2.5   TEMPORAL AVERAGES

Averages taken along a single sample record are called the *temporal averages*. Consider a sample record of a typical random process shown in Fig. 8.2. The probability of $X(t)$ to take values within the range x and $x + \Delta x$ may be obtained by dividing the sum of all time intervals Δt_i which Δx occupies, by the total time period of the sample, T; i.e.,

$$P[x \leq X(t) \leq x + \Delta x] = \lim_{T \to \infty} \frac{1}{T} \sum_{i=1}^{n} \Delta t_i. \tag{8.6}$$

Note that the sample length, T, should be large enough to exhibit all dynamic characteristics of the system.

Let $g[x(t)]$ be a given function of x. Then, the temporal average of $g(x)$ may be determined by taking the time average over the sample interval T; i.e.,

$$E[g(x)] = \lim_{T \to \infty} \frac{1}{T} \int_0^T g[x(t)]\, dt. \tag{8.7}$$

When $g(x)$ is simply x itself, the temporal average given by Eq. (8.7) becomes

$$E[x] = \lim_{T \to \infty} \frac{1}{T} \int_0^T x(t)\, dt, \tag{8.8}$$

which defines the *temporal mean* of x, or the *temporal expected value* of x. When $g(x)$ is the function x^2, Eq. (8.7) becomes

$$E[x^2] = \lim_{T\to\infty} \frac{1}{T} \int_0^T [x(t)]^2 \, dt, \tag{8.9}$$

which defines the *temporal mean square value* of x. The square root of Eq. (8.9) is called the *temporal root mean square value*, or the *temporal rms value*. Similar to the ensemble average case, the temporal mean square value, $E[x^2]$, represents the average energy of an ergodic random process.

The temporal average of the square of the deviation of x from its temporal mean, $E[x]$, is the *temporal variance* of x, σ^2; i.e.,

$$\sigma^2 = E\left[(x - E[x])^2\right],$$
$$= \lim_{T\to\infty} \frac{1}{T} \int_0^\infty \{x(t) - E[x]\}^2 \, dt,$$

or

$$\sigma^2 = E[x^2] - (E[x])^2.$$

For an ergodic zero-mean process, or for $E[x] = 0$,

$$\sigma^2 = E[x^2] = \text{average energy.}$$

8.2.6 AUTO-CORRELATION FUNCTION (ACF)

The dependence of the values of a random process at time t_1 on its values at t_2 is given by the *autocorrelation function*

$$AC_x(t_1, t_2) = E[x(t_1)\, x(t_2)],$$

or

$$AC_x(t_1, t_2) = \int_{-\infty}^\infty \int_{-\infty}^\infty x_1\, x_2\, f(x_1, t_1; x_2, t_2)\, dx_1\, dx_2. \tag{8.10}$$

In general, $AC_x(t_1, t_2)$ depends on t_1 and t_2. But if the process is *stationary*, or the process statistics are invariant to a time shift, $AC_x(t_1, t_2)$ depends only on the time lag between t_1 and t_2, or $\tau = |t_2 - t_1|$; i.e.,

$$AC_x(t_1, t_2) = AC_x(t_1, t_1 + \tau) = E[x(t)\, x(t + \tau)] = AC_x(\tau),$$

and

$$AC_x(\tau) = AC_x(-\tau). \tag{8.11}$$

The autocorrelation function provides, indirectly, the information on the frequency content of a random process. This may be seen later in the definition of the *spectral density function*.

A closely related function is the *autocovariance function*

$$AV_x(t_1, t_2) = E\left\{ \left[x(t_1) - E[x_1]\right]\left[x(t_2) - E[x_2]\right] \right\},$$

or

$$AV_x(t_1, t_2) = \int_{-\infty}^{\infty} \int_{-\infty}^{\infty} \{x(t_1) - E[x_1]\} \{x(t_2) - E[x_2]\}$$
$$\cdot f(x_1, t_1; x_2, t_2) \, dx_1 \, dx_2.$$

$$(8.12)$$

8.2.7   CROSS-CORRELATION FUNCTION (CCF)

The dependence of the values of one random process $X(t)$ at time t_1 on the values of another random process $Y(t)$ at time t_2 is given by the *cross-correlation function*

$$CC_{xy}(t_1, t_2) = E[x(t_1) \, y(t_2)],$$

or

$$CC_{xy}(t_1, t_2) = \int_{-\infty}^{\infty} \int_{-\infty}^{\infty} x_1 \, y_2 f(x_1, t_1; y_2, t_2) \, dx_1 \, dy_2. \qquad (8.13)$$

In general, $CC_{xy}(t_1, t_2)$ depends on t_1 and t_2. But if the two processes are stationary, $CC_{xy}(t_1, t_2)$ depends only on the time lag between t_1 and t_2, or $\tau = |t_2 - t_1|$; i.e.,

$$CC_{xy}(t_1, t_2) = CC_{xy}(\tau) = E[x(t) \, y(t + \tau)].$$

Note that $CC_{xy}(\tau) \neq CC_{yx}(\tau)$, but they follow the following relationship:

$$CC_{xy}(\tau) = CC_{yx}(-\tau),$$

and

$$CC_{yx}(\tau) = CC_{xy}(-\tau).$$

A closely related function is the *cross-covariance function*

$$CV_{xy}(t_1, t_2) = E\left\{ \left[x(t_1) - E[x_1]\right]\left[y(t_2) - E[y_2]\right] \right\},$$

or

$$CV_{xy}(t_1, t_2) = \int_{-\infty}^{\infty} \int_{-\infty}^{\infty} \{x(t_1) - E[x_1]\}\{y(t_2) - E[y_2]\}$$
$$\cdot f(x_1, t_1; y_2, t_2) \, dx_1 \, dy_2. \qquad (8.14)$$

8.2.8   STATIONARITY OF A RANDOM PROCESS

A random process is said to be *stationary* if its probability structure is invariant under arbitrary shifts of the time scale, in other words, the process is invariant to a time shift and all statistical properties, such as the autocorrelation function given by Eq. (8.10), are the same for a time shift, τ. Thus, $X(t)$ is stationary for all $t_1, t_2, \cdots, t_n$, and an arbitrary constant τ if for all n the following equality exits:

$$f(x_1, t_1; x_2, t_2; \cdots; x_n, t_n) = f(x_1, t_1 + \tau; x_2, t_2 + \tau; \cdots; x_n, t_n + \tau).$$

(8.15)

If this property holds for a particular n, the random process is called *stationary of order n*. If a process is stationary of order n, then it is also stationary of all orders less than n.

Physically, stationarity implies a measure of temporal uniformity in the characteristics of the factors contributing to the randomness. In particular the first-order probability density, $f[x(t)]$, becomes a universal function independent of time t. This implies that all the averages based on $f(x)$ (e.g., the mean $E[x]$ and the variance σ^2) are constants independent of time. If the second-order probability density $f[x(t_1), x(t_2)]$ is to be invariant under any time scale shift (stationary of order $n \geq 2$), then $f[x(t_1), x(t_2)]$ and therefore the autocorrelation function $AC_x(t_1, t_2) = E[x(t_1)\ x(t_2)]$ are only dependent on the time lag $|t_2 - t_1|$. Therefore, it is possible to partially verify the stationarity experimentally by obtaining a large family of sample functions and then calculating averages, such as the mean and the autocorrelation for many different times. If there is a substantial agreement among the results at different times, then the stationarity is warranted.

In practice, no random process can be truly stationary. However, long segments of random process realizations exhibiting uniform characteristics can be treated as stationary.

8.2.9   ERGODICITY OF A RANDOM PROCESS

An *ergodic process* is one for which ensemble averages are equal to the corresponding temporal averages taken along any representative sample function. Mathematically, the ergodic process may be stated according to the strong law of large numbers by

$$P\left\{ \lim_{T \to \infty} \frac{1}{T} \int_0^T g[x(t)]\ dt = \int_{\text{ensemble}} g(x)\ f(x)\ dx \right\} \equiv 1. \qquad (8.16)$$

An ergodic process is necessarily stationary. However, a random process can be stationary without being ergodic.

It is possible to verify experimentally whether a particular process is or is not ergodic by processing a large number of samples, but this is very time consuming. However, if it can be assumed ahead of time that a particular pro- cess is ergodic, then all statistical information can be obtained from a single, sufficiently long sample. In situations where statistical estimates are desired but only one sample of a stationary process is available, it is a common prac- tice to proceed on the *assumption* that the process is ergodic. These initial estimates can then be updated when and if further data become available.

8.2.10 SPECTRAL DENSITY FUNCTION

The *spectral density function*, also called the *power spectral density function* (*PSD*) or the *mean square spectral density function*, of a stationary random process $X(t)$, $S_x(\omega)$, is defined as the Fourier transform of its autocorrelation function; i.e.,

$$S_x(\omega) = \frac{1}{2\pi} \int_{-\infty}^{\infty} AC_x(\tau) e^{-i\omega\tau} d\tau, \qquad (8.17)$$

provided that

$$AC_x(\tau \to \infty) = 0,$$

and

$$\int_{-\infty}^{\infty} |AC_x(\tau)| \, d\tau < \infty.$$

If $S_x(\omega)$ is known, the autocorrelation function $AC_x(\tau)$ can be recovered by taking the inverse Fourier transform; i.e.,

$$AC_x(\tau) = \int_{-\infty}^{\infty} S_x(\omega) \, e^{i\omega\tau} \, d\omega. \qquad (8.18)$$

The pair of Fourier transforms given by Eqs. (8.17) and (8.18) is known as the Wiener-Khinchin relations between the PSD and the ACF of a stationary random process. They provide a transformation from the time domain into the frequency domain and vice versa.

When $\tau = 0$, $AC_x(0)$ becomes the mean square value which represents the average energy of a stationary random process. On the other hand, when $\tau = 0$ Eq. (8.18) implies that $AC_x(0)$ is equal to the area under the spectral density curve; i.e.,

$$AC_x(0) = E[x^2] = \int_{-\infty}^{\infty} S_x(\omega) \, d\omega = \text{average energy.}$$

Therefore, the power spectral density function $S_x(\omega)$ gives the spectral dis- tribution of the average energy of a stationary random process.

In experimental work a different unit of spectral density is widely used. The difference arises due to the use of frequency, f, in cycles per second, or Hz, in place of ω in radians per second, or rps, and due to counting only positive frequencies, or $f > 0$, instead of counting both positive and negative frequencies, or $-\infty < \omega < \infty$. The spectral density function in terms of f is called the *experimental spectral density function* and is denoted by $W_x(f)$, where f is the frequency in Hz. The relationship between $W_x(f)$ and $S_x(\omega)$ is so established that the area under $W_x(f)$ is equal to the area under $S_x(\omega)$ and equal to the mean square value, or

$$\int_0^\infty W_x(f)\, df = \int_{-\infty}^\infty S_x(\omega)\, d\omega = E[x^2]. \tag{8.19}$$

Note that $S_x(\omega)$ is an even function of ω because

$$S_x(\omega) = \frac{1}{2\pi} \int_{-\infty}^\infty AC_x(\tau)\, e^{-i\omega\tau}\, d\tau,$$

$$= \frac{1}{2\pi} \int_{-\infty}^\infty AC_x(\tau)\, (\cos\omega\tau - i\sin\omega\tau)\, d\tau,$$

or

$$S_x(\omega) = \frac{1}{2\pi} \left[\int_{-\infty}^\infty AC_x(\tau)\, \cos\omega\tau\, d\tau - i \int_{-\infty}^\infty AC_x(\tau)\, \sin\omega\tau\, d\tau \right].$$

From Eq. (8.11) we know that $AC_x(\tau)$ is an even function of τ, or

$$AC_x(\tau) = AC_x(-\tau).$$

Since $\sin\omega\tau$ is an odd function of τ and $\cos\omega\tau$ is an even function of τ, then $[AC_x(\tau)\, \cos\omega\tau]$ is even and $[AC_x(\tau)\, \sin\omega\tau]$ is odd. Therefore,

$$S_x(\omega) = \frac{1}{2\pi} \left[2 \int_0^\infty AC_x(\tau)\, \cos\omega\tau\, d\tau - 0 \right],$$

or

$$S_x(\omega) = \frac{1}{\pi} \int_0^\infty AC_x(\tau)\, \cos\omega\tau\, d\tau, \tag{8.20}$$

which is an even function of ω.

Since $\omega = 2\pi f$, then,

$$\int_{-\infty}^\infty S_x(\omega)\, d\omega = 2 \int_0^\infty S_x(\omega)\, d\omega,$$

$$= 2 \int_0^\infty S_x(\omega)\, 2\pi df,$$

or

$$\int_{-\infty}^{\infty} S_x(\omega) \, d\omega = \int_0^{\infty} [4 \, \pi \, S_x(\omega)] \, df. \qquad (8.21)$$

Comparing Eq. (8.19) with Eq. (8.21) yields

$$\int_0^{\infty} [4 \, \pi \, S_x(\omega)] \, df = \int_0^{\infty} W_x(f) \, df,$$

or

$$W_x(f) = 4 \, \pi \, S_x(\omega). \qquad (8.22)$$

8.2.11   CROSS-SPECTRAL DENSITY FUNCTION

The *cross-spectral density function* of two stationary random processes, $X(t)$ and $Y(t)$, is defined as the Fourier transform of their cross-correlation function $CC_{xy}(\tau)$; i.e.,

$$S_{xy}(\omega) = \frac{1}{2\pi} \int_{-\infty}^{\infty} CC_{xy}(\tau) \, e^{-i\omega\tau} \, d\tau, \qquad (8.23)$$

provided that

$$\int_{-\infty}^{\infty} |CC_{xy}(\tau)| \, d\tau < \infty.$$

The inverse Fourier transform on $S_{xy}(\omega)$ recovers the cross-correlation function $CC_{xy}(\tau)$; i.e.,

$$CC_{xy}(\tau) = \int_{-\infty}^{\infty} S_{xy}(\omega) \, e^{i\omega\tau} \, d\omega. \qquad (8.24)$$

8.2.12   SPECTRAL CLASSIFICATION OF STATIONARY RANDOM PROCESSES

A stationary random process can be classified as a *wide-band* or a *narrow-band process* depending on the nature of its power spectral density (*PSD*).

A *wide-band process* is a stationary random process whose power spectral density has significant values over a wide range of frequencies, roughly of the same order of magnitude as the central frequency of the band. An example of a wide-band spectrum and a possible sample function are displayed in Fig. 8.3.

A common idealization of a wide-band random process in analytical investigation is the assumption of a uniform spectral density, S_0, as shown in Fig. 8.4. A process with such a spectrum is called *white noise* in analogy with white light which spans the visible spectrum more or less uniformly.

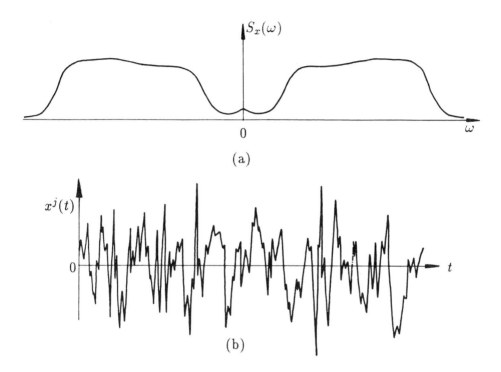

Fig. 8.3– A wide-band stationary random process. (a) Spectral density function and (b) a time history of sample function.

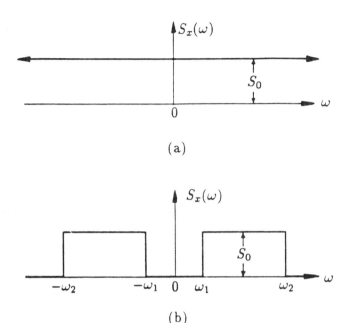

Fig. 8.4– White noise spectra. (a) Ideal white noise and (b) Band-limited white noise.

The *ideal white noise* shown in Fig. 8.4 (a) is supposed to have a uniform density $S_x(\omega) = S_0$ over all frequencies. Thus, the autocorrelation function of white noise is a delta function, or

$$AC_x(\tau) = \int_{-\infty}^{\infty} S_0 \, e^{i\omega\tau} \, d\omega = 2\pi \, S_0 \, \delta(\tau), \qquad (8.25)$$

where $\delta(\tau)$ is the Dirac delta function, or unit impulse, and

$$\delta = \begin{cases} \infty, \text{ for } \tau = 0, \\ 0, \quad \text{for } \tau \neq 0. \end{cases}$$

This is a physically unrealizable concept since the mean square value (average energy) of such a process would be infinite because the area under the spectrum is infinite. The *band-limited white noise* spectrum in Fig. 8.4 (b), however, is a close approximation to many physically realizable random processes. For band-limited white noise, the mean square value is simply

$$2S_0(\omega_2 - \omega_1),$$

or the total area under the spectrum. The autocorrelation function is

$$AC_x(\tau) = 2S_0 \frac{\sin \omega_2\tau - \sin \omega_1\tau}{\tau}. \qquad (8.26)$$

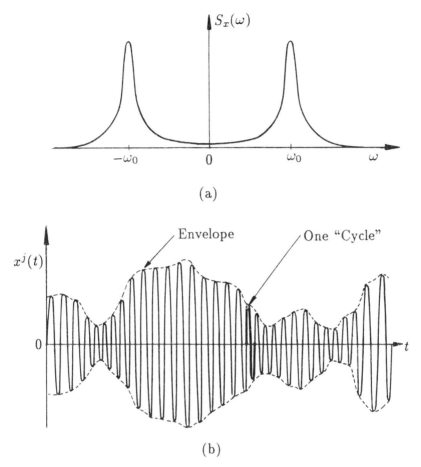

Fig. 8.5– A narrow-band stationary process. (a) Spectral density and (b) a time history of sample function.

A *narrow-band process* is a stationary random process whose power spectral density function has significant values over a narrow frequency band around a central frequency. The band width is small compared with the magnitude of the central frequency. An example of a narrow-band spectrum and a possible sample function are displayed in Fig. 8.5.

A narrow-band random process resembles a harmonic function with randomly varying amplitudes, which is typically encountered as the response in a strongly resonant vibratory system when the excitation is a wide-band process.

8.2.13 THE NORMAL OR GAUSSIAN RANDOM PROCESS

Consider a random process $X(t)$, $t \in T$. Let $X_1 = X(t_1)$, $X_2 = X(t_2)$, $\cdots$, $X_n = X(t_n)$, be n random variables for any integer n and any subset $t_1, t_2, \cdots, t_n \in T$. The random process $X(t)$ is said to be *normal* or *Gaussian* if the random variables $X_1, X_2, \cdots, X_n$ are jointly normal.

Many random processes in nature which play the role of excitations to vibratory systems are at least approximately normal. This observation is made plausible by the central limit theorem, which says in this case that a random process will be approximately normal if each of its sample functions can be considered to have been generated by the superposition of a large number of independent random sources, no single one of which contributed significantly. Furthermore when the excitation of a linear system is a normal process, the response will still be normal but with different distribution parameters. The normal, or Gaussian, process is one of the very few processes for which the higher-order probability distributions have been explicitly written and studied [1; 2].

8.2.14 SOME IMPORTANT STATISTICS OF A RANDOM PROCESS

In the analysis and design of systems, within a probability framework, the following statistics play an important role:

1. The number of crossings at level $X(t) = \alpha$ by $X(t)$ in the time interval $(t_0, t_0 + T)$.

2. The number of peaks of $X(t)$ above a given level $X(t) = \alpha$ in the time interval T.

3. The fraction of time spent by $X(t)$ above a level $X(t) = \alpha$ in the time interval T.

4. The envelope of $X(t)$, its *pdf* and level crossing rate.

5. The first passage time – the time at which $X(t)$ crosses a level $X(t) = \alpha$ for the first time.

6. The maximum of $X(t)$ in the time interval T.

These concepts and the associated equations are summarized next.

8.2.14.1 LEVEL CROSSINGS

For any stationary process, the expected number per unit time (the *expected rate*) of crossing level $X(t) = \alpha$ *with a positive slope* at time t, which is defined as the probability density function of crossing the level $X(t) = \alpha$ with a positive slope at time t, is given by

$$n_x^+(\alpha, t) = \int_0^\infty \dot{x} \, f(\alpha, \dot{x}, t) \, d\dot{x}, \tag{8.27}$$

where

$$\dot{x} = \frac{d[x(t)]}{dt},$$

and

$$f(\alpha, \dot{x}, t) = \text{joint } pdf \text{ of } x \text{ and } \dot{x} \text{ at time } t \text{ evaluated at } x = \alpha.$$

Similarly, the *expected rate of crossing* level $X(t) = \alpha$ with a negative slope at time t, which is defined as the probability density function of the time of crossing level $X(t) = \alpha$ with a negative slope at time t, is given by

$$n_x^-(\alpha, t) = \int_{-\infty}^0 (-\dot{x}) \, f(\alpha, \dot{x}, t) \, d\dot{x}. \tag{8.28}$$

The expected rate of crossing level $X(t) = 0$ at time t with a positive or a negative slope is called the *expected zero-up-crossing rate* or the *expected zero-down-crossing rate*, respectively. Substituting $\alpha = 0$ into Eqs. (8.27) and (8.28) yields

$$n_x^+(0, t) = \int_0^\infty \dot{x} \, f(0, \dot{x}, t) \, d\dot{x}, \tag{8.29}$$

and

$$n_x^-(0, t) = \int_{-\infty}^0 (-\dot{x}) \, f(0, \dot{x}, t) \, d\dot{x}, \tag{8.30}$$

respectively, where

$$f(0, \dot{x}, t) = \text{joint } pdf \text{ of } x \text{ and } \dot{x} \text{ at time } t \text{ evaluated at } x = 0.$$

The *expected rate of crossing* level $X(t) = \alpha$ at time t, with either a positive or a negative slope, which is defined as the probability density function of the time of crossing level $X(t) = \alpha$ at time t, is given by

$$n_x(\alpha, t) = n_x^+(\alpha, t) + n_x^-(\alpha, t) = \int_{-\infty}^\infty |\dot{x}| \, f(\alpha, \dot{x}, t) \, d\dot{x}. \tag{8.31}$$

The *expected number of crossings*, $N_x(\alpha; T)$, at level $X(t) = \alpha$ in time interval $(t_0, t_0 + T)$ is given by

$$N_x(\alpha; T) = \int_{t_0}^{t_0 + T} n_x(\alpha, t) \, dt. \tag{8.32}$$

If $X(t)$ is a stationary process, then

$$f(x, \dot{x}, t) = f(x, \dot{x}), \tag{8.33}$$

and Eqs. (8.27) through (8.32) become

$$n_x^+(\alpha, t) = n_x^+(\alpha) = \int_0^\infty \dot{x}\, f(\alpha, \dot{x})\, d\dot{x}, \tag{8.34}$$

$$n_x^-(\alpha, t) = n_x^-(\alpha) = \int_{-\infty}^0 (-\dot{x})\, f(\alpha, \dot{x})\, d\dot{x}, \tag{8.35}$$

$$n_x^+(0, t) = n_x^+(0) = \int_0^\infty \dot{x}\, f(0, \dot{x})\, d\dot{x}, \tag{8.36}$$

$$n_x^-(0, t) = n_x^-(0) = \int_{-\infty}^0 (-\dot{x})\, f(0, \dot{x})\, d\dot{x}, \tag{8.37}$$

$$n_x(\alpha, t) = n_x(\alpha) = \int_{-\infty}^\infty |\dot{x}|\, f(\alpha, \dot{x})\, d\dot{x}, \tag{8.38}$$

and

$$N_x(\alpha; T) = n_x(\alpha)\, T, \tag{8.39}$$

respectively.

As a special case where $X(t)$ is a *stationary Gaussian process with zero mean*, we have the following results:

$$n_x^+(\alpha, t) = n_x^+(\alpha) = n_x^-(\alpha, t) = n_x^-(\alpha),$$

$$= \frac{1}{2\pi} \left(\frac{\sigma_{\dot{x}}}{\sigma_x} \right) e^{-\frac{1}{2}\left(\frac{\alpha}{\sigma_x}\right)^2}, \tag{8.40}$$

$$n_x(\alpha, t) = n_x(\alpha) = n_x^+(\alpha) + n_x^-(\alpha),$$

$$= \frac{1}{\pi} \left(\frac{\sigma_{\dot{x}}}{\sigma_x} \right) e^{-\frac{1}{2}\left(\frac{\alpha}{\sigma_x}\right)^2}, \tag{8.41}$$

$$n_x^+(0, t) = n_x^+(0) = n_x^-(0, t) = n_x^-(0),$$

$$= \frac{1}{2\pi} \left(\frac{\sigma_{\dot{x}}}{\sigma_x} \right), \tag{8.42}$$

and

$$n_x(0, t) = n_x(0) = n_x^+(0) + n_x^-(0) = \frac{1}{\pi} \left(\frac{\sigma_{\dot{x}}}{\sigma_x} \right), \tag{8.43}$$

where

$$\sigma_x \quad = \text{standard deviation, or } rms \text{ value, of } X(t),$$

$$= \left[\int_{-\infty}^\infty S_x(\omega)\, d\omega \right]^{1/2}, \tag{8.44}$$

$\sigma_{\dot{x}}$ = standard deviation, or *rms* value, of $\dot{X}(t)$,

$$= \left[\int_{-\infty}^{\infty} \omega^2 \, S_x(\omega) \, d\omega \right]^{1/2}, \tag{8.45}$$

and

$S_x(\omega)$ = power spectral density function of $X(t)$.

8.2.14.2 PEAKS

The *expected rate of the occurrence of peaks above the level* $X(t) = \alpha$ is given by

$$m(\alpha, t) = \int_{\alpha}^{\infty} dx \int_{-\infty}^{0} |\ddot{x}| \, f(x, 0, \ddot{x}, t) \, d\ddot{x}, \tag{8.46}$$

where

$$\ddot{x} = \frac{d^2 [x(t)]}{dt^2},$$

and

$f(x, 0, \ddot{x}, t)$ = joint *pdf* of x, $\dot{x}$ and $\ddot{x}$ at time t, evaluated at $\dot{x} = 0$.

The *expected total rate of peaks per unit time* without any restriction on the peak level is given by

$$m_T(t) = \int_{-\infty}^{0} |\ddot{x}| \, f(0, \ddot{x}, t) \, d\ddot{x}, \tag{8.47}$$

where

$f(0, \ddot{x}, t)$ = joint *pdf* of $\dot{x}$ and $\ddot{x}$ at time t, evaluated at $\dot{x} = 0$.

The *cumulative distribution function of the peaks* at time t is given by

$$F_p(\alpha, t) = P[(\text{peak hight at time } t) \leq \alpha],$$

$$= 1 - \frac{m(\alpha, t)}{m_T(t)},$$

or, from Eq. (8.46),

$$F_p(\alpha, t) = 1 - \frac{1}{m_T(t)} \int_{\alpha}^{0} dx \int_{-\infty}^{0} |\ddot{x}| \, f(x, 0, \ddot{x}, t) \, d\ddot{x}. \tag{8.48}$$

The *probability density function of the peaks* at time t is obtained from

$$f_p(\alpha, t) = \frac{\partial}{\partial \alpha} [F_p(\alpha, t)],$$

or

$$f_p(\alpha, t) = \frac{1}{m_T(t)} \int_{-\infty}^{0} |\ddot{x}| \, f(\alpha, 0, \ddot{x}, t) \, d\ddot{x}, \tag{8.49}$$

where

$$f(\alpha, 0, \ddot{x}, t) = \text{joint } pdf \text{ of } x, \dot{x} \text{ and } \ddot{x} \text{ at time } t, \text{ evaluated}$$
$$\text{at } x = \alpha \text{ and } \dot{x} = 0.$$

The *expected number of peaks above the level* $X(t) = \alpha$ *in the time interval* T is obtained from

$$M(\alpha; T) = \int_{t_0}^{t_0+T} m(\alpha, t) dt. \tag{8.50}$$

If $X(t)$ is a stationary process, then Eqs. (8.46) through (8.50) become

$$m(\alpha, t) = m(\alpha) = \int_{\alpha}^{\infty} dx \int_{-\infty}^{0} |\ddot{x}| \, f(x, 0, \ddot{x}) \, d\ddot{x},$$

$$m_T(t) = m_T = \int_{-\infty}^{0} |\ddot{x}| \, f(0, \ddot{x}) \, d\ddot{x},$$

$$F_p(\alpha, t) = F_p(\alpha) = 1 - \frac{1}{m_T} \int_{\alpha}^{0} dx \int_{-\infty}^{0} |\ddot{x}| \, f(x, 0, \ddot{x}) \, d\ddot{x},$$

$$f_p(\alpha, t) = f_p(\alpha) = \frac{1}{m_T} \int_{-\infty}^{0} |\ddot{x}| \, f(\alpha, 0, \ddot{x}) \, d\ddot{x},$$

and

$$M(\alpha; T) = m(\alpha) \, T,$$

respectively.

The ratio of the expected zero-up-crossing rate to the expected total rate of peaks is called the *irregularity factor* of the random process $X(t)$ and is denoted by ξ; i.e.,

$$\xi = \frac{n_x^+(0, t)}{m_T(t)} \leq 1. \tag{8.51}$$

The square root of $(1 - \xi^2)$ is called the *spectral width parameter* of $X(t)$ and is denoted by ϵ; i.e.,

$$\epsilon = \sqrt{1 - \xi^2} \leq 1. \tag{8.52}$$

The irregularity factor, ξ, and the spectral width parameter, ϵ, are measures of the bandwidth of a random process:

1. If $X(t)$ is a narrow-band process, there are equal number of peaks and zero-up-crossings as shown in Fig. 8.5, or

$$n_x^+(0,t) = m_T(t).$$

Then,

$$\xi = \frac{n_x^+(0,t)}{m_T(t)} = 1,$$

and

$$\epsilon = \sqrt{1 - \xi^2} = 0.$$

2. If $X(t)$ is a wide-band process, the number of peaks is much higher than the number of zero-up-crossings as shown in Fig. 8.3, or

$$m_T(t) \gg n_x^+(0,t).$$

Then,

$$\xi = \frac{n_x^+(0,t)}{m_T(t)} = 0,$$

and

$$\epsilon = \sqrt{1 - \xi^2} = 1.$$

3. Practically, a process may be considered as a narrow-band process if

$$0 \le \epsilon < 0.14,$$

or

$$0.99 < \xi \le 1,$$

and a wide-band process if

$$0.14 < \epsilon \le 1,$$

or

$$0 \le \xi < 0.99.$$

It has been shown [2, p. 115] that if $X(t)$ is a stationary Gaussian process with zero mean, then

$$m_T(t) = m_T = \frac{1}{2\pi}\left(\frac{\sigma_{\ddot{x}}}{\sigma_{\dot{x}}}\right), \tag{8.53}$$

$$\xi_N = \frac{n_x^+(0)}{m_T} = \frac{\frac{1}{2\pi}\left(\frac{\sigma_{\dot{x}}}{\sigma_x}\right)}{\frac{1}{2\pi}\left(\frac{\sigma_{\ddot{x}}}{\sigma_{\dot{x}}}\right)} = \frac{\sigma_{\dot{x}}^2}{\sigma_x \sigma_{\ddot{x}}}, \tag{8.54}$$

and

$$\epsilon_N = \sqrt{1 - \xi^2} = \sqrt{1 - \frac{\sigma_{\ddot{x}}^4}{\sigma_x^2 \sigma_{\ddot{x}}^2}}, \tag{8.55}$$

where the subscript N implies a Gaussian process,

$\sigma_{\ddot{x}}$ = standard deviation, or *rms* value of $\ddot{X}(t)$,

or

$$\sigma_{\ddot{x}} = \left[\int_{-\infty}^{\infty} \omega^4 \, S_x(\omega) d\omega\right]^{1/2}. \tag{8.56}$$

Let Z be a random variable denoting peak heights of a stationary Gaussian random process $X(t)$ with zero mean. Then, the general expression for the probability density function of Z is given by [1; 2]

$$f(Z) = \epsilon_N \left[\frac{1}{\sqrt{2\pi}\sigma_x} e^{-\frac{1}{2}\left(\frac{Z}{\epsilon_N \sigma_x}\right)^2}\right]$$
$$+ \left[\xi_N \, \Phi\left(\frac{\xi_N Z}{\epsilon_N \sigma_x}\right)\right]\left[\frac{Z}{\sigma_x^2} e^{-\frac{1}{2}\left(\frac{Z}{\sigma_x}\right)^2}\right], \tag{8.57}$$

where

$\Phi(\cdot)$ = *cdf* of the standardized normal distribution.

It may be observed that the general peak height distribution is a quasi-mixed distribution of the normal and the Rayleigh.

From Eq. (8.57) the following may be concluded:

1. If $X(t)$ is a *stationary narrow-band Gaussian process with zero mean*, then its peak follows a *Rayleigh distribution*. This is so because if the process is narrow-band then $\xi_N = 1$ and $\epsilon_N = 0$. Substituting these into Eq. (8.57) yields

$$f(Z) = \frac{Z}{\sigma_x^2} e^{-\frac{1}{2}\left(\frac{Z}{\sigma_x}\right)^2}, \tag{8.58}$$

which is a *Rayleigh distribution* with a scale parameter of σ_x.

2. If $X(t)$ is a *stationary wide-band Gaussian process with zero mean*, then its peak follows a *normal distribution*. This is so because if the process is wide-band then $\xi_N = 0$ and $\epsilon_N = 1$. Substituting these into Eq. (8.57) yields

$$f(Z) = \frac{1}{\sqrt{2\pi}\sigma_x} e^{-\frac{1}{2}\left(\frac{Z}{\sigma_x}\right)^2}, \tag{8.59}$$

which is a *normal distribution* with zero mean and a standard deviation of σ_x.

If $X(t)$ is a *nonstationary (wide-band or narrow-band) random process*, then the distribution of its peak can be obtained by substituting the expression for the joint probability density function $f(x, 0, \ddot{x}, t)$ in Eq. (8.48). It has been shown [2, p. 117] that the Weibull distribution, expressed as

$$F_p(\alpha, T) = P[(\text{peak hight at time T}) \le \alpha] = 1 - e^{-\frac{1}{\beta}\left(\frac{\alpha}{\sigma}\right)^{\beta}}, \qquad (8.60)$$

fits the peak distribution very closely, where parameters β and σ depend on the characteristics of the nonstationarity and the process duration, T. Note that if $\beta = 2$ and $\sigma = \sigma_x$, Eq. (8.60) reduces to the Rayleigh distribution.

8.2.14.3   FRACTIONAL OCCUPATION TIME

Consider a random process $X(t)$, $t \in T$. The *fractional occupation time* is defined as the ratio of the time spent by the random process above the level $X(t) = \alpha$ to the duration of the process, T, and is denoted $Y_\alpha(T)$.

The distribution of $Y_\alpha(T)$ is very difficult to determine. However, its mean and variance can be readily obtained by the following equations [2, p. 117]:

$$E[Y_\alpha(T)] = \frac{1}{T} \int_{t=t_0}^{t_0+T} \int_{x=\alpha}^{\infty} f(x, t) \; dx \; dt, \qquad (8.61)$$

and

$$
\begin{aligned}
Var[Y_\alpha(T)] = \frac{1}{T^2} \int_{t_1=t_0}^{t_0+T} \int_{t_2=t_0}^{t_0+T} \int_{x_1=\alpha}^{\infty} \int_{x_2=\alpha}^{\infty} \\
\cdot f(x_1, t_1; x_2, t_2) \; dx_1 \; dx_2 \; dt_1 \; dt_2 \\
- \{E[Y_\alpha(T)]\}^2,
\end{aligned}
\qquad (8.62)
$$

where

$t_0 = $ starting moment of the random process $X(t)$.

8.2.14.4   ENVELOPES

Geometrically, an envelope $A(t)$ of the random process $X(t)$ is conceived to be a curved tangent to the peaks of $X(t)$, as shown in Fig. 8.5b. Thus, the envelope of a random process is also a random process. A well-accepted definition for the envelope of a narrow-band random process is due to Rice [1; 3]. Let a narrow-band random process $X(t)$ be represented by

$$X(t) = A(t) \; \cos[\omega_m t + \phi(t)], \qquad (8.63)$$

where ω_m is a representative midband frequency of the narrow band, and $A(t)$, the amplitude, and $\phi(t)$, the phase angle, are random processes which vary much more slowly than $X(t)$ with respect to t. According to Rice's definition the random process $A(t)$ is the envelope process of $X(t)$.

If $X(t)$ is a stationary Gaussian with zero mean, then the *pdf* of $A(t)$, $f(a)$, is given by [1; 2]

$$f(a) = \frac{a}{\sigma_x^2}\, e^{-\frac{1}{2}\left(\frac{a}{\sigma_x}\right)^2}, \quad 0 \le a < \infty, \tag{8.64}$$

which is the Rayleigh distribution and has already been shown to be the *pdf* of the peak height of a narrow-band Gaussian process in the preceding section. It is plausible that for a stationary narrow-band random process, the distribution of the envelope and that of the peak heights ought to have certain similarities.

The expected rate of crossing the level $X(t) = \alpha$ with a positive slope by the envelope of a stationary Gaussian random process $X(t)$, $n_A^+(\alpha)$, is given by [2]

$$n_A^+(\alpha) = \frac{\Delta\,\alpha}{\sqrt{2\pi}\,\sigma_x^2}\, e^{-\frac{1}{2}\left(\frac{\alpha}{\sigma_x}\right)^2}, \tag{8.65}$$

where

$$\Delta = \sqrt{\sigma_{\dot{x}}^2 - \frac{\lambda_1^2}{\sigma_x^2}}, \tag{8.66}$$

$$\lambda_1 = \int_{-\infty}^{\infty} \omega\, S_x(\omega)\, d\omega, \tag{8.67}$$

$S_x(\omega)$ = spectral density function of $X(t)$,

σ_x = *rms* value of $X(t)$ as given by Eq. (8.44) ,

and

$\sigma_{\dot{x}}$ = *rms* value of $\dot{X}(t)$ as given by Eq. (8.45) ,

For a more detailed study of the behavior of the envelope process $A(t)$, the readers are referred to [1; 2].

8.2.14.5 FIRST-PASSAGE TIME

The *first-passage time*, T_f, of a random process $X(t)$ is defined as the time at which $X(t)$ crosses a level α for the first time, as shown in Fig. 8.6. Obviously T_f is a random variable associated with the random process $X(t)$.

First-passage time represents an important characteristic of random processes from the viewpoint of safety, or performance, of a system. For a reliability-based analysis and design of a system, it is often necessary to establish the distribution of the first-passage time. In general, such problems are very difficult. An exact solution of this classical problem has not been

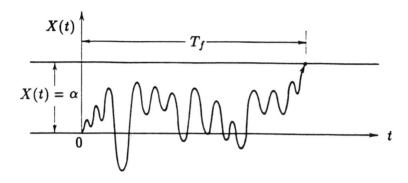

Fig. 8.6– The first-passage crossing of the level $X(t) = \alpha$ and the first-passage time.

available. Presented next are two approximate solutions for stationary random processes based on the Poisson process and the two-state Markov process assumptions, respectively.

1. The Poisson process assumption

Let $X(t)$ be a stationary narrow-band process with zero mean and $N(t)$ be the number of up-crossings of level $X(t) = \alpha$ in the time interval $(0, t]$. If the level $X(t) = \alpha$ is sufficiently high, it is reasonable to assume that a crossing at that level is a rare event. Therefore, the crossings may be assumed to be independent. Since $X(t)$ is stationary, $N(t)$ is a Poisson process with stationary increments and an arrival rate of

$$\lambda_P(\alpha) = n_x^+(\alpha). \tag{8.68}$$

Then, the underlying distribution of the first-passage time is the exponential with parameter $\lambda_P(\alpha)$; i.e.,

$$f(T_f) = \lambda_P(\alpha)\, e^{-\lambda_P(\alpha)\, T_f} = n_x^+(\alpha)\, e^{-n_x^+(\alpha)\, T_f}. \tag{8.69}$$

The corresponding reliability and unreliability functions are

$$R(T_f) = e^{-\lambda_P(\alpha)\, T_f} = e^{-n_x^+(\alpha)\, T_f}, \tag{8.70}$$

and

$$Q(T_f) = 1 - e^{-\lambda_P(\alpha)\, T_f} = 1 - e^{-n_x^+(\alpha)\, T_f}, \tag{8.71}$$

respectively, which for small values of $Q(T_f)$ is given approximately by

$$Q(T_f) \cong \lambda_P(\alpha)\, T_f = n_x^+(\alpha)\, T_f.$$

The mean and the standard deviation of the first-passage time is

$$E[T_f] = \sigma_{T_f} = \frac{1}{\lambda_P(\alpha)} = \frac{1}{n_x^+(\alpha)}. \tag{8.72}$$

The *independent-level-crossing assumption* is very poor for a narrow-band random process, where crossings tend to occur in clumps, and for low crossing levels. For the narrow-band process, it is more reasonable to assume that the *independent envelope crossings* occur with a Poisson arrival rate of

$$\lambda_P(\alpha) = n_A^+(\alpha) < n_x^+(\alpha).$$

Then, Eqs. (8.69) through (8.72) become

$$f(T_f) = n_A^+(\alpha)\, e^{-n_A^+(\alpha)\, T_f}, \tag{8.73}$$

$$R(T_f) = e^{-n_A^+(\alpha)\, T_f}, \tag{8.74}$$

$$Q(T_f) = 1 - e^{-n_A^+(\alpha)\, T_f}, \tag{8.75}$$

and

$$E[T_f] = \sigma_{T_f} = \frac{1}{n_A^+(\alpha)}. \tag{8.76}$$

Note that the actual values of the first-passage probabilities may deviate significantly from the estimates based on the Poisson level and envelope crossing assumptions. Also note that the exponential first-passage-time distribution implies zero initial probability of failure, or,

$$Q(T_f = 0) = 1 - e^0 = 0,$$

which may not be the case for a random start.

If $X(t)$ is a stationary, Gaussian and narrow-band random process, the level crossing rates of the process and its envelope, $n_x^+(\alpha)$ and $n_A^+(\alpha)$, are given by Eqs. (8.40) and (8.65). Then, the Poisson arrival rate, assuming an independent level crossing of the process, is given by

$$\lambda_P(\alpha) = n_x^+(\alpha) = \frac{1}{2\pi} \left(\frac{\sigma_{\dot{x}}}{\sigma_x} \right) e^{-\frac{1}{2}\left(\frac{\alpha}{\sigma_x} \right)^2}. \tag{8.77}$$

The Poisson arrival rate, assuming an independent level crossing of the envelope process, is

$$\lambda_P(\alpha) = n_A^+(\alpha) = \frac{\Delta\, \alpha}{\sqrt{2\pi}\, \sigma_x^2} e^{-\frac{1}{2}\left(\frac{\alpha}{\sigma_x} \right)^2}. \tag{8.78}$$

Substituting Eq. (8.77) into Eq. (8.71), and Eq. (8.78) into Eq. (8.75), yields the cumulative distribution functions of the first-passage time; i.e.,

$$Q(T_f) = F(T_f) = 1 - e^{-\frac{1}{2\pi}\left(\frac{\sigma_{\dot{x}}}{\sigma_x}\right)e^{-\left(\frac{\eta^2}{2}\right)}T_f}$$

(8.79)

assuming an independent level crossing of the process, and

$$Q(T_f) = F(T_f) = 1 - e^{-\frac{\Delta}{\sqrt{2\pi}}\frac{\eta}{\sigma_x}e^{-\left(\frac{\eta^2}{2}\right)}T_f}$$

(8.80)

assuming an independent level crossing of the envelope, where

$$\eta = \alpha/\sigma_x = \text{normalized level.}$$

2. The two-state Markov process assumption

Define a *two-state, zero-one process*, $Z_\alpha(t)$, with the following properties:

$$Z_\alpha(t) = \begin{cases} 1, & \text{if } X(t) \geq \alpha, \\ 0, & \text{if } X(t) < \alpha. \end{cases}$$

The random process $Z_\alpha(t)$ can be expressed in terms of a unit step function $U(\ )$:

$$Z_\alpha(t) = U[X(t) - \alpha] = \begin{cases} 1, & \text{if } X(t) \geq \alpha, \\ 0, & \text{if } X(t) < \alpha, \end{cases}$$

as shown in Fig. 8.7(b).

The sequence of times $t_0^{(i)}$, $t_1^{(i)}$, $i = 1, 2, 3, ...$, are sample values of random variables T_0 and T_1 corresponding to states 0 and 1. It is assumed that

(1) T_0 and T_1 are independent,

(2) $(T_0 + T_1)$ has a continuous distribution, and

(3) $E[T_0] < \infty$ and $E[T_1] < \infty$.

Then, the level crossing of the process can be described as a *two-state, Markovian renewal process*.

According to renewal theory, it can be shown [2, p. 133] that the initial failure probability and reliability are given by

$$Q(T_f = 0) = P[Z_\alpha(0) = 1] = \frac{E[T_1]}{E[T_0] + E[T_1]} = \frac{n_x^+(\alpha)}{n_x^+(0)},$$

(8.81)

and

$$R(T_f = 0) = P[Z_\alpha(0) = 0] = \frac{E[T_0]}{E[T_0] + E[T_1]} = 1 - \frac{n_x^+(\alpha)}{n_x^+(0)},$$

(8.82)

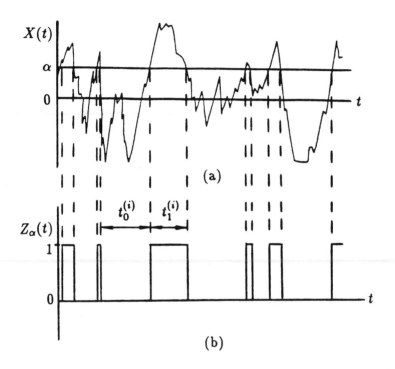

Fig. 8.7– (a) Up- and down-crossings of a level $X(t) = \alpha$;
(b) Sample function of a two-state, zero-one process
showing sample values of random variables T_0 and T_1.

respectively. The probability density function of T_f is given by

$$f(T_f) = R(T_f = 0) \left[\frac{1}{E[T_0]} \int_{T_f}^{\infty} f(T_0) \, dT_0 \right]. \tag{8.83}$$

Assume that T_0 is exponentially distributed with *pdf*

$$f(T_0) = \lambda_M \, e^{-\lambda_M \, T_0}, \tag{8.84}$$

where

$$\lambda_M = \frac{1}{E[T_0]}. \tag{8.85}$$

Then, substituting Eqs. (8.84) and (8.85) into Eq. (8.83) yields

$$f(T_f) = R(T_f = 0) \, \lambda_M \, e^{-\lambda_M T_f}. \tag{8.86}$$

When $X(t)$ is a stationary Gaussian process, it has been shown [2, p. 134] that

$$\lambda_M = \frac{1}{2\pi} \left(\frac{\sigma_{\dot{x}}}{\sigma_x} \right) \left(1 - e^{-K \, \eta} \right) \Big/ \left(e^{\frac{\eta^2}{2}} - 1 \right), \tag{8.87}$$

and

$$R(T_f = 0) = 1 - e^{\frac{\eta^2}{2}}, \tag{8.88}$$

where

$$K = \frac{\Delta \sqrt{2\pi}}{\sigma_{\dot{x}}},$$

and

$$\eta = \alpha / \sigma_x.$$

It was shown [2, pp. 138-139] that the results based on the assumption of a two-state Markov crossing are the closest to simulation results and converge asymptotically to those based on the assumption of a Poisson crossing as the level α becomes high.

The distributions of the first-passage times derived for a stationary process, as presented previously, can be extended to nonstationary processes by assuming that the level-crossing rate, $\lambda(t)$, is a continuous function of time. The readers are referred to [2] for details.

8.2.14.6 MAXIMUM VALUE OF A RANDOM PROCESS IN A TIME INTERVAL

The maximum value of a random process, $X(t)$, in a specified time interval is a random variable. Let

$$X_{max} = MAX\{X(t)\}, \quad \text{for} \quad t_0 < t \le t_0 + T.$$

The probability distribution of X_{max} is closely related to the distribution of the first-passage time T_f and to the distribution of the peak values in $(t_0, t_0 + T]$. For example,

$$F_{X_{max}}(\alpha) = P[X_{max} \leq \alpha] = P[T_f > T], \tag{8.89}$$

and

$$F_{X_{max}}(\alpha) = P[X_{max} \leq \alpha] = [F_p(\alpha)]^{N_p(T)}, \tag{8.90}$$

where

$$F_p(\alpha) = cdf \text{ of the peaks evaluated at level } \alpha,$$

and

$$N_p(T) = \text{total number of peaks in } (t_0, t_0 + T].$$

For a *stationary, narrow-band Gaussian process*, $X(t)$, we have the following closed form for $F_{X_{max}}(\alpha)$ [2, p. 141]:

$$F_{X_{max}}(\alpha) = e^{-n_x^+(0)\, T\, e^{-\frac{1}{2}\left(\frac{\alpha}{\sigma_x}\right)^2}}, \tag{8.91}$$

or

$$F_{X_{max}}(\alpha) \cong e^{-e^{-C_1(\eta - C_1)}}, \tag{8.92}$$

where

$$C_1 = \sqrt{2\, \log_e[n_x^+(0)\, T]}. \tag{8.93}$$

The mean and variance for X_{max} are given by [2, p. 141]

$$E[X_{max}] = \mu_{X_{max}} = C\,\sigma_x, \tag{8.94}$$

and

$$Var[X_{max}] = \sigma_{X_{max}}^2 = \frac{\pi^2}{6}\left(\frac{\sigma_x}{C_1}\right)^2 = 1.638\left(\frac{\sigma_x}{C_1}\right)^2, \tag{8.95}$$

where

$$C = C_1 + \frac{0.5772}{C_1}. \tag{8.96}$$

For a nonstationary, narrow-band Gaussian process, $X(t)$, if

$$T \gg T_0 = \frac{2\pi}{\omega_0},$$

and

$$\frac{\alpha}{\sigma} \gg 1,$$

where

ω_0 = central frequency of the narrow band,

and

σ = Weibull scale parameter of the peak
distribution, as in Eq. (8.60),

then the asymptotic distribution of X_{max} is given by [2, p. 143]

$$F_{X_{max}}(\alpha) = e^{-N_x^+(0;T)} \, e^{-\frac{1}{\beta}\left(\frac{\alpha}{\sigma}\right)^\beta}, \tag{8.97}$$

or

$$F_{X_{max}}(\alpha) \cong e^{-e^{-D_1^{\beta-1}\left(\frac{\alpha}{\sigma}-D_1\right)}}, \tag{8.98}$$

where

β = Weibull shape parameter of the peak
distribution, as in Eq. (8.60),

and

$$D_1 = \left\{\beta \, \log_e[N_x^+(0;T)]\right\}^{1/\beta}. \tag{8.99}$$

The mean and variance for X_{max} are given by [2, p. 143]

$$E[X_{max}] = \mu_{X_{max}} = D \, \sigma, \tag{8.100}$$

and

$$Var[X_{max}] = \sigma_{X_{max}}^2 = 1.638 \left(\frac{\sigma^2}{D_1^{\beta-1}}\right), \tag{8.101}$$

respectively, where

$$D = D_1 + \frac{0.5772}{D_1}. \tag{8.102}$$

Note that a stationary random process can be regarded as a special case
of the foregoing nonstationary random process where

$$N_x^+(0;T) = n_x^+(0) \, T,$$
$$\beta = 2,$$

and

$$\sigma = \sigma_x.$$

Then, Eq. (8.97) reduces to Eq. (8.91).

8.3 FAILURE DUE TO RANDOM VIBRATION AND STRUCTURAL RELIABILITY EVALUATION

8.3.1 FAILURE MECHANISMS

The response process of a vibratory system may be that of a displacement, or of an acceleration, or of a stress, or of some other dynamic quantity. It would be highly desirable if the statistics on such dynamic response quantities would be converted simply into statistics on the reliability of the structure or equipment in the random vibration environment. This would then provide information directly useful to the designer or test engineer. Even though many failure mechanisms of structures are basically governed by the vibration environment, it should not be forgotten that other environmental factors such as temperature, pressure, load, radiation, etc. can often interact with the vibration environment to alter the nature of the failure mechanisms; e.g., the strength of a vibrating member can be sharply temperature dependent. Here, only random vibration caused failures are considered. Reliability quantification for the temperature-vibration combined environment will be discussed in the next section.

Let $Y(t)$ be a dynamic response quantity in a vibratory system. At least three different failure mechanisms can be envisioned:

1. Failure can occur the first time that $Y(t)$ reaches a certain fixed level Y_F. For example, $Y(t)$ might be the relative displacement necessary to cause a short circuit; or $Y(t)$ might be the relative displacement of a resiliently supported package which fails when the resilient support bottoms out after traveling through the clearance space Y_F; or $Y(t)$ might be the maximum stress at a critical location on the structure with Y_F being the allowable yield strength.

2. Failure can occur when the fraction of time for which $Y > Y_F$ is greater than some predetermined fraction ε. This might be the case, for example, in which $Y(t)$ is the relative displacement of an electronic component, and Y_F is the threshold displacement for malfunction, but the communication link of which this is a part is not considered to fail until the malfunction occurs for more than a certain percent of the time.

3. Failure can be due to an accumulation of damage . Each excursion of $Y(t)$ does a small but definite amount of damage which depends on the amplitude of the excursion. Failure occurs when the accumulation of these small damages reaches a fixed total. For example, the fracture of metal elements due to *fatigue* can be assumed to be due to such a mechanism.

8.3.2 FAILURE PROBABILITY DUE TO THE FIRST EXCURSION UP TO A CERTAIN LEVEL

Failures due to the relative displacements exceeding the allowable space, or the maximum stress exceeding the allowable yield strength, are first-passage failures in nature. Let $Y(t)$ be the associated dynamic response in the vibratory system. The random time to structure failure (the first exceedance), T_f, is the so-called first-passage time. It has been shown in Section 8.2 that under the Poisson assumption the *pdf* of T_f, and the reliability and unreliability, are given by Eqs. (8.69) through (8.71), assuming independent level crossings of $Y(t)$, or

$$f(T_f) = n_y^+(Y_F) \, e^{-n_y^+(Y_F) \, T_f}, \tag{8.103}$$

$$R(T_f) = e^{-n_y^+(Y_F) \, T_f}, \tag{8.104}$$

and

$$Q(T_f) = 1 - e^{-n_y^+(Y_F) \, T_f}, \tag{8.105}$$

and by Eqs. (8.73) through (8.75) assuming independent envelope crossings,

$$f(T_f) = n_A^+(Y_F) \, e^{-n_A^+(Y_F) \, T_f}, \tag{8.106}$$

$$R(T_f) = e^{-n_A^+(Y_F) \, T_f}, \tag{8.107}$$

and

$$Q(T_f) = 1 - e^{-n_A^+(Y_F) \, T_f}, \tag{8.108}$$

where

Y_F = critical value of $Y(t)$, the first exceedance of which leads to a failure,

$n_y^+(Y_F)$ = expected rate of crossing level $Y(t) = Y_F$ by $Y(t)$ itself, as given by Eq. (8.77),

and

$n_A^+(Y_F)$ = expected rate of crossing the level $Y(t) = Y_F$ by the envelope of $Y(t)$, as given by Eq. (8.78).

Under the two-state Markov crossing assumption, the *pdf* of T_f when $Y(t)$ is a stationary Gaussian process is given by Eq. (8.86), or

$$f(T_f) = R(T_f = 0) \, \lambda_M \, e^{-\lambda_M T_f}, \tag{8.109}$$

where

$$\lambda_M = \frac{1}{2\pi}\left(\frac{\sigma_{\dot{y}}}{\sigma_y}\right)\left(1 - e^{-K\,\eta}\right)\Big/\left(e^{\frac{\eta^2}{2}} - 1\right),$$ (8.110)

$$R(T_f = 0) = 1 - e^{\frac{\eta^2}{2}},$$ (8.111)

$$K = \frac{\Delta\sqrt{2\pi}}{\sigma_{\dot{y}}},$$ (8.112)

$$\eta = \alpha/\sigma_y,$$ (8.113)

$$\Delta = \sqrt{\sigma_{\dot{y}}^2 - \frac{\lambda_1^2}{\sigma_y^2}}, \quad \text{as given by Eq. (8.66)},$$

and

$$\lambda_1 = \int_{-\infty}^{\infty} \omega\, S_y(\omega)\, d\omega, \quad \text{as given by Eq. (8.67)}.$$

EXAMPLE 8–1

A single-degree-of-freedom ($SDOF$) system is subjected to a stationary Gaussian force excitation $x(t)$ with zero mean and uniform spectral density S_0 (white noise). The governing equation of motion for the $SDOF$ system is given by [2, p. 176]

$$\ddot{y} + 2\,\xi\,\omega_n\,\dot{y} + \omega_n^2\,y = \frac{x(t)}{m},$$ (8.114)

where

y = relative displacement response of mass m,

ω_n = natural frequency of the system, in radians per second (rps),

$\omega_n = \sqrt{\dfrac{k}{m}}$,

k = spring stiffness,

ξ = damping ratio = $\dfrac{C}{C_c}$,

C = damping coefficient of the system,

C_c = critical damping coefficient of the system,

and

$C_c = 2\,m\,\omega_n$.

The dynamic analysis yields the following *rms* values of $Y(t)$ and $\dot{Y}(t)$ [2, p. 187]:

$$\sigma_Y = \sqrt{\frac{\pi\,S_0}{2\,\xi\,(\omega_n)^3}},$$ (8.115)

and

$$\sigma_{\dot{Y}} = \sqrt{\frac{\pi S_0}{2 \xi \omega_n}}, \tag{8.116}$$

respectively, and the value of Δ in Eq. (8.66) as

$$\Delta = \sqrt{\frac{2 S_0}{\omega_n}}, \quad \text{for } \xi \ll 1. \tag{8.117}$$

Do the following:

1. Determine the reliability, $R(t)$, which is defined as the probability that the response displacement $Y(t)$ has not crossed a prespecified barrier $Y_F = \alpha = b \sigma_y$ after time t with b being a specified constant, assuming

   (a) an independent level crossing and a Poisson distribution for the first-passage time,

   (b) an independent envelope crossing and a Poisson distribution for the first-passage time,

   (c) a two-state Markovian level crossing.

2. If the natural frequency is $\omega_n = 100$ rps and $\xi = 0.004$ (light damping), determine the allowable operating time T_0 for which we can be 99% sure that $Y(t)$ will not reach the level $Y_F = \alpha = 5 \sigma_y$, assuming

   (a) an independent level crossing and a Poisson distribution for the first-passage time,

   (b) an independent envelope crossing and a Poisson distribution for the first-passage time, or

   (c) a two-state Markovian level crossing.

3. Compare the results obtained in Case 2 under the three different assumptions.

SOLUTIONS TO EXAMPLE 8–1

1. Since given is a $SDOF$ system with natural frequency ω_n and a stationary Gaussian excitation $x(t)$, then the response $y(t)$ will be a stationary Gaussian narrow-band process with central frequency ω_n.

   (a) Under the Poisson process assumption for the independent level crossings, the first-passage time is exponentially distributed. The reliability function is given by Eq. (8.104), or

   $$R(t) = e^{-n_y^+(Y_F) t}, \tag{8.118}$$

where, from Eq. (8.77),

$$n_y^+(Y_F) = \frac{1}{2\pi} \left(\frac{\sigma_{\dot{y}}}{\sigma_y}\right) e^{-\frac{1}{2}\left(\frac{Y_F}{\sigma_y}\right)^2},$$

$$= \frac{1}{2\pi} \left(\frac{\sigma_{\dot{y}}}{\sigma_y}\right) e^{-\frac{1}{2}\left(\frac{b\,\sigma_y}{\sigma_y}\right)^2},$$

or

$$n_y^+(Y_F) = \frac{1}{2\pi} \left(\frac{\sigma_{\dot{y}}}{\sigma_y}\right) e^{-\frac{b^2}{2}}. \tag{8.119}$$

Substituting Eqs. (8.115) and (8.116) into Eq. (8.119) yields

$$n_y^+(Y_F) = \frac{1}{2\pi} \left(\frac{\sqrt{\pi\,S_0/2\,\xi\,\omega_n}}{\sqrt{\pi\,S_0/2\,\xi\,(\omega_n)^3}}\right) e^{-\frac{b^2}{2}},$$

or

$$n_y^+(Y_F) = \frac{\omega_n}{2\,\pi} e^{-\frac{b^2}{2}}. \tag{8.120}$$

Then, the reliability function, given by Eq. (8.118), becomes

$$R(t) = e^{-\left(\frac{\omega_n}{2\,\pi} e^{-\frac{b^2}{2}}\right) t}. \tag{8.121}$$

(b) Under the Poisson process assumption for the independent envelope crossings, the first-passage time is also exponentially distributed. The reliability function is then given by Eq. (8.107), or

$$R(t) = e^{-n_A^+(Y_F)\,t}, \tag{8.122}$$

where, from Eq. (8.78),

$$n_A^+(Y_F) = \frac{\Delta\,Y_F}{\sqrt{2\pi}\,\sigma_y^2} e^{-\frac{1}{2}\left(\frac{Y_F}{\sigma_y}\right)^2},$$

and from Eq. (8.117) $\Delta = \sqrt{\frac{2\,S_0}{\omega_n}}$, then

$$n_A^+(Y_F) = \frac{\sqrt{2\,S_0/\omega_n}\,(b\,\sigma_y)}{\sqrt{2\pi}\,\sigma_y^2} e^{-\frac{1}{2}\left(\frac{b\,\sigma_y}{\sigma_y}\right)^2},$$

or

$$n_A^+(Y_F) = \left(\frac{b}{\sigma_y}\right)\sqrt{\frac{S_0}{\omega_n\,\pi}}\, e^{-\frac{b^2}{2}}. \tag{8.123}$$

Substituting Eq. (8.115) into Eq. (8.123) yields

$$n_A^+(Y_F) = \left(\frac{b}{\sqrt{\pi\,S_0/2\,\xi\,(\omega_n)^3}}\right)\sqrt{\frac{S_0}{\omega_n\,\pi}}\, e^{-\frac{b^2}{2}},$$

or

$$n_A^+(Y_F) = \frac{b\,\omega_n\,\sqrt{2\,\xi}}{\pi} e^{-\frac{b^2}{2}}. \tag{8.124}$$

Then, the reliability function of Eq. (8.122) becomes

$$R(t) = e^{-\left(\frac{b\ \omega_n\ \sqrt{2\,\xi}}{\pi}\ e^{-\frac{b^2}{2}}\right)t}.$$

(8.125)

(c) Under the two-state-Markovian-level-crossing assumption, the *pdf* of T_f is given by Eq. (8.109). The reliability function is then given by

$$R(t) = \int_t^\infty f(T_f)\,dT_f,$$

or

$$R(t) = R(T_f = 0)\ e^{-\lambda_M t},$$

(8.126)

where λ_M and $R(T_f = 0)$ are given by Eqs. (8.110) and (8.111), respectively.

Substituting Eqs. (8.115), (8.116) and (8.117) into Eqs. (8.112), (8.113), and then into Eqs. (8.110) and (8.111), yields

$$K = \frac{\Delta\ \sqrt{2\pi}}{\sigma_{\dot{y}}},$$

$$= \frac{\sqrt{2\ S_0/\omega_n}\ \sqrt{2\pi}}{\sqrt{\pi\ S_0/(2\ \xi\ \omega_n)}},$$

or

$$K = 2\sqrt{2\ \xi},$$

(8.127)

$$\eta = \alpha/\sigma_y = b\ \sigma_y/\sigma_y = b,$$

(8.128)

$$\lambda_M = \frac{1}{2\pi}\left(\frac{\sqrt{\pi\ S_0/2\ \xi\ \omega_n}}{\sqrt{\pi\ S_0/2\ \xi\ (\omega_n)^3}}\right)$$

$$\cdot\left(1 - e^{-2\sqrt{2\ \xi}\ b}\right)\Big/\left(e^{\frac{b^2}{2}} - 1\right),$$

or

$$\lambda_M = \frac{\omega_n}{2\pi}\left(1 - e^{-2\sqrt{2\ \xi}\ b}\right)\Big/\left(e^{\frac{b^2}{2}} - 1\right),$$

(8.129)

and

$$R(T_f = 0) = 1 - e^{-\frac{\eta^2}{2}} = 1 - e^{-\frac{b^2}{2}}.$$

(8.130)

Substituting Eqs. (8.129) and (8.130) into Eq. (8.126) yields

$$R(t) = \left(1 - e^{-\frac{b^2}{2}}\right)\ e^{-\left[\frac{\omega_n}{2\pi}\left(1 - e^{-2\sqrt{2\ \xi}\ b}\right)\Big/\left(e^{\frac{b^2}{2}} - 1\right)\right]t}.$$

(8.131)

2. (a) Substituting $\omega_n = 100$ rps, $b = 5$ and $R(T_0) = 0.99$ into Eq. (8.121) yields

$$0.99 = e^{-\left(\frac{100}{2\pi} e^{-\frac{5^2}{2}}\right) T_0}.$$

Solving for T_0 yields

$$T_0 = \frac{2\pi}{100} e^{\frac{25}{2}} \log_e \left(\frac{1}{0.99}\right),$$

$T_0 = 169.45$ seconds, or 2.82 minutes.

(b) Substituting $\omega_n = 100$ rps, $b = 5$, $\xi = 0.004$, and $R(T_0) = 0.99$ into Eq. (8.125) yields

$$0.99 = e^{-\left(\frac{5\ (100)\ \sqrt{2\ (0.004)}}{\pi} e^{-\frac{5^2}{2}}\right) T_0}.$$

Solving for T_0 yields

$$T_0 = \frac{\pi}{500\sqrt{0.008}} e^{\frac{25}{2}} \log_e \left(\frac{1}{0.99}\right),$$

$T_0 = 189.45$ seconds, or 3.15 minutes.

(c) Substituting $\omega_n = 100$ rps, $b = 5$, $\xi = 0.004$, and $R(T_0) = 0.99$ into Eq. (8.131) yields

$$0.99 = \left(1 - e^{-\frac{5^2}{2}}\right) e^{-\left[\frac{100}{2\pi}\left(1 - e^{-2\sqrt{2\ (0.004)}\ 5}\right)\right] \big/ \left(e^{\frac{5^2}{2}} - 1\right)} T_0.$$

Solving for T_0 yields

$$T_0 = \left[\frac{2\pi}{100}\left(e^{\frac{25}{2}} - 1\right) \big/ \left(1 - e^{-2\sqrt{0.008}\ 5}\right)\right] \log_e \frac{\left(1 - e^{-\frac{25}{2}}\right)}{0.99},$$

$T_0 = 286.53$ seconds, or 4.78 minutes.

3. Comparing the results under the three different assumptions, it may be seen that the first two assumptions give conservative estimates of the allowable operating time T_0 which are lower than that under the two-state Markovian process assumption; i.e., $2.82 < 3.15 < 4.78$. Since $Y(t)$ is a narrow-band process in this example, crossings will tend to occur in clumps. Therefore, the first assumption, or the independent level crossing assumption, is very poor in this case. The result under the second assumption, or independent envelope crossing assumption, is superior to that under the first assumption. Meanwhile, the result under the two-state Markovian process assumption is closest to the simulation results and converges asymptotically to those based on the first two assumptions [2, p. 135]. Then, the order of superiority for these three assumptions, when $Y(t)$ is a narrow-band random process,

is the following: *Two-state Markov > Poisson with independent envelope crossing > Poisson with independent level crossing.*

8.3.3 FAILURE PROBABILITY DUE TO RESPONSE REMAINING ABOVE A FIXED LEVEL FOR TOO GREAT A FRACTION OF THE TIME

For systems which operate intermittently, or which handle highly redundant information, so that an occasional isolated malfunction is not too serious, it may be assumed that an occasional crossing of $Y(t)$ over a fixed threshold level $Y_F = \alpha$ does not in itself cause a failure. The failure occurs only when the fraction of time in $(t_0, t_0 + T)$ for which $Y(t) > \alpha$, $Y_\alpha(T)$, becomes greater than a fixed fraction ϵ. It is very difficult to determine the distribution for $Y_\alpha(T)$. But it has been shown in Section 8.2 that the mean of $Y_\alpha(T)$ is given by Eq. (8.61), or

$$E[Y_\alpha(T)] = \frac{1}{T} \int_{t=t_0}^{t_0+T} \int_{y=\alpha}^{\infty} f(y,t)\, dy\, dt.$$

If $Y(t)$ is a stationary ergodic process, then

$$f(y,t) = f(y),$$

and

$$Y_\alpha(T) = Y_\alpha.$$

Therefore,

$$E[Y_\alpha(T)] = \frac{1}{T} \int_{t=t_0}^{t_0+T} \int_{y=\alpha}^{\infty} f(y)\, dy\, dt,$$

or

$$E[Y_\alpha] = \int_{y=\alpha}^{\infty} f(y)\, dy,$$

which is independent of process length T.

If the exponential distribution can be assumed for Y_α, then the reliability function is given by

$$R(\epsilon, \alpha) = P[Y_\alpha < \epsilon] = 1 - e^{-\frac{\epsilon}{E[Y_\alpha]}} = 1 - e^{-\frac{\epsilon}{\int_\alpha^\infty f(y)\, dy}}. \tag{8.132}$$

EXAMPLE 8–2

Let $Y(t)$ represent the dynamic response of a system, which is an ergodic Gaussian process with a zero mean and an *rms* value of σ_y. Determine the reliability of this system under $\epsilon = 1\%$ and $\alpha = 3\sigma_y$.

SOLUTION TO EXAMPLE 8–2

Since $Y(t)$ is ergodic Gaussian with $\mu_y = 0$ and σ_y, then

$$f(y) = \frac{1}{\sqrt{2\pi}\,\sigma_y}\, e^{-\frac{1}{2}\left(\frac{y}{\sigma_y}\right)^2},$$

and

$$\int_\alpha^\infty f(y)\,dy = 1 - \Phi\left(\frac{\alpha}{\sigma_y}\right),$$

where

$\Phi(\ ) = cdf$ of the standardized normal distribution $N(0,1)$.

Therefore,

$$
\begin{aligned}
R(\epsilon, \alpha) &= R(0.01, 3\sigma_y), \\
&= 1 - e^{-\frac{0.01}{[1-\Phi(3\sigma_y/\sigma_y)]}}, \\
&= 1 - e^{-\frac{0.01}{[1-\Phi(3)]}}, \\
&= 1 - e^{-\frac{0.01}{[1-0.99865]}},
\end{aligned}
$$

or

$$R(0.01, 3\sigma_y) = 0.99939.$$

8.3.4   FAILURE PROBABILITY DUE TO FATIGUE

8.3.4.1   PALMGREN AND MINER'S LINEAR DAMAGE ACCUMULATION HYPOTHESIS

In standard fatigue tests of structural elements where the stress varies with time, as shown in Fig. 8.8, but its amplitude, S, is held constant, the number of cycles until fracture, $N(S)$, is counted. When a large number of identical samples are tested in a range of stress amplitudes, a linear approximation between $N(S)$ and S on a log-log scale, as shown in Fig. 8.9, can be obtained from test data; i.e.,

$$\log_{10} N(S) = a - b\,\log_{10} S, \quad S_E \le S \le S_Y, \tag{8.133}$$

or

$$N(S)\,S^b = A, \quad S_E \le S \le S_Y, \tag{8.134}$$

where constants b and c depend on the material's fatigue properties,

$A = 10^a,$

$S_E = $ endurance limit,

and

$$S_Y = \text{yielding strength.}$$

Equations (8.133) and (8.134) are the so-called $S - N$ curves or "fatigue curves."

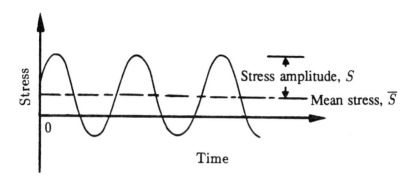

Fig. 8.8– Stress history during a constant amplitude fatigue test.

Strictly speaking, the actual fatigue curve is not a single curve but a family of distributions scattered over a two-dimensional plane of $S(N)$ and N. For the sake of simplicity, it is assumed here that there is a single well-defined $S - N$ curve for the structural material involved.

When the stress in the structure does not have a fixed amplitude but is a random process, the prediction of the probability of fatigue failure becomes difficult. However, the fixed amplitude fatigue data can be approximately, but "rationally," extrapolated to the random stress amplitude case by means of the *linear damage accumulation hypothesis* proposed by Palmgren and Miner [3; 4]. According to this hypothesis when n_i cycles of stress amplitude S_i have been experienced, the structure has "used up" a fraction of $\frac{n_i}{N(S_i)}$ of its fatigue life. The damage, D, due to loading with groups of different stress amplitudes will then accumulate linearly; i.e.,

$$D = \frac{n_1}{N(S_1)} + \frac{n_2}{N(S_2)} + \cdots = \sum_i \frac{n_i}{N(S_i)}. \qquad (8.135)$$

The Palmgren-Miner hypothesis states that the structure will undergo fatigue failure when $D \geq 1$. Note that under this hypothesis there is no restriction on the order of application of the different stress levels. However, the experimental values of D at failure have been found to range between 0.18 and 23.0 [5; 6], depending on the material, the test conditions and the order of loading history. It was observed [6] that a low-to-high loading sequence ($S_1 < S_2 < S_3 < \cdots$) resulted in high D values ($\overline{D} > 1$). A high-to-low loading

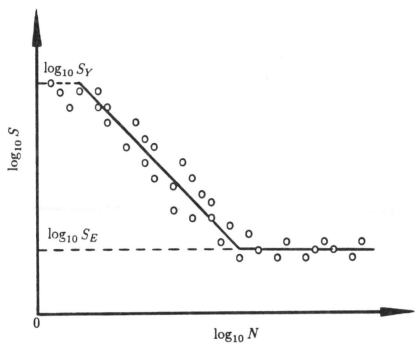

Fig. 8.9– A typical $S - N$ curve.

sequence $(S_1 > S_2 > S_3 > \cdots)$ gave low D values $(\overline{D} < 1)$. This behavior applies particularly to strain-hardening materials. For a loading history with representative high, low, and medium stress levels in random order, the value of D appears to be close to unity. Therefore, the Palmgren-Miner hypothesis can be applied to a random process where the stress amplitudes change from cycle to cycle randomly [3; 7].

Equation (8.135) may be extended to a continuous loading process where the stress amplitude varies continuously; i.e.,

$$D = \int_{S=0}^{\infty} \frac{n(S)}{N(S)} \, dS, \tag{8.136}$$

where

$\quad\quad n(S)dS$ = number of cyclic loadings with amplitude
$\quad\quad\quad\quad$ in the stress interval $(S, S + dS)$,

and

$\quad\quad N(S)$ = fatigue life at stress amplitude S.

8.3.4.2 STATIONARY NARROW-BAND RANDOM STRESSING

When the stressed element is part of a sharply resonant vibratory system, the response stress, $Y(t)$, will be a narrow-band process with the central frequency equal to the natural frequency of the structure. It is assumed that $Y(t)$ is a stationary, random process. Therefore, $Y(t)$ may have a sample function like that shown in Fig. 8.5(b), where individual "cycles" are obvious although their amplitudes may vary in a random fashion. An incremental "damage" can be ascribed to each cycle using the Palmgren-Miner hypothesis. Accumulating these damages over a time interval T leads to a total damage $D(T)$ which is a monotonically increasing function of T. At a certain time point, T_F, where $D(T_F) \geq 1$, fatigue failure occurs according to the Palmgren-Miner criterion.

8.3.4.2.1 Mean Value of the Cumulative Damage

Let $n_y^+(0)$ be the expected zero-up-crossing rate of the narrow-band random stress response $Y(t)$. Then, there will be on the average $[n_y^+(0)\ T]$ stress cycles in time interval T. The expected fraction of these cycles whose stress amplitudes lie between S and $S + dS$ is $[f_p(S)\ dS]$, where $f_p(S)$ is the *pdf* of the stress peaks. The random number of such peaks in time interval T with amplitudes in the interval $(S, S + dS)$ is given by

$$n(S)\ dS = [n_y^+(0)\ T][f_p(S)\ dS]. \tag{8.137}$$

One single peak of amplitude S will cause an incremental damage of $1/N(S)$. The elementary fractional damage in time T due to all cycles with peaks ranging from S to $S + dS$ is

$$\frac{n(S)}{N(S)}\ dS = n_y^+(0)\ T\ \frac{f_p(S)}{N(S)}\ dS. \tag{8.138}$$

The expected cumulative damage in time interval T resulting from the cycles of all stress amplitudes is given by

$$\overline{D}(T) = \int_0^\infty \frac{n(S)}{N(S)}\ dS = n_y^+(0)\ T \int_0^\infty \frac{f_p(S)}{N(S)}\ dS, \tag{8.139}$$

which is a linear function of T. This expression may be used to predict the mean time to fatigue failure, $\overline{T}_F$; i.e.,

$$\overline{T}_F = T\Big|_{\overline{D}(T)=1} = \frac{1}{n_y^+(0)\ \int_0^\infty \frac{f_p(S)}{N(S)}\ dS}, \tag{8.140}$$

where $T\Big|_{\overline{D}(T)=1}$ implies the time during which the average cumulative damage is equal to one (1).

If $Y(t)$ is a stationary, narrow-band Gaussian process with zero mean, its peaks will follow a Rayleigh distribution with a *pdf* given by Eq. (8.58), or

$$f_p(S) = \frac{S}{\sigma_y^2} \, e^{-\frac{1}{2}\left(\frac{S}{\sigma_y}\right)^2}. \tag{8.141}$$

The expected zero-up-crossing rate, $n_y^+(0)$, can be evaluated using Eq. (8.120) by letting $Y_F = b\,\sigma_y = 0$; i.e.,

$$n_y^+(0) = n_y^+(Y_F = 0) = \frac{\omega_n}{2\,\pi}\, e^0,$$

or

$$n_y^+(0) = \frac{\omega_n}{2\,\pi}, \tag{8.142}$$

where

ω_n = natural frequency of the structure, in radians per second (rps).

Substituting Eqs. (8.134), (8.141) and (8.142) into Eq. (8.139) yields

$$\begin{aligned}
\overline{D}(T) &= \left(\frac{\omega_n}{2\,\pi}\right) T \int_0^\infty \frac{(S/\sigma_y^2)\, e^{-\frac{1}{2}(S/\sigma_y)^2}}{A/S^b}\, dS, \\
&= 2\left(\frac{\omega_n}{2\,\pi}\right)\left(\frac{T}{A}\right)(\sqrt{2}\,\sigma_y)^b \\
&\quad \cdot \int_0^\infty \left(S/\sqrt{2}\,\sigma_y\right)^{b+1} e^{-(S/\sqrt{2}\,\sigma_y)^2}\, d\left(\frac{S}{\sqrt{2}\,\sigma_y}\right).
\end{aligned} \tag{8.143}$$

Let $Z = (S/\sqrt{2}\,\sigma_y)^2$, then $S/\sqrt{2}\,\sigma_y = Z^{1/2}$ and $d(S/\sqrt{2}\,\sigma_y) = d\sqrt{Z} = \frac{1}{2} Z^{-1/2}\, dZ$. Equation (8.143) becomes

$$\begin{aligned}
\overline{D}(T) &= 2\left(\frac{\omega_n}{2\,\pi}\right)\left(\frac{T}{A}\right)(\sqrt{2}\,\sigma_y)^b \\
&\quad \cdot \int_0^\infty Z^{\frac{b+1}{2}}\, e^{-Z}\left(\frac{1}{2}\right) Z^{-\frac{1}{2}}\, dZ, \\
&= \left(\frac{\omega_n}{2\,\pi}\right)\left(\frac{T}{A}\right)(\sqrt{2}\,\sigma_y)^b \int_0^\infty Z^{\frac{b}{2}}\, e^{-Z}\, dZ,
\end{aligned}$$

or

$$\overline{D}(T) = \left(\frac{\omega_n}{2\,\pi}\right)\left(\frac{T}{A}\right)(\sqrt{2}\,\sigma_y)^b\, \Gamma\left(\frac{b}{2}+1\right), \tag{8.144}$$

where

$$\Gamma(n) = \text{Gamma function of } n = \int_0^\infty Z^{n-1} e^{-Z} dZ.$$

Equation (8.144) states that the expected cumulative damage is linearly proportional to the expected number of stress cycles in time T, $n_y^+(0) \, T = \left(\frac{\omega_n}{2\pi}\right) T$, and also nonlinearly proportional to the *rms* value of $Y(t)$, σ_y.

Correspondingly, the mean time to fatigue failure, given by Eq. (8.140), becomes

$$\overline{T}_F = T\Big|_{\overline{D}(T)=1} = \frac{A}{\left(\frac{\omega_n}{2\pi}\right) \left(\sqrt{2}\,\sigma_y\right)^b \Gamma\left(\frac{b}{2}+1\right)}. \tag{8.145}$$

8.3.4.2.2 Standard Deviation of the Cumulative Damage

Mark [8] did a comprehensive study on the inherent variation in fatigue damage due to random vibration. Under the assumptions that the random stress process $Y(t)$ is stationary Gaussian and the damping is light ($\xi \leq 0.05$), he derived the expression for the ratio of the standard deviation of the damage, $\sigma_{D(T)}$, to the mean value of the damage, $\overline{D}(T)$, at the end of an interval of duration T as follows [3; 8]:

$$\frac{\sigma_{D(T)}}{\overline{D}(T)} = \frac{1}{\sqrt{n_y^+(0)\,T}} \sqrt{\frac{\psi_1(b)}{\xi} - \frac{\psi_2(b)}{\xi^2\,n_y^+(0)\,T} + \frac{\psi_3(b)}{n_y^+(0)\,T}}, \tag{8.146}$$

or

$$\frac{\sigma_{D(T)}}{\overline{D}(T)} \simeq \sqrt{\frac{\psi_1(b)}{\xi\,n_y^+(0)\,T}}, \tag{8.147}$$

where

$n_y^+(0)\,T$ = expected total number of cycles in time interval T,

ξ = damping ratio as introduced in Example 8–1,

b = exponent in the fatigue law of Eqs. (8.133) and (8.134),

and the quantities $\psi_1(b)$, $\psi_2(b)$ and $\psi_3(b)$ are functions of b and are tabulated in Table 8.1 for odd integer values of b [3]. The values of $\psi_1(b)$, $\psi_2(b)$ and $\psi_3(b)$ for any other value of b can be calculated by semilog linear interpolation; i.e., interpolate with b versus $\log_{10} \psi_i(b)$, $i = 1, 2, 3$.

Note that the approximate expression given by Eq. (8.147) is valid because in most applications the expected cycle number, $n_y^+(0)\,T$, is large enough to make $n_y^+(0)\,T \gg 1$ and $\xi\,n_y^+(0)\,T \gg 1$ so that both the second and the third terms under the square root of Eq. (8.146) will be negligible as compared with the first term.

TABLE 8.1– Values of $\psi_1(b)$, $\psi_2(b)$ and $\psi_3(b)$ as functions of the fatigue law exponent b [3].

b	$\psi_1(b)$	$\psi_2(b)$	$\psi_3(b)$
1	0.0414	0.00323	0.0796
3	0.3690	0.02900	0.2120
5	1.2800	0.09040	0.6790
7	3.7200	0.22300	2.3300
9	10.7000	0.51800	8.2800
11	31.5000	1.23000	30.0000
13	96.7000	3.06000	111.2000
15	308.0000	8.11000	415.0000

Similar to the case of $\overline{D}(T)$, the damage standard deviation, $\sigma_{D(T)}$, when $Y(t)$ is a stationary, narrow-band, Gaussian process with zero mean, can be obtained by substituting Eqs. (8.142) and (8.144) into Eq. (8.147); i.e.,

$$\sigma_{D(T)} \cong \overline{D}(T) \sqrt{\frac{\psi_1(b)}{\xi \, n_y^+(0) \, T}},$$

$$= \left(\frac{\omega_n}{2\pi}\right) \left(\frac{T}{A}\right) (\sqrt{2}\,\sigma_y)^b \, \Gamma\left(\frac{b}{2}+1\right)$$

$$\cdot \sqrt{\frac{\psi_1(b)}{\xi \left(\frac{\omega_n}{2\pi}\right) T}},$$

or

$$\sigma_{D(T)} \cong \frac{(\sqrt{2}\,\sigma_y)^b}{A} \, \Gamma\left(\frac{b}{2}+1\right) \sqrt{\left(\frac{\omega_n}{2\pi}\right) \frac{\psi_1(b) \, T}{\xi}}, \qquad (8.148)$$

which is proportional to $\sqrt{T}$ and inversely proportional to $\sqrt{\xi}$. That is, the longer the random vibration, or the lighter the damping, the larger the damage variability.

8.3.4.2.3 Distributions of the Cumulative Damage and the Fatigue Life

Equations (8.144) and (8.148) indicate that there is a distribution for the actual damage in interval T with mean, $\overline{D}(T)$, and standard deviation, $\sigma_{D(T)}$. Correspondingly, the time to fatigue failure (or when $D = 1$), T_F, is also a random variable. The central limit theorem can be invoked to show that both the distribution of D and the distribution of T_F approach the normal [3]; i.e.,

$$D \sim N\left[\overline{D}(T), \sigma_{D(T)}^2\right], \tag{8.149}$$

and

$$T_F \sim N\left(\overline{T}_F, \sigma_{T_F}^2\right), \tag{8.150}$$

where $\overline{D}(T)$, $\sigma_{D(T)}$ and $\overline{T}_F$ are given by Eqs. (8.144), (8.148) and (8.145), respectively. The value of σ_{T_F} can be estimated as follows:

(a) Determine the "3 σ" upper limit of $D(T)$, $D_U(T)$; i.e.,

$$D_U(T) = \overline{D}(T) + 3\,\sigma_{D(T)},$$
$$= \left(\frac{\omega_n}{2\pi}\right)\left(\frac{T}{A}\right)(\sqrt{2}\,\sigma_y)^b\,\Gamma\left(\frac{b}{2}+1\right)$$
$$+\frac{3\,(\sqrt{2}\,\sigma_y)^b}{A}\,\Gamma\left(\frac{b}{2}+1\right)\sqrt{\left(\frac{\omega_n}{2\pi}\right)\frac{\psi_1(b)\,T}{\xi}},$$

or

$$D_U(T) = h_1\,T + h_2\,\sqrt{T}, \tag{8.151}$$

where

$$h_1 = \left(\frac{\omega_n}{2\pi}\right)(\sqrt{2}\,\sigma_y)^b\,\Gamma\left(\frac{b}{2}+1\right)\bigg/A, \tag{8.152}$$

and

$$h_2 = \frac{3\,(\sqrt{2}\,\sigma_y)^b}{A}\,\Gamma\left(\frac{b}{2}+1\right)\sqrt{\left(\frac{\omega_n}{2\pi}\right)\frac{\psi_1(b)}{\xi}}. \tag{8.153}$$

(b) Determine the "3 σ" lower limit of $D(T)$, $D_L(T)$, as follows:

$$D_L(T) = \overline{D}(T) - 3\,\sigma_{D(T)} = h_1\,T - h_2\,\sqrt{T}. \tag{8.154}$$

(c) Determine the "3 σ" lower limit of T_F, T_{FL}, by letting $D_U(T) = 1$ and solving for T; i.e.,

$$h_1\,T + h_2\,\sqrt{T} = 1. \tag{8.155}$$

This is so because when fatigue failure occurs, or when $D(T) = 1$, the lower limit of the fatigue life, T_{FL}, corresponds to the intersection of $D(T) = 1$ and the upper bound of the fatigue curve, $D_U(T)$, as shown in Fig. 8.10.

Let $\sqrt{T} = Z$, then $T = Z^2$ and Eq. (8.155) becomes

$$h_1\,Z^2 + h_2\,Z - 1 = 0.$$

Solving for Z yields

$$Z = \frac{-h_2 + \sqrt{h_2^2 + 4\,h_1}}{2\,h_1},$$

where only the "+" sign is taken because $Z = \sqrt{T}$ is always non-negative. Consequently, the "3 σ" lower limit for T_F, T_{FL}, is given by

$$T_{FL} = Z^2 = \left(\frac{-h_2 + \sqrt{h_2^2 + 4\,h_1}}{2\,h_1}\right)^2. \tag{8.156}$$

(d) Determine the "3 σ" upper limit of T_F, T_{FU}, by letting $D_L(T) = 1$ and solving for T; i.e.,

$$h_1\,T - h_2\,\sqrt{T} = 1. \tag{8.157}$$

This is so because when fatigue failure occurs, or when $D(T) = 1$, the upper limit of fatigue life, T_{FU}, corresponds to the intersection of $D(T) = 1$ and the lower bound of the fatigue curve $D_L(T)$, as shown in Fig. 8.10.

Let $\sqrt{T} = Z$, then $T = Z^2$ and Eq. (8.157) becomes

$$h_1\,Z^2 - h_2\,Z - 1 = 0.$$

Solving for Z yields

$$Z = \frac{h_2 + \sqrt{h_2^2 + 4\,h_1}}{2\,h_1},$$

where only the "+" sign is taken because $Z = \sqrt{T}$ is always non-negative. Consequently, the "3 σ" upper limit for T_F, T_{FU}, is given by

$$T_{FU} = Z^2 = \left(\frac{h_2 + \sqrt{h_2^2 + 4\,h_1}}{2\,h_1}\right)^2. \tag{8.158}$$

(e) Determine the standard deviation of the fatigue life distribution, σ_{T_F}, by letting

$$T_{FL} = \overline{T}_F - 3\,\sigma_{T_F}, \tag{8.159}$$

and

$$T_{FU} = \overline{T}_F + 3\,\sigma_{T_F}. \tag{8.160}$$

Subtracting Eq. (8.159) from Eq. (8.160) and solving for σ_{T_F} yields

$$\sigma_{T_F} = \frac{T_{FU} - T_{FL}}{6},$$

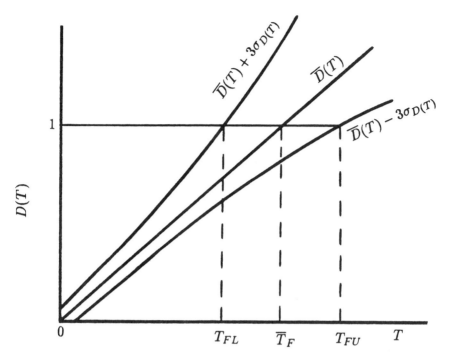

Fig. 8.10– Determination of the standard deviation of the fatigue
life distribution.

$$= \frac{1}{6} \left[\left(\frac{h_2 + \sqrt{h_2^2 + 4\, h_1}}{2\, h_1} \right)^2 \right.$$

$$\left. - \left(\frac{-h_2 + \sqrt{h_2^2 + 4\, h_1}}{2\, h_1} \right)^2 \right].$$

Simplifying yields

$$\sigma_{T_F} = \frac{h_2\, \sqrt{h_2^2 + 4\, h_1}}{6\, h_1^2}. \tag{8.161}$$

The procedures **(a)** through **(e)** can be visualized through Fig. 8.10.

8.3.4.2.4   Reliability Evaluation for Fatigue Failure

Having derived the parameter estimates for cumulative damage and fatigue
life, the fatigue reliability can be readily calculated from

$$R(T) = P[D(T) < 1] = \Phi \left[\frac{1 - \overline{D}(T)}{\sigma_{D(T)}} \right], \tag{8.162}$$

or, equivalently, from

$$R(T) = P(T_F > T) = 1 - \Phi \left(\frac{T - \overline{T}_F}{\sigma_{T_F}} \right) = \Phi \left(\frac{\overline{T}_F - T}{\sigma_{T_F}} \right), \tag{8.163}$$

where

$\Phi(\ ) = cdf$ of the standardized normal distribution, $N(0,1)$.

Equations (8.162) and (8.163) should yield very close results.

Correspondingly, the probability of fatigue failure due to random vibration for time T, is given by

$$Q(T) = P[D(T) \geq 1] = 1 - \Phi \left[\frac{1 - \overline{D}(T)}{\sigma_{D(T)}} \right] = \Phi \left[\frac{\overline{D}(T) - 1}{\sigma_{D(T)}} \right], \tag{8.164}$$

or by

$$Q(T) = P(T_F \leq T) = \Phi \left(\frac{T - \overline{T}_F}{\sigma_{T_F}} \right). \tag{8.165}$$

EXAMPLE 8–3

During *ESS* an aircraft line replaceable unit (LRU), which is a 20-lb
rectangular box of sheet metal construction containing edge-supported PC
boards, is randomly vibrated by small electrodynamic thrusters to produce
an exitation vector directed diagonally through the LRU's center of mass,
as shown in Fig. 8.11. The 6-$grms$, 20 to 2,000 Hz spectrum recommended
by NAVMAT P-9492 [9] is used. The diagonal force vector has been found
superior to the usual orthogonal approach (using a conventional slip-plate
vibration fixture in the axis perpendicular to the PC board plane, as well
as in the two remaining orthogonal axes) in exiting the PC board [9]. This
is so because some degree of PC board bending is required to stress solder
joints and other board metallization, whereas none of the three conventional
orthogonal axis tests exited the PC board's fundamental bending resonance
better than the diagonal exitation test.

The bending stress response spectra in the engineering form, $W_S(f)$, at
the center of a PC board, for LRU exitation in the X (longitudinal) axis and
in the diagonal V (longitudinal–vertical) axis, are shown in Fig. 8.12 where

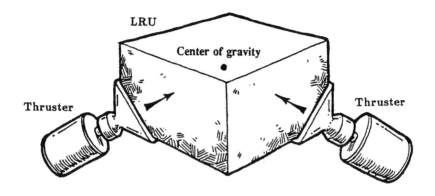

Fig. 8.11– Diagonal force vector vibration applied to a typical *LRU* with either electrodynamic or pneumatic thrusters.

only part of the response in the 20–250 Hz range is presented. It may be seen that the unconventional diagonal vibration test clearly exited the fundamental resonance frequency at 125 Hz. Therefore, the mounted resonant frequency of the PC board is determined to be 125 Hz. This process is the so-called *Frequency Range Influence Test* . The *rms* value of the response stress, $S(t)$, at the solder joint is measured to be $\sigma_S = 5.4 \times 10^3$ psi. The damping ratio of the *LRU* is 0.01. The equivalent $S - N$ fatigue curve of the solder joint is given by

$$N(S)\ S^4 = \left(2 \times 10^5\right)^4 .$$

Do the following:

1. Determine the mean and standard deviation of the cumulative damage due to such a random vibration for a time interval of 30 minutes.

2. Determine the mean and standard deviation of the time to fatigue failure of this *LRU*.

3. Determine the reliability of the *LRU* under the given random vibration for 30 minutes assuming the normal distribution for both of the cumulative damage and the fatigue life.

4. Determine how long should this random vibration last so that we can be 99% sure that the fatigue failure will occur?

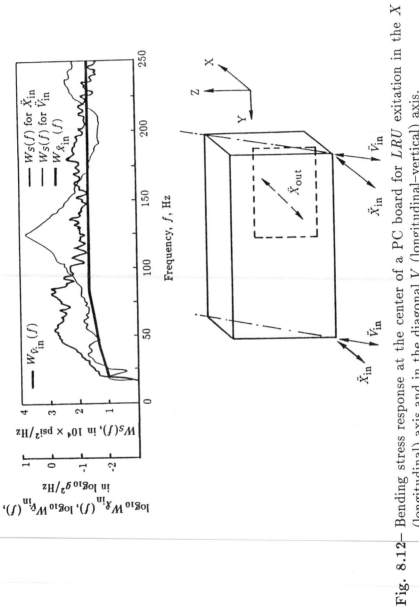

Fig. 8.12— Bending stress response at the center of a PC board for *LRU* excitation in the *X* (longitudinal) axis and in the diagonal *V* (longitudinal–vertical) axis.

266

SOLUTIONS TO EXAMPLE 8–3

1. From Fig. 8.12 it may be seen that the bending stress response under the diagonal exitation is a narrow-band process. The natural or fundamental frequency for this PC board in the *LRU* is 125 Hz. Then,

$$\omega_n = 125 \, (2\,\pi) \text{ rps},$$

or

$$\frac{\omega_n}{2\,\pi} = 125 \text{ Hz.}$$

The constant parameters for the solder joint fatigue curve are the following:

$$b = 4,$$

and

$$A = \left(2 \times 10^5\right)^4.$$

Substituting these and $\sigma_S = 5.4 \times 10^3$ psi into Eqs. (8.144) and (8.148) yields

$$\overline{D}(T = 30 \text{ min}) = (125) \left[\frac{30 \times 60}{\left(2 \times 10^5\right)^4} \right]$$
$$\cdot \left(\sqrt{2} \times 5.4 \times 10^3\right)^4 \, \Gamma\left(\frac{4}{2} + 1\right),$$

or

$$\overline{D}(T = 30 \text{ min}) = 0.95659,$$

and

$$\sigma_{D(T=30 \text{ min})} \cong \frac{\left(\sqrt{2} \times 5.4 \times 10^3\right)^4}{\left(2 \times 10^5\right)^4} \, \Gamma\left(\frac{4}{2} + 1\right)$$
$$\cdot \sqrt{(125) \frac{\psi_1(4) \, (30 \times 60)}{0.01}}.$$

The value for $\psi_1(4)$ can be obtained by semilog linear interpolation using Table 8.1; i.e.,

$$\frac{\log_{10} \psi_1(4) - \log_{10} \psi_1(3)}{\log_{10} \psi_1(5) - \log_{10} \psi_1(3)} = \frac{4 - 3}{5 - 3},$$

or

$$\frac{\log_{10} \psi_1(4) - \log_{10} 0.3690}{\log_{10} 1.2800 - \log_{10} 0.3690} = \frac{1}{2},$$

or

$$\psi_1(4) = 0.6873.$$

Then,

$$\sigma_{D(T=30 \text{ min})} \cong \frac{\left(\sqrt{2} \times 5.4 \times 10^3\right)^4}{\left(2 \times 10^5\right)^4} \Gamma\left(\frac{4}{2}+1\right)$$
$$\cdot \sqrt{(125) \frac{(0.6873)(30 \times 60)}{0.01}},$$

or

$$\sigma_{D(T=30 \text{ min})} \cong 0.01672.$$

Therefore, under the normal distribution assumption for the cumulative damage, we have

$$D(T = 30 \text{ min}) \sim N\left(\overline{D}(30 \text{ min}),\ \sigma^2_{D(T=30 \text{ min})}\right),$$

or

$$D(T = 30 \text{ min}) \sim N\left(0.95659,\ 0.01672^2\right).$$

2. Substitution of $b = 4$, $A = \left(2 \times 10^5\right)^4$, $\frac{\omega_n}{2\pi} = 125$ Hz, and $\sigma_S = 5.4 \times 10^3$ psi into Eqs. (8.145) and (8.161) yields

$$\overline{T}_F = \frac{\left(2 \times 10^5\right)^4}{(125)\left(\sqrt{2} \times 5.4 \times 10^3\right)^4 \Gamma\left(\frac{4}{2}+1\right)},$$

or

$$\overline{T}_F = 1881.6764 \text{ seconds, or } 31.3613 \text{ minutes},$$

and

$$\sigma_{T_F} = \frac{h_2 \sqrt{h_2^2 + 4 h_1}}{6 h_1^2},$$

where from Eqs. (8.152) and (8.153)

$$h_1 = \left(\frac{\omega_n}{2\pi}\right)\left(\sqrt{2}\,\sigma_S\right)^b \Gamma\left(\frac{b}{2}+1\right)\Big/ A,$$
$$= (125)\left(\sqrt{2} \times 5.4 \times 10^3\right)^4$$
$$\cdot \Gamma\left(\frac{4}{2}+1\right)\Big/\left(2 \times 10^5\right)^4,$$

or

$$h_1 = 0.00053144,$$

and

$$h_2 = \frac{3\left(\sqrt{2}\,\sigma_S\right)^b}{A} \Gamma\left(\frac{b}{2}+1\right)\sqrt{\left(\frac{\omega_n}{2\pi}\right)\frac{\psi_1(b)}{\xi}},$$

$$= \frac{3 \left(\sqrt{2} \times 5.4 \times 10^3\right)^4}{(2 \times 10^5)^4} \, \Gamma \left(\frac{4}{2} + 1\right)$$
$$\cdot \sqrt{(125) \frac{0.6873}{0.01}},$$

or

$$h_2 = 0.0011820.$$

Then,

$$\sigma_{T_F} = \frac{0.0011820 \, \sqrt{0.0011820^2 + 4 \, (0.00053144)}}{6 \, (0.00053144)^2},$$

or

$$\sigma_{T_F} = 32.1705 \text{ seconds, or } 0.5362 \text{ minutes.}$$

Therefore,

$$T_F \sim N \left(\overline{T}_F, \sigma^2_{T_F}\right) = N \left(31.3613, 0.5362^2\right).$$

3. Under the normal distribution assumption for the cumulative damage, the reliability of the *LRU* for 30 minutes of random vibration is given by Eq. (8.162), or

$$R(T = 30 \text{ min}) = \Phi \left[\frac{1 - \overline{D}(T = 30 \text{ min})}{\sigma_{D(T=30 \text{ min})}}\right],$$
$$= \Phi \left(\frac{1 - 0.95659}{0.01672}\right),$$
$$= \Phi(2.5963),$$

or

$$R(T = 30 \text{ min}) = 0.995288.$$

Under the normal distribution assumption for the fatigue life, the reliability of the *LRU* for 30 minutes of random vibration is given by Eq. (8.163), or

$$R(T = 30 \text{ min}) = \Phi \left(\frac{\overline{T}_F - 30}{\sigma_{T_F}}\right),$$
$$= \Phi \left(\frac{31.3613 - 30}{0.5362}\right),$$
$$= \Phi(2.5388),$$

or

$$R(T = 30 \text{ min}) = 0.994438.$$

By comparing the result obtained using the cumulative damage distribution, or $R(T = 30$ min$) = 0.995288$, with that obtained using the
fatigue life distribution, or $R(T = 30$ min$) = 0.994438$, respectively, it
may be seen that they are very close with an absolute difference of

$$\left|0.995288 - 0.994438\right| = 0.00085.$$

4. The desired random vibration length for a 99% failure probability can
be obtained by letting $Q(T) = 0.99$ in Eq. (8.165) and solving for T;
i.e.,

$$0.99 = \Phi\left(\frac{T - \overline{T}_F}{\sigma_{T_F}}\right),$$

or

$$0.99 = \Phi\left(\frac{T - 31.3613}{0.5362}\right).$$

Then, from the standardized normal *cdf* table,

$$\frac{T - 31.3613}{0.5362} = 2.3267,$$

or

$$T = 32.6089 \text{ minutes, or 33 minutes.}$$

8.3.4.3 STATIONARY WIDE-BAND RANDOM STRESSING

If the response stress process of a vibrated element, $Y(t)$, is not a narrow-band
process but a wide-band process, then we can not apply the results based on
the narrow-band stressing assumption directly. However, these results can
be modified by introducing an *equivalent narrow-band process* to the original
wide-band process.

8.3.4.3.1 Equivalent Narrow-Band Process and the Equivalent Damage

Let $Y_{WB}(t)$ be a stationary wide-band Gaussian process with a *rms* value
of σ_y and an expected zero-up-crossing rate of $n_y^+(0)$. A stationary narrow-
band process, $Y_{NB}(t)$, which has the same *rms* value and the same expected
zero-up-crossing rate as those of $Y_{WB}(t)$, is called the *equivalent narrow-band
process* of $Y_{WB}(t)$. The general expression for fatigue damage under wide-
band stressing of $Y_{WB}(t)$, is given by [10]

$$D = \varphi(\epsilon; b) \, D_{NB}, \tag{8.166}$$

where

$$D_{NB} = \text{cumulative damage due to the equivalent}$$
$$\text{narrow-band process } Y_{NB}(t),$$

and

$\varphi(\epsilon; b)$ = modification factor which is a function of the spectral width parameter, ϵ, given by Eq. (8.52) and the fatigue law exponent b in Eqs. (8.133) and (8.134).

Wirsching et al [11; 12] derived the empirical modification factor by simulating processes having a variety of spectral shapes and using the rainflow method to measure the magnitudes. The resultant expression for $\varphi(\epsilon; b)$ is given by

$$\varphi(\epsilon; b) = K_1(b) + [1 - K_1(b)](1 - \epsilon)^{K_2(b)}, \tag{8.167}$$

where

$$K_1(b) = 0.926 - 0.033\, b, \tag{8.168}$$

and

$$K_2(b) = 1.587\, b - 2.323. \tag{8.169}$$

8.3.4.3.2 Mean Value of the Cumulative Damage and Fatigue Life

The mean value of the cumulative damage under the equivalent narrow-band stress $Y_{NB}(t)$, for a time period T, $\overline{D}_{NB}(T)$, is calculated by Eq. (8.144); i.e.,

$$\overline{D}_{NB}(T) = n_y^+(0)\ \left(\frac{T}{A}\right)\ (\sqrt{2}\ \sigma_y)^b\ \Gamma\left(\frac{b}{2} + 1\right), \tag{8.170}$$

where

$n_y^+(0)$ = expected zero-up-crossing rate of the equivalent narrow-band process, which corresponds to term $\left(\frac{\omega_n}{2\,\pi}\right)$ in Eq. (8.144) for the real narrow-band stress case,

A = fatigue law constant in Eq. (8.134),

and

b = fatigue law exponent in Eqs. (8.133) and (8.134).

Therefore, the mean value of the cumulative damage under the actual wide-band stressing of $Y_{WB}(t)$, for a time period T, can be obtained by taking the expectation of both sides of Eq. (8.166) and substituting Eq. (8.170) into it; i.e.,

$$\begin{aligned}
\overline{D}(T) &= E[D(T)] = E[\varphi(\epsilon; b)\ D_{NB}(T)] \\
&= \varphi(\epsilon; b)\ E[D_{NB}(T)] = \varphi(\epsilon; b)\ \overline{D}_{NB}(T),
\end{aligned}$$

or

$$\overline{D}(T) = \varphi(\epsilon; b) \, n_y^+(0) \, \left(\frac{T}{A}\right) \, (\sqrt{2}\,\sigma_y)^b \, \Gamma\left(\frac{b}{2}+1\right), \tag{8.171}$$

where $\varphi(\epsilon; b)$ is given by Eq. (8.167).

Correspondingly, the mean time to fatigue failure is given by

$$T_F = T\Big|_{\overline{D}(T)=1} = \frac{A}{\varphi(\epsilon; b) \, n_y^+(0) \, (\sqrt{2}\,\sigma_y)^b \, \Gamma\left(\frac{b}{2}+1\right)}. \tag{8.172}$$

8.3.4.3.3 Standard Deviation of the Cumulative Damage

According to Eq. (8.147), the standard deviation of the cumulative damage under wide-band stressing, using Eq. (8.171), is given by

$$\sigma_{D(T)} \cong \overline{D}(T) \, \sqrt{\frac{\psi_1(b)}{\xi \, n_y^+(0) \, T}},$$

$$= \varphi(\epsilon; b) \, n_y^+(0) \, \left(\frac{T}{A}\right) \, (\sqrt{2}\,\sigma_y)^b \, \Gamma\left(\frac{b}{2}+1\right)$$

$$\cdot \sqrt{\frac{\psi_1(b)}{\xi \, n_y^+(0) \, T}},$$

or

$$\sigma_{D(T)} \cong \frac{\varphi(\epsilon; b) \, (\sqrt{2}\,\sigma_y)^b \, \Gamma\left(\frac{b}{2}+1\right)}{A} \, \sqrt{\frac{\psi_1(b) \, n_y^+(0) \, T}{\xi}}, \tag{8.173}$$

where

ξ = damping ratio as introduced in Example 8–1,

and

$\psi_1(b)$ = function of b as given in Table 8.1.

8.3.4.3.4 Distributions of the Cumulative Damage and the Fatigue Life

Similar to the case of narrow-band stressing, the central limit theorem can be invoked to show that the cumulative damage, D, and the fatigue life, T_F, under wide-band stressing both approach the normal distribution; i.e.,

$$D \sim N\left[\overline{D}(T), \sigma_{D(T)}^2\right],$$

and

$$T_F \sim N\left(\overline{T}_F, \sigma_{T_F}^2\right),$$

where $\overline{D}(T)$, $\sigma_{D(T)}$, and $\overline{T}_F$ are given by Eqs. (8.171), (8.173) and (8.172), respectively. The value of σ_{T_F} in this case can be obtained, similarly, by following Steps (a) through (e) presented in the narrow-band stressing case covered in Section 8.3.4.2.3. It turns out that σ_{T_F} under wide-band stressing has a similar expression to σ_{T_F} under narrow-band stressing given by Eq. (8.161); i.e.,

$$\sigma_{T_F} = \frac{h_2' \sqrt{(h_2')^2 + 4\,h_1'}}{6\,(h_1')^2}. \tag{8.174}$$

where

$$h_1' = \varphi(\epsilon;b)\,n_y^+(0)\,(\sqrt{2}\,\sigma_y)^b\,\Gamma\left(\frac{b}{2}+1\right)\Big/A, \tag{8.175}$$

and

$$h_2' = \frac{3\,\varphi(\epsilon;b)\,(\sqrt{2}\,\sigma_y)^b}{A}\,\Gamma\left(\frac{b}{2}+1\right)\,\sqrt{n_y^+(0)\left[\frac{\psi_1(b)}{\xi}\right]}. \tag{8.176}$$

8.3.4.3.5   Reliability Evaluation for Fatigue Failure.

Having derived the parameter estimates for cumulative damage and fatigue life under wide-band stressing, the fatigue reliability can be readily calculated from

$$R(T) = P[D(T) < 1] = \Phi\left[\frac{1 - \overline{D}(T)}{\sigma_{D(T)}}\right], \tag{8.177}$$

or, equivalently, from

$$R(T) = P(T_F > T) = 1 - \Phi\left(\frac{T - \overline{T}_F}{\sigma_{T_F}}\right) = \Phi\left(\frac{\overline{T}_F - T}{\sigma_{T_F}}\right), \tag{8.178}$$

where

$\Phi(\ )$ = cdf of the standardized normal distribution, $N(0,1)$,

and $\overline{D}(T)$, $\sigma_{D(T)}$, $\overline{T}_F$ and σ_{T_F} are given by Eqs. (8.171), (8.173), (8.172) and (8.174), respectively. Equations (8.177) and (8.178) should yield very close results.

Correspondingly, the probability of fatigue failure due to wide-band random vibration for time T, is given by

$$Q(T) = P[D(T) \geq 1] = 1 - \Phi\left[\frac{1 - \overline{D}(T)}{\sigma_{D(T)}}\right] = \Phi\left[\frac{\overline{D}(T) - 1}{\sigma_{D(T)}}\right], \tag{8.179}$$

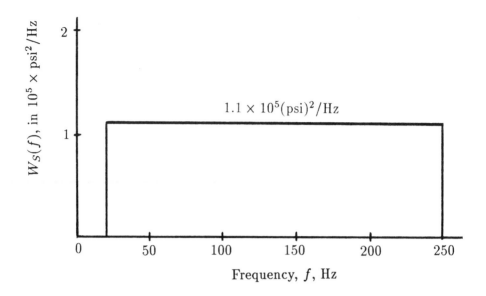

Fig. 8.13– Bending stress response at the center of a PC board for *LRU* exitation perpendicular to the PC board plane, for Example 8–4.

or by

$$Q(T) = P(T_F \leq T) = \Phi \left(\frac{T - \overline{T}_F}{\sigma_{T_F}} \right).$$ (8.180)

EXAMPLE 8–4

The bending stress response spectrum in the engineering form, $W_S(f)$, at the center of a PC board in a line replaceable unit (*LRU*) under random exitation perpendicular to the PC board plane is given in Fig. 8.13, which is a wide-band process and has been approximated by a rectangular spectrum with a magnitude of 1.1×10^5 (psi)^2/Hz. The damping ratio of the *LRU* is 0.01. The equivalent $S - N$ fatigue curve of the solder joint is given by

$$N(S) \, S^4 = \left(2 \times 10^5 \right)^4 .$$

Do the following:

1. Find the equivalent narrow-band process to the given stress response spectrum.

2. Determine the mean and standard deviation of the cumulative damage due to such a random vibration for a time interval of 40 minutes.

3. Determine the mean and standard deviation of the time to fatigue failure of this *LRU*.

4. Determine the reliability of the *LRU* under the given random vibration for 40 minutes, assuming that the cumulative damage and the fatigue life are both normally distributed.

5. How long should this random vibration last so that we can be 99% sure that the fatigue failure will occur?

SOLUTIONS TO EXAMPLE 8–4

1. The bending stress response is a wide-band process in this case. Therefore, we need to find its equivalent narrow-band process which has the same *rms* value and the same zero-up-crossing rate.

The *rms* value of this stress response is given by Eqs. (8.44) and (8.19), or

$$\sigma_S = \sqrt{\int_{-\infty}^{\infty} S_S(\omega)\, d\omega} = \sqrt{\int_{0}^{\infty} W_S(f)\, df}. \tag{8.181}$$

Substituting

$$W_S(f) = \begin{cases} 1.1 \times 10^5 (\text{psi})^2/\text{Hz}, & \text{for } 20 \text{ Hz } \leq f \leq 250 \text{ Hz}, \\ 0, & \text{otherwise}, \end{cases}$$

$$\tag{8.182}$$

into Eq. (8.181) yields

$$\sigma_S = \sqrt{\int_{20}^{250} 1.1 \times 10^5 \, df} = \sqrt{1.1 \times 10^5 \, (250 - 20)},$$

or

$$\sigma_S = 5.0299 \times 10^3 \text{ psi}.$$

The zero-up-crossing rate of the bending stress response is given by Eq. (8.42), or

$$n_S^+(0) = \frac{1}{2\pi} \left(\frac{\sigma_{\dot{S}}}{\sigma_S} \right),$$

where $\sigma_{\dot{S}}$ is the *rms* value of $\dot{S}(t)$ and given by Eq. (8.45), or

$$\sigma_{\dot{S}} = \left[\int_{-\infty}^{\infty} \omega^2 \, S_S(\omega)\, d\omega \right]^{1/2}.$$

Then,

$$n_S^+(0) = \frac{1}{2\pi} \sqrt{\frac{\int_{-\infty}^{\infty} \omega^2 \, S_S(\omega) \, d\omega}{\int_{-\infty}^{\infty} S_S(\omega) \, d\omega}}.$$

From Eq. (8.22), $S_S(\omega) = \frac{1}{4\pi} W_S(f)$. Substituting this and $\omega = 2\pi f$ into the above equation yields

$$n_S^+(0) = \frac{1}{2\pi} \sqrt{\frac{2 \int_0^{\infty} (2\pi f)^2 \left(\frac{1}{4\pi}\right) W_S(f) \, (2\pi) \, df}{2 \int_0^{\infty} \left(\frac{1}{4\pi}\right) W_S(f) \, (2\pi) \, d\omega \, df}},$$

or

$$n_S^+(0) = \sqrt{\frac{\int_0^{\infty} f^2 \, W_S(f) \, df}{\int_0^{\infty} W_S(f) \, df}}. \tag{8.183}$$

Substituting $W_S(f)$, given by Eq. (8.182), into Eq. (8.183) yields

$$n_S^+(0) = \sqrt{\frac{\int_{20}^{250} f^2 \, (1.1 \times 10^5) \, df}{\int_{20}^{250} 1.1 \times 10^5 \, df}},$$

$$= \sqrt{\frac{\frac{1}{3}(250^3 - 20^3)}{(250 - 20)}},$$

or

$$n_S^+(0) = 150.4438 \text{ Hz}.$$

Therefore, the equivalent narrow-band process to $W_S(f)$ has a *rms* value of $\sigma_S = 5.0299 \times 10^3$ psi and a central frequency of $n_y^+(0) = 150.4438$ Hz.

Under the stationary Gaussian and zero mean assumption for the bending stress process, the spectral width parameter, ϵ, is given by Eq. (8.55), or

$$\epsilon = \sqrt{1 - \frac{\sigma_{\ddot{S}}^4}{\sigma_S^2 \, \sigma_{\ddot{S}}^2}},$$

where $\sigma_{\ddot{S}}$ is the *rms* value of $\ddot{S}(t)$ and is given by Eq. (8.56), or

$$\sigma_{\ddot{S}} = \left[\int_{-\infty}^{\infty} \omega^4 \, S_S(\omega) \, d\omega\right]^{1/2}.$$

Therefore,

$$\epsilon = \sqrt{1 - \frac{\left[\int_{-\infty}^{\infty} \omega^2 \, S_S(\omega) \, d\omega\right]^2}{\left[\int_{-\infty}^{\infty} S_S(\omega) \, d\omega\right] \left[\int_{-\infty}^{\infty} \omega^4 \, S_S(\omega) \, d\omega\right]}}. \tag{8.184}$$

Substituting $\omega = 2\pi f$ and $S_S(\omega) = \frac{1}{4\pi}W_S(f)$ into Eq. (8.184) yields

$$\epsilon = \sqrt{1 - \frac{\left[2\int_0^\infty (2\pi f)^2 \frac{1}{4\pi} W_S(f) (2\pi) df\right]^2}{\left[2\int_0^\infty \frac{1}{4\pi}W_S(f)(2\pi)df\right]\left[2\int_0^\infty (2\pi f)^4 \frac{1}{4\pi}W_S(f)(2\pi)df\right]}},$$

or

$$\epsilon = \sqrt{1 - \frac{\left[\int_0^\infty f^2 W_S(f) df\right]^2}{\left[\int_0^\infty W_S(f) df\right]\left[\int_0^\infty f^4 W_S(f) df\right]}}. \qquad (8.185)$$

Substituting $W_S(f)$ given by Eq. (8.182) into Eq. (8.185) yields

$$\epsilon = \sqrt{1 - \frac{\left[\int_{20}^{250} f^2 (1.1 \times 10^5) df\right]^2}{\left[\int_{20}^{250} 1.1 \times 10^5 df\right]\left[\int_{20}^{250} f^4 (1.1 \times 10^5) df\right]}},$$

$$= \sqrt{1 - \frac{\left[\frac{1}{3}(250^3 - 20^3)\right]^2}{[(250 - 20)]\left[\frac{1}{5}(250^5 - 20^5)\right]}},$$

or

$$\epsilon = 0.6299.$$

2. The parameters for the solder joint fatigue curve are

$$b = 4,$$

and

$$A = \left(2 \times 10^5\right)^4.$$

The modification factor $\varphi(\epsilon; b)$ is given by Eq. (8.167), or

$$\varphi(\epsilon; b) = \varphi(0.6299; 4),$$
$$= K_1(4) + [1 - K_1(4)](1 - 0.6299)^{K_2(4)},$$

where from Eqs. (8.168) and (8.169)

$$K_1(4) = 0.926 - 0.033 (4) = 0.794,$$

and

$$K_2(4) = 1.587 (4) - 2.323 = 4.025.$$

Then,

$$\varphi(0.6299; 4) = 0.794 + (1 - 0.794)(1 - 0.6299)^{4.025},$$

or

$$\varphi(0.6299; 4) = 0.7978.$$

Substituting $b = 4$, $A = (2 \times 10^5)^4$, $\varphi(\epsilon; b) = 0.7978$, $n_y^+(0) = 150.4438$ Hz and $\sigma_S = 5.0299 \times 10^3$ psi into Eqs. (8.171) and (8.173) yields

$$\overline{D}(T = 40 \text{ min}) = 0.7978 \, (150.4438) \left[\frac{40 \times 60}{(2 \times 10^5)^4} \right]$$

$$\cdot \left(\sqrt{2} \times 5.0299 \times 10^3 \right)^4$$

$$\cdot \Gamma \left(\frac{4}{2} + 1 \right),$$

or

$$\overline{D}(T = 40 \text{ min}) = 0.9219,$$

and

$$\sigma_{D(T=40 \text{ min})} \cong \frac{0.7978 \, \left(\sqrt{2} \times 5.0299 \times 10^3 \right)^4}{(2 \times 10^5)^4}$$

$$\cdot \Gamma \left(\frac{4}{2} + 1 \right)$$

$$\cdot \sqrt{(150.4438) \frac{\psi_1(4) \, (40 \times 60)}{0.01}}.$$

The value for $\psi_1(4)$ has been found in Example 8–3 to be 0.6873. Then,

$$\sigma_{D(T=40 \text{ min})} \cong \frac{0.7978 \, \left(\sqrt{2} \times 5.0299 \times 10^3 \right)^4}{(2 \times 10^5)^4}$$

$$\cdot \Gamma \left(\frac{4}{2} + 1 \right)$$

$$\cdot \sqrt{(150.4438) \frac{0.6873 \, (40 \times 60)}{0.01}},$$

or

$$\sigma_{D(T=40 \text{ min})} \cong 0.0127.$$

Therefore, under the normal distribution assumption for the cumulative damage,

$$D(T = 40 \text{ min}) \sim N \left(\overline{D}(40 \text{ min}), \, \sigma^2_{D(T=40 \text{ min})} \right),$$

or

$$D(T = 40 \text{ min}) \sim N \left(0.9219, \, 0.0127^2 \right).$$

3. Substituting $b = 4$, $A = (2 \times 10^5)^4$, $\varphi(\epsilon; b) = 0.7978$, $n_y^+(0) = 150.4438$ Hz and $\sigma_S = 5.0299 \times 10^3$ psi into Eqs. (8.172) and (8.174) yields

$$\overline{T}_F = \frac{\left(2 \times 10^5 \right)^4}{0.7978 \, (150.4438) \, \left(\sqrt{2} \times 5.0299 \times 10^3 \right)^4 \, \Gamma \left(\frac{4}{2} + 1 \right)},$$

or

$$\overline{T}_F = 2603.3003 \text{ seconds, or } 43.3883 \text{ minutes,}$$

and

$$\sigma_{T_F} = \frac{h_2' \sqrt{(h_2')^2 + 4\, h_1'}}{6\, (h_1')^2},$$

where from Eqs. (8.175) and (8.176)

$$h_1' = 0.7978\, (150.4438)\, (\sqrt{2} \times 5.0299 \times 10^3)^4$$
$$\cdot \Gamma\left(\frac{4}{2} + 1\right) \Big/ (2 \times 10^5)^4,$$

or

$$h_1' = 0.0003841,$$

and

$$h_2' = \frac{3\,(0.7978)\,(\sqrt{2} \times 5.0299 \times 10^3)^4}{(2 \times 10^5)^4}$$
$$\cdot \Gamma\left(\frac{4}{2} + 1\right)\, \sqrt{150.4438\left(\frac{0.6873}{0.01}\right)},$$

or

$$h_2' = 0.0007789.$$

Then,

$$\sigma_{T_F} = \frac{0.0007789 \sqrt{0.0007789^2 + 4\,(0.0003841)}}{6\,(0.0003841)^2},$$

or

$$\sigma_{T_F} = 34.4969 \text{ seconds, or } 0.5749 \text{ minutes.}$$

Therefore,

$$T_F \sim N\left(\overline{T}_F, \sigma_{T_F}^2\right) = N\left(34.4969, 0.5749^2\right).$$

4. Under the normal distribution assumption for the cumulative damage, the reliability of the *LRU* for 40 minutes of random vibration is given by Eq. (8.177), or

$$R(T = 40 \text{ min}) = \Phi\left[\frac{1 - \overline{D}(T = 40 \text{ min})}{\sigma_{D(T=40 \text{ min})}}\right],$$
$$= \Phi\left(\frac{1 - 0.9219}{0.0127}\right),$$
$$= \Phi(6.1496),$$

or

$$R(T = 40 \text{ min}) \cong 1.0.$$

Under the normal distribution assumption for the fatigue life, the reliability of the LRU for 40 minutes of random vibration is given by Eq. (8.178), or

$$R(T = 40 \text{ min}) = \Phi\left(\frac{\overline{T}_F - 40}{\sigma_{T_F}}\right),$$

$$= \Phi\left(\frac{43.3883 - 40}{0.5749}\right),$$

$$= \Phi(5.8937),$$

or

$$R(T = 40 \text{ min}) \cong 1.0.$$

5. The desired random vibration length for a 99% screening probability can be obtained by letting $Q(T) = 0.99$ in Eq. (8.165) and solving for T; i.e.,

$$0.99 = \Phi\left(\frac{T - \overline{T}_F}{\sigma_{T_F}}\right),$$

or

$$0.99 = \Phi\left(\frac{T - 43.3883}{0.5749}\right).$$

Then, from the standardized normal cdf table,

$$\frac{T - 43.3883}{0.5749} = 2.3267,$$

or

$$T = 44.7259 \text{ minutes.}$$

8.3.4.4   DETERMINATION OF THE OPTIMUM SCREENING LENGTH – BIMODALLY-DISTRIBUTED $S - N$ CURVE APPROACH

The fitted $S - N$ curve presented in Fig. 8.9 is, in general, the mean value curve or the median curve (under normal distribution assumption) of Eq. (8.133); i.e.,

$$E[\log_{10} N(S)] = \hat{a} - \hat{b}\, E[\log_{10} S], \tag{8.186}$$

where both the stress amplitude, S, and the number of cycles to fatigue failure, $N(S)$, are random variables in nature. As stated in Section 8.3.4.1, the actual fatigue curve is not a single curve but a family of distributions

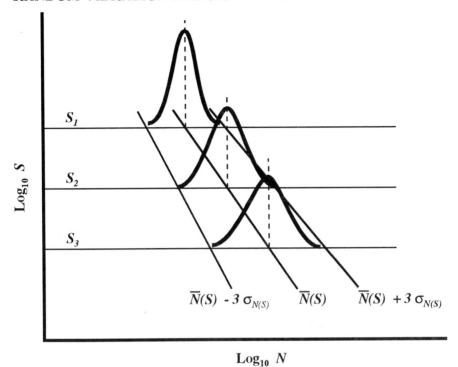

Fig. 8.14– The distributed fatigue life under various stress amplitudes.

scattered over a two-dimensional plane of $S(N)$ and N. Figure 8.14 gives an example of the distributed fatigue lives, N, under various stress amplitudes S. Figure 8.15 gives another example of the distributed strength (the allowable operating stress amplitude), S, under various specified operating cycles, N. Therefore, the *confidence level* for the reliability, unreliability, and fatigue life under narrow-band or wide-band stressing, which are based on the mean $S - N$ curve parameters, such as those obtained in Examples 8–3 and 8–4, will be 50%.

It should be noted that the fatigue life for the solder joint in Examples 8–3 and 8–4 is not the screening time which will be much smaller than the real fatigue life. For example, if we narrow-band stress the PC board given in Example 8–3 for 33 minutes, or wide-band stress the PC board given in Example 8–4 for 45 minutes, we will exhaust almost 50% of the fatigue life of those PC boards; i.e., we are over-damaging the PC boards by screening them for too long a time period. Therefore, determination of the optimum screening time requires the use of an appropriate $S - N$ curve among its distributed family, which represents the inherent fatigue curve of the defects.

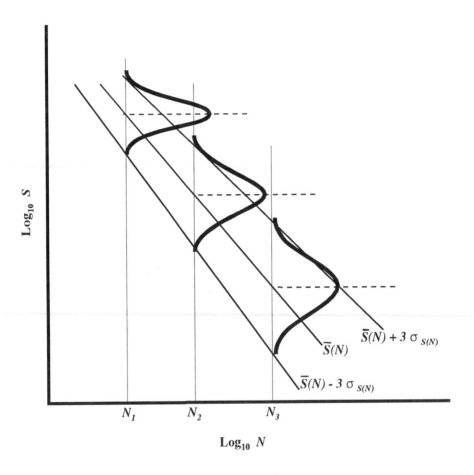

Fig. 8.15– The distributed strength (allowable operating stress amplitude) for various specified operating cycles.

Here we propose, according to the physics of failure, a bimodally-distributed $S-N$ (*P-S-N*) diagram, for the fatigue strength of non-screened units including electronic structural elements, such as solder joints, etc. This proposed $S - N$ diagram is shown in Fig. 8.16 and can be mathematically expressed by

$$f_{N(S)}(N) = \alpha\, f_{N_b(S)}(N) + (1 - \alpha)\, f_{N_g(S)}(N), \quad S \geq 0, \quad N \geq 0,$$

$$(8.187)$$

where

$$S = \text{applied cyclic stress amplitude,}$$
$$N(S) = \text{number of cycles until fracture for the}$$
$$\text{whole population under a fixed stress}$$
$$\text{amplitude, } S,$$
$$\alpha = \text{proportion of the substandard subpopulation,}$$
$$N_b(S) = \text{number of cycles until fracture for the}$$
$$\text{substandard (bad) subpopulation under}$$
$$\text{a fixed stress amplitude, } S,$$
$$f_{N_b(S)}(N) = pdf \text{ of } N_b(S),$$
$$1 - \alpha = \text{proportion of the standard subpopulation,}$$
$$N_g(S) = \text{number of cycles until fracture for the}$$
$$\text{standard (good) subpopulation under}$$
$$\text{a fixed stress amplitude, } S,$$

and

$$f_{N_g(S)}(N) = pdf \text{ of } N_g(S).$$

It may be seen from Eq. (8.187) and Fig. 8.16 that the number of cycles to fracture failure of a structural element, which is subjected to a constant stress application, follows a bimodal mixed distribution. This mixed fatigue life distribution is composed of a substandard (bad) subpopulation of 100 α%, which is to be eliminated by random vibration screen, and a standard (good) subpopulation of 100 $(1 - \alpha)$%, which is to be preserved after screen and is expected to dominate the residual fatigue strength of the screened equipment in the field.

Let 100 α% be the average defect percentage of the equipment population in solder joints and other metalization aspects, which may be obtained from historical screening tests, material fatigue tests, or field operation records. Let

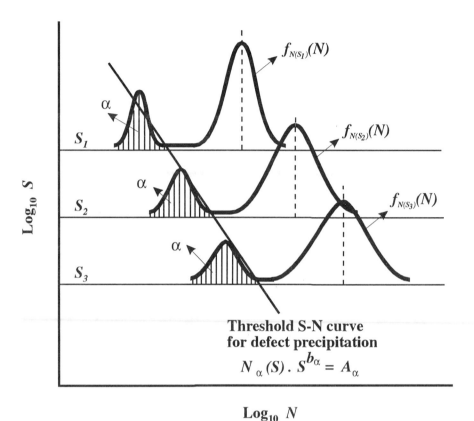

Fig. 8.16– A typical bimodally-distributed $S - N$ (P-S-N) diagram, for the fatigue strength of non-screened electronic structural elements, and its threshold $S - N$ curve for fatigue defect precipitation.

$N_\alpha(S)$ be the 100 $\alpha\%$ confidence limit of fatigue life under stress amplitude S. Then, $N_\alpha(S)$ cycles of stressing at amplitude S will yield 100 $\alpha\%$ fatigue failures, as shown in Fig. 8.16. These 100 $\alpha\%$ fatigue-failed equipment are among those that have the weakest fatigue strength.

Following the same reasoning, screening the equipment, by either a narrow-band or an equivalent wide-band stressing, for a time period which is calculated using the parameters of the 100 $\alpha\%$ confidence limit of the $S-N$ curve will evoke 100 $\alpha\%$ fatigue failures; i.e., *the "threshold $S-N$ curve" for fatigue defect precipitation should be the 100 $\alpha\%$ confidence limit of the $P-S-N$ diagram.*

Having chosen the threshold $S-N$ curve for fatigue defect precipitation, the mean and standard deviation for the cumulative damage and the screening time, or fatigue defect precipitation time, or fatigue life of the defects, under both narrow-band and wide-band stressing can be obtained by substituting the corresponding $S-N$ curve parameters into Eqs. (8.144), (8.148), (8.145), (8.161); and (8.171), (8.173), (8.172) and (8.174), respectively.

Let b_α and A_α be the parameters of the threshold $S-N$ curve; i.e., of the 100 $\alpha\%$ confidence limit of the $S-N$ curve family, or

$$N_\alpha(S)\, S^{b_\alpha} = A_\alpha. \tag{8.188}$$

Then, according to Eqs. (8.144) and (8.148), the mean and standard deviation of the cumulative damage under a narrow-band stressing for time period T are

$$\overline{D}(T) = \left(\frac{\omega_n}{2\,\pi}\right)\,\left(\frac{T}{A_\alpha}\right)\,(\sqrt{2}\,\sigma_y)^{b_\alpha}\,\Gamma\left(\frac{b_\alpha}{2}+1\right), \tag{8.189}$$

and

$$\sigma_{D(T)} \cong \frac{(\sqrt{2}\,\sigma_y)^{b_\alpha}}{A_\alpha}\,\Gamma\left(\frac{b_\alpha}{2}+1\right)\,\sqrt{\left(\frac{\omega_n}{2\,\pi}\right)\frac{\psi_1(b_\alpha)\,T}{\xi}}. \tag{8.190}$$

According to Eqs. (8.145) and (8.161), the mean and standard deviation of the screening time, or the fatigue defect precipitation time, or the fatigue life of the defects, under a narrow-band stressing, are

$$\overline{T}_F = \frac{A_\alpha}{\left(\frac{\omega_n}{2\,\pi}\right)\,(\sqrt{2}\,\sigma_y)^{b_\alpha}\,\Gamma\left(\frac{b_\alpha}{2}+1\right)}, \tag{8.191}$$

and

$$\sigma_{T_F} = \frac{h_2\,\sqrt{h_2^2 + 4\,h_1}}{6\,h_1^2}, \tag{8.192}$$

where from Eqs. (8.152) and (8.153)

$$h_1 = \left(\frac{\omega_n}{2\,\pi}\right)\,(\sqrt{2}\,\sigma_y)^{b_\alpha}\,\Gamma\left(\frac{b_\alpha}{2}+1\right)\Big/A_\alpha, \tag{8.193}$$

and

$$h_2 = \frac{3\left(\sqrt{2}\,\sigma_y\right)^{b_\alpha}}{A_\alpha}\,\Gamma\left(\frac{b_\alpha}{2}+1\right)\sqrt{\left(\frac{\omega_n}{2\,\pi}\right)\frac{\psi_1(b_\alpha)}{\xi}}.$$ (8.194)

According to Eqs. (8.171) and (8.173), the mean and standard deviation of the cumulative damage under a wide-band stressing for a time period T are

$$\overline{D}(T) = \varphi(\epsilon;b_\alpha)\,n_y^+(0)\,\left(\frac{T}{A_\alpha}\right)\,\left(\sqrt{2}\,\sigma_y\right)^{b_\alpha}\,\Gamma\left(\frac{b_\alpha}{2}+1\right),$$ (8.195)

and

$$\sigma_{D(T)} \cong \frac{\varphi(\epsilon;b_\alpha)\,\left(\sqrt{2}\,\sigma_y\right)^{b_\alpha}\,\Gamma\left(\frac{b_\alpha}{2}+1\right)}{A_\alpha}\,\sqrt{n_y^+(0)\,\frac{\psi_1(b_\alpha)\,T}{\xi}},$$ (8.196)

where from Eqs. (8.167), (8.168) and (8.169)

$$\varphi(\epsilon;b_\alpha) = K_1(b_\alpha) + [1 - K_1(b_\alpha)](1-\epsilon)^{K_2(b_\alpha)},$$ (8.197)

where

$$K_1(b_\alpha) = 0.926 - 0.033\,b_\alpha,$$ (8.198)

and

$$K_2(b_\alpha) = 1.587\,b_\alpha - 2.323.$$ (8.199)

According to Eqs. (8.172) and (8.174), the mean and standard deviation of the screening time, or fatigue defect precipitation time, or fatigue life of the defects, under a wide-band stressing, are the following:

$$\overline{T}_F = \frac{A_\alpha}{\varphi(\epsilon;b_\alpha)\,n_y^+(0)\,\left(\sqrt{2}\,\sigma_y\right)^{b_\alpha}\,\Gamma\left(\frac{b_\alpha}{2}+1\right)},$$ (8.200)

and

$$\sigma_{T_F} = \frac{h_2'\,\sqrt{(h_2')^2 + 4\,h_1'}}{6\,(h_1')^2},$$ (8.201)

where from Eqs. (8.175) and (8.176)

$$h_1' = \varphi(\epsilon;b_\alpha)\,n_y^+(0)\,\left(\sqrt{2}\,\sigma_y\right)^{b_\alpha}\,\Gamma\left(\frac{b_\alpha}{2}+1\right)\Big/A_\alpha,$$ (8.202)

and

$$h_2' = \frac{3\,\varphi(\epsilon;b_\alpha)\,\left(\sqrt{2}\,\sigma_y\right)^{b_\alpha}}{A_\alpha}\,\Gamma\left(\frac{b_\alpha}{2}+1\right)\,\sqrt{n_y^+(0)\,\left[\frac{\psi_1(b_\alpha)}{\xi}\right]},$$ (8.203)

and $\varphi(\epsilon;b_\alpha)$ is given by Eq. (8.197).

With the above mean and standard deviation values, the following can be readily determined:

1. Reliability or the anti-screen (screen resistance) probability of the fatigue defects, $R(T)$, for a specified screening time period T.

2. Unreliability, or screening probability, $Q(T)$, for a specified screening time period T.

3. The optimum screening length, T_S, for a specified screening probability P_S.

The anti-screen and screen probabilities are given by Eqs. (8.162) and (8.164), or (8.177) and (8.179); i.e.,

$$R(T) = \Phi \left[\frac{1 - \overline{D}(T)}{\sigma_{D(T)}} \right], \tag{8.204}$$

and

$$Q(T) = \Phi \left[\frac{\overline{D}(T) - 1}{\sigma_{D(T)}} \right], \tag{8.205}$$

or, equivalently, by Eqs. (8.163) and (8.165), or (8.178) and (8.180); i.e.,

$$R(T) = \Phi \left(\frac{\overline{T}_F - T}{\sigma_{T_F}} \right), \tag{8.206}$$

and

$$Q(T) = \Phi \left(\frac{T - \overline{T}_F}{\sigma_{T_F}} \right), \tag{8.207}$$

where $\overline{D}(T)$, $\sigma_{D(T)}$, $\overline{T}_F$ and σ_{T_F} are given by Eqs. (8.189) through (8.192) for narrow-band stressing, and by Eqs. (8.195), (8.196), (8.200) and (8.201) for wide-band stressing.

The optimum screening length, T_S, for a specified screening probability P_S, is given by

$$T_S = T \Big|_{Q(T) = P_s} = \overline{T}_F + \Phi^{-1}(P_S)\, \sigma_{T_F}, \tag{8.208}$$

where

$$\Phi^{-1}(P_S) = \text{inverse function of the } cdf \text{ of the standardized}$$
$$\text{normal distribution,}$$

and $\overline{T}_F$ and σ_{T_F} are given by Eqs. (8.191) and (8.192) for narrow-band stressing, and by Eqs. (8.200) and (8.201) for wide-band stressing.

It may be seen that determination of $R(T)$, $Q(T)$ and particularly T_S requires the availability of the values of percent fatigue defectives, α, and the distributed $S - N$ diagram such that the $100\,\alpha\%$ confidence limit $S - N$ curve can be obtained.

EXAMPLE 8-5

Determine the optimum screening length, T_S, for the LRU described in Example 8-3, for a specified screening probability of 99%, given that the solder joint defect percentage is 10% and the 10% failed life $S - N$ curve for the solder joint is

$$N(S)\ S^4 = \left(5 \times 10^{4.5}\right)^4,$$

where

$$b_\alpha = 4,$$

and

$$A_\alpha = \left(5 \times 10^{4.5}\right)^4.$$

SOLUTION TO EXAMPLE 8-5

First, the mean and the standard deviation for the fatigue life of the defective solder joint of the PC board under the given stressing conditions are determined. From Example 8-3, it is known that the stress response under diagonal force excitation is a narrow-band process and

$$\frac{\omega_n}{2\,\pi} = 125 \text{ Hz},$$
$$\xi = 0.01,$$

and

$$\sigma_S = 5.4 \times 10^3 \text{ psi}.$$

Substituting these, $b_\alpha = 4$ and $A_\alpha = \left(5 \times 10^{4.5}\right)^4$ into Eqs. (8.191), (8.193), (8.194) and (8.192) yields

$$\overline{T}_F = \frac{\left(5 \times 10^{4.5}\right)^4}{(125)\ \left(\sqrt{2} \times 5.4 \times 10^3\right)^4\ \Gamma\left(\frac{4}{2} + 1\right)},$$

or

$$\overline{T}_F = 735.02985 \text{ seconds},$$

and

$$h_1 = (125)\ \left(\sqrt{2} \times 5.4 \times 10^3\right)^4\ \Gamma\left(\frac{4}{2} + 1\right) \Big/ \left(5 \times 10^{4.5}\right)^4,$$

or

$$h_1 = 0.00136,$$

and

$$h_2 = \frac{3\left(\sqrt{2}\times 5.4\times 10^3\right)^4}{(5\times 10^{4.5})^4}\ \Gamma\left(\frac{4}{2}+1\right)\ \sqrt{(125)\ \frac{0.6873}{0.01}},$$

or

$$h_2 = 0.00303,$$

and

$$\sigma_{T_F} = \frac{0.00303\ \sqrt{0.00303^2 + 4\ (0.00136)}}{6\ (0.00136)^2},$$

or

$$\sigma_{T_F} = 20.12036 \text{ seconds,}$$

where $\psi_1(b_\alpha) = \psi(4) = 0.6873$ has been calculated in Example 8–3 using semi-log linear interpolation.

Therefore, the fatigue life of the defective solder joint under the given narrow-band stressing, T_F, follows a normal distribution with a mean of 735.02985 seconds and a standard deviation of 20.12036 seconds, or

$$T_F \sim N\left(\overline{T}_F, \sigma_{T_F}^2\right) = N\left(735.02985, 20.12036^2\right).$$

Correspondingly, the optimum screening length for the specified screening probability of $P_S = 99\%$, T_S, is given by Eq. (8.208), or

$$\begin{aligned} T_S &= \overline{T}_F + \Phi^{-1}(P_S)\ \sigma_{T_F}, \\ &= 735.02985 + 20.12036\ \Phi^{-1}(0.99), \\ &= 735.02985 + 20.12036 \times 2.3267, \end{aligned}$$

or

$$T_S = 781.84390 \text{ seconds, or 13 minutes.}$$

EXAMPLE 8–6

Determine the optimum screening length, T_S, for the *LRU* described in Example 8–4, for a specified screening probability of 99%, given that the solder joint defect percentage is 10% and the 10% confidence limit $S - N$ curve for the solder joint is

$$N(S)\ S^4 = \left(5\times 10^{4.5}\right)^4,$$

where

$$b_\alpha = 4,$$

and

$$A_\alpha = \left(5 \times 10^{4.5}\right)^4 .$$

SOLUTION TO EXAMPLE 8-6

First, the mean and standard deviation for the fatigue life of the defective solder joint of the PC board under the given stressing conditions are determined. From Example 8–4, it is known that the stress response under the orthogonal force excitation is a wide-band process, and

$$n_S^+(0) = 150.4438 \text{ Hz},$$
$$\epsilon = 0.6299,$$
$$\xi = 0.01,$$

and

$$\sigma_S = 5.0299 \times 10^3 \text{ psi}.$$

For $\epsilon = 0.6299$ and $b_\alpha = 4$, $\varphi(\epsilon; b_\alpha) = \varphi(0.6299; 4) = 0.7978$ has been calculated in Example 8–4.

Substituting these, and $b_\alpha = 4$ and $A_\alpha = \left(5 \times 10^{4.5}\right)^4$ into Eqs. (8.200), (8.202), (8.203) and (8.201) yields

$$\overline{T}_F = \frac{\left(5 \times 10^{4.5}\right)^4}{0.7978 \,(150.4438) \,\left(\sqrt{2} \times 5.0299 \times 10^3\right)^4 \,\Gamma\left(\frac{4}{2}+1\right)},$$

or

$$\overline{T}_F = 1016.91417 \text{ seconds},$$

and

$$h_1' = 0.7978 \,(150.4438) \left(\sqrt{2} \times 5.0299 \times 10^3\right)^4$$
$$\cdot \Gamma\left(\frac{4}{2}+1\right) \Big/ \left(5 \times 10^{4.5}\right)^4 ,$$

or

$$h_1' = 0.00098,$$

and

$$h_2' = \frac{3\,(0.7978)\,\left(\sqrt{2} \times 5.0299 \times 10^3\right)^4}{\left(5 \times 10^{4.5}\right)^4} \,\Gamma\left(\frac{4}{2}+1\right)$$
$$\cdot \sqrt{(150.4438)\,\frac{0.6873}{0.01}},$$

or

$$h_2' = 0.00199,$$

and

$$\sigma_{T_F} = \frac{0.00199 \sqrt{0.00199^2 + 4 \,(0.00098)}}{6 \,(0.00098)^2},$$

or

$$\sigma_{T_F} = 21.56491 \text{ seconds,}$$

where $\psi_1(b_\alpha) = \psi(4) = 0.6873$ has been calculated in Example 8–3 using semi-log linear interpolation.

Therefore, the fatigue life of the defective solder joint under the given wide-band stressing, T_F, follows a normal distribution with a mean of 1016.91417 seconds and a standard deviation of 21.56491 seconds, or

$$T_F \sim N\left(\overline{T}_F, \sigma_{T_F}^2\right) = N\left(1016.91417, 21.56491^2\right).$$

Correspondingly, the optimum screening length for the specified screening probability of $P_S = 99\%$, T_S, is given by Eq. (8.208), or

$$\begin{aligned} T_S &= \overline{T}_F + \Phi^{-1}(P_S)\, \sigma_{T_F}, \\ &= 1016.91417 + 21.56491\, \Phi^{-1}(0.99), \\ &= 1016.91417 + 21.56491 \times 2.3267, \end{aligned}$$

or

$$T_S = 1067.08924 \text{ seconds, or 18 minutes.}$$

8.4 OBTAIN THE STRESS RESPONSE *PSD* FUNCTION FROM FINITE ELEMENT ANALYSIS

The techniques and procedures, presented in this chapter, for random vibration damage evaluation, fatigue life prediction, screening probability calculation and optimum screen time determination, require the availability of the stress response *PSD* function in the structural element. As discussed previously, the *PSD* function directly reflects energy distribution in the frequency domain, and can be related to the number of zero-crossings and the peak distribution of a random stress process. Now the question is how one can get this *PSD* function for a particular randomly-vibrated equipment.

Actually measuring and recording the stress-response time history requires prototypes, measurement equipment and testing time. A very effective way of obtaining the stress response *PSD* function is to conduct a steady-state dynamic response analysis using the finite element analysis (FEA) technique such that the need for prototypes, test equipment and testing time will be eliminated or reduced.

There are commercial software for FEA of equipment subjected to random vibration stressing, such as *ABAQUS* by HKS Inc. [28] and *PATRAN/P3* by PDA Engineering [29; 30]. Recently, Pacific Numerix Corp. at Scottsdale, Arizona presented a new software named *PCB Vibration Plus* [31]. PCB Vibration Plus is the first specialized 3-D FEA software that allows complete modeling of printed circuit boards (*PCB*) along with components and lead/solder joint interconnects. *PCB* Vibration Plus can predict *PCB*'s stress responses to various environments, including static loading, transient loading, frequency spectrum loading and random vibration loading.

REFERENCES

1. Lin, Y. K., *Probabilistic Theory of Structural Dynamics*, McGraw-Hill, New York, 366 pp., 1967.

2. Nigam, N. C., *Introduction to Random Vibrations*, The MIT Press, Cambridge, Massachusetts, 341 pp., 1983.

3. Crandall, S. H. and Mark, W. D., *Random Vibration in Mechanical Systems*, Academic Press, New York, 166 pp., 1963.

4. Yang, C. Y., *Random Vibration of Structures*, John Wiley, New York, 295 pp., 1986.

5. Aerojet Nuclear System Company, "Simplified PRE-PDR Techniques for Assessing Component Reliability – NERVA," *Reliability Calculations for the Cases of Combined Stress and Fatigue Loading*, 62 pp., 1970.

6. Kececioglu, Dimitri B., Chester, L. B. and Gardner, E. O., "Sequential Cumulative Fatigue Reliability," *Proceedings of Annual Reliability and Maintainability Symposium*, pp. 463-469, 1974.

7. Lambert, R. G., "Accelerated Test Rationale for Fracture Mechanics Effects," *Proceedings of the Institute of Environmental Sciences*, pp. 642-648, 1990.

8. Mark, W. D., *The Inherent Variation in Fatigue Damage Resulting from Random Vibration*, Ph.D. Dissertation, Department of Mechanical Engineering, MIT, Cambridge, Massachusetts, August 1961.

9. Silver, W. and Caruso, H., "A Rational Approach to Stress Screening Vibration," *Proceedings of Annual Reliability and Maintainability Symposium*, pp. 384-388, 1981.

10. Wirsching, P. H. and Light, M. C., "Probability Based Fatigue Design Criteria for Ocean Structure," *Final Report, American Petroleum Institute, PRAC Project No. 15,* 1979.

11. Wirsching, P. H. and Light, M. C., "Digital Simulation of Fatigue Damage in Offshore Structures," *Computational Methods in Offshore Structures, ASME,* 1980.

12. Wirsching, P. H., *Random Vibration,* Lecture Notes, The University of Arizona, Tucson, Arizona, 380 pp., 1992.

13. Rice, S. O., "Mathematical Analysis of Random Noise," *Bell System Technical Journal,* 23: pp. 282-332; 24: pp. 46-156; also in Wax, N. ed., *Selected Papers on Noise and Stochastic Processes,* Dover, New York, pp. 133-249, 1954.

14. Shinozuka, M. and Yang, J.-N., "Peak Structural Response to Nonstationary Random Excitations," *Journal of Sound Vibration,* Vol. 14, No. 4, pp. 505-517, 1971.

15. Ibrahim, R. A., *Parametric Random Vibration,* Research Studies Press Ltd., 342 pp., 1985.

16. Newland, D. E., *Random Vibrations and Spectral Analysis,* Longman Group Limited, 377 pp., 1984.

17. Yang, J.-N. and Heer, E., "Reliability of Randomly Excited Structures," *AIAA Journal,* Vol. 9, No. 7, pp. 1262-1268, July 1971.

18. Miner, M. A., "Cumulative Damage in Fatigue," *Journal of Applied Mechanics,* Vol. 12, pp. A159-164, September 1945.

19. Gatts, R., R., "Applications of a Cumulative Damage Concept to Fatigue," *ASME Publication Paper 60-WA-144,* 1960.

20. Bendat, J. et al, "Advanced Concepts of Stochastic Processes and Statistics for Flight Vehicle Vibration Estimation and Measurement," *ASD-TDR-62-973,* Aeronautical Systems Division, Wright-Patterson Air Force Base, Ohio, December 1962.

21. Curtis, A. J. and McKain, R. D., *A Quantified Method of Tailoring Input Spectra for Random Vibration Screens,* Final Report for Period March 1983 to March 1987 prepared for Office of Assistant Secretary of the Navy, 83 pp., June 1987.

22. Arnola, L, E., "Selecting Random Vibration Screening Levels," *Proceedings of the Institute of Environmental Sciences,* Orlando, Florida, pp. 312-315, 1984.

23. Lambert, R. G., "Random Vibration *ESS* Adequacy Prediction Method," *Proceedings of the Institute of Environmental Sciences,* San Diego, California, pp. 374-383, 1991.

24. Bastien, G. J., "Random Vibration of Printed Wiring Assemblies," *Proceedings of the Institute of Environmental Sciences,* Anaheim, California, pp. 204-206, 1989.

25. Freeman, M. T., "3-axis Vibration Test System Simulates Real World," *Test Engineering & Management,* pp. 10-16, December/January 1990-91.

26. Markstein, H. W., "Dealing with Vibration in Electronics," *Electronic Packaging & Production,* pp. 30-34, June 1989.

27. Kindig, W. G. and McGrath, J. D., "Vibrations, Random Required," *Proceedings of the Annual Reliability and Maintainability Symposium,* San Francisco, pp. 143-147, 1984.

28. *ABAQUS User's Manual,* Hibbitt, Karlsson and Sorensen (HKS) Inc., Pawtucket, Rhode Island 02860, 1994.

29. *P/FATIGUE User's Manual,* Release 2.5, Publication No. 2190089, PDA Engineering, PATRAN Division, 1993.

30. Hu, J. M., "Life Prediction and Damage Acceleration Based on the Power Spectral Density of Random Vibration," *Journal of the IES,* pp. 34-40, January/February 1995.

31. *PCB Vibration Plus,* Pacific Numerix Corp., Scottsdale, Arizona 85258, 1994.

Chapter 9

ESS BY COMBINED ENVIRONMENTS – THERMAL CYCLING AND RANDOM VIBRATION

9.1 INTRODUCTION

In the last two chapters, we discussed the quantification of *ESS* by thermal cycling and by random vibration, separately. However, the combined thermal cycling and random vibration scenarios are often used in *ESS* to ensure more effective screening. These two stresses have been found to be the two most effective stresses among all kinds of screening stresses. They are two different stresses, but they are not independent during combined *ESS*. The vibration resisting stress capability (fatigue strength) of hardware has been shown to be successively and incrementally degraded by thermal action [1; 2].

Thermal stresses during thermal cycling are produced by restrictions on the natural expansions and contractions of a material. These restrictions can be an external constraint that prevents free expansion and compression of the entire item, such as the interaction of two or more material interfaces in a structure which have different coefficients of thermal expansion. A more common occurrence is that of varying the temperature throughout the volume of the item. This will cause a nonuniform distribution of stresses. Thermal cycling as a life accelerating process, has been quantified in Chapter 7 using the modified Arrhenius model.

From the fracture mechanics point of view, thermal cycling reliability analysis can be based on an application of the Coffin-Manson model [3; 4; 5;

295

6; 7] which treats thermal expansions and compressions, and their associated stress-strain relations under temperature cycling [8; 9], as a low-cycle fatigue problem. Therefore, coupled thermal-vibration screening can be treated as a combined fatigue problem.

The mechanics of the fatigue process is divided into three stages:

1. Flaw or crack initiation.

2. Flaw or crack propagation.

3. Catastrophic failure as the final result.

The fatigue-induced flaw propagation and catastrophic failure mechanics and dynamics, and the resulting reliability have been treated extensively in the literature [10; 3; 4; 5; 6; 7; 8]. Two general approaches exist to the solution of the reliability of structures subjected to fatigue:

1. The Birnbaum-Saunders damage potential probability model [11; 12].

2. The fracture mechanics probability model, such as that used in the random vibration screening section.

However, a mature quantitative physical analysis of the coupled thermal-vibration screening problem has not as yet been available in the literature. In 1982 Chenoweth [2] presented a combined reliability model which is based on the coupling of a thermal fatigue model (fatigue life distribution) with a random vibration fatigue model (fatigue life distribution). This model is introduced next.

Figure 9.1 [2] indicates the defect population which thermal cycling and random vibration screenings are designed to disclose. These defects are usually of much lower strength than the design levels of strength. Also characterized in Fig. 9.1 is the thermal cycling and random vibration induced stress ranges. The objective of screening is to position this range precisely in the region where it will most likely excite and disclose the defective population with minimal effect on the designed-in strength population.

The combined thermal-cycling and random-vibration screening is expected to precipitate defects by virtue of shorter failure precipitation times and in greater numbers, and a greater variety of incipient failure modes. These failure modes include flaws, improperly annealed metal, inadequate stress relief (wires and structures) and other defects. Some failure modes are strictly thermal and others are strictly vibrational. Other failure modes are initially precipitated by thermal stress and subsequently precipitated by vibration, and vice versa. In addition, there are failure modes which require both vibration and thermal input to precipitate the failure. Because of these coupled-environment impacts, the failure statistics requires correlation among all stress variables.

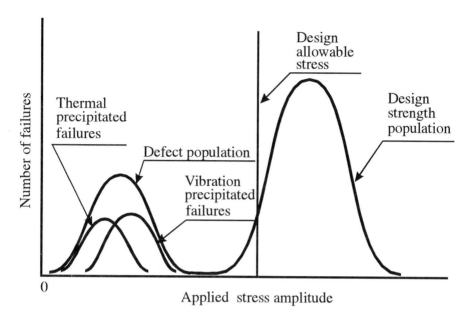

Fig. 9.1– Idealized failure distributions precipitated by random vibration and thermal cycling [2].

9.2   THE THERMAL-FATIGUE LIFE MODEL

It has been shown that thermal stress and strain can be explicitly related to cyclic stress-strain relations [13]. Lambert [8] and Mohler [9] have shown, through a number of data sets of thermal cycling, that the number of thermal cycles to failure, N_{TC}, is lognormally distributed with a *pdf* of

$$f_{N_{TC}}(n) = \frac{1}{\sqrt{2\pi}\,\sigma\,n}\; exp\left[-\frac{1}{2}\left(\frac{\log_e n - \mu}{\sigma}\right)^2\right],$$ (9.1)

where

$$n = \text{value of } N_{TC},$$
$$\mu = \text{mean of } \log_e N_{TC},$$

and

$$\sigma = \text{standard deviation of } \log_e N_{TC}.$$

The *cdf* is

$$F_{N_{TC}}(n) = \int_0^n f_{N_{TC}}(\tau)d\tau = \Phi\left(\frac{\log_e n - \mu}{\sigma}\right),$$ (9.2)

where

$$\Phi(\ ) = cdf \text{ of the standardized normal distribution, } N(0,1).$$

9.3   THE RANDOM VIBRATION FATIGUE LIFE MODEL

Several data sets indicate [2] that the number of random vibration cycles to failure, N_{RV}, follows the *pdf*

$$f_{N_{RV}}(n) = \begin{cases} \frac{\beta}{M_0}\left(\frac{n}{M_0}\right)^{\beta-1}, & \text{for } 0 \leq n \leq M_0, \\ 0, & \text{for } n > M_0, \end{cases} \tag{9.3}$$

which is the so-called *asymptotically truncated Weibull distribution* with a right-truncation point at $n = M_0$, where

$\quad\quad n = $ value of N_{RV},

$\quad\quad \beta = $ shape parameter,

and

$\quad\quad M_0 = $ maximum value of N_{RV} when all (or nearly all) of the defective units in the population have failed.

The number of random vibrations, n, in terms of vibration time, t, is given by [2]

$$n = \left[\frac{\int_0^\infty f^2\, W(f)\, df}{\int_0^\infty W(f)\, df}\right]^{\frac{1}{2}} t, \tag{9.4}$$

where

$\quad\quad f = $ frequency in Hz,

and

$\quad\quad W(f) = $ power spectral density function of the input acceleration spectrum in g^2/Hz.

The *cdf* of N_{RV} is

$$F_{N_{RV}}(n) = \int_0^n f_{N_{RV}}(\tau)d\tau = \begin{cases} \left(\frac{n}{M_0}\right)^{\beta}, & \text{for } 0 \leq n \leq M_0, \\ 1, & \text{for } n > M_0. \end{cases} \tag{9.5}$$

The vibration screening stress levels are planned to be so low, such as 6 *grms* for the acceleration power spectral density function, that the elastic stresses in the vicinity of cracks are very low relative to the plastic stress.

9.4   COMBINED RELIABILITY ANALYSIS

To accomplish the combined reliability analysis, the following variable transformations [2] are needed:

$$X = \frac{N_{RV}}{M_0} \leq 1, \tag{9.6}$$

and

$$Y = \frac{N_{TC}}{N_0} \leq 1, \tag{9.7}$$

where

$$N_0 = 99th \text{ percentile of } f_{N_{TC}}(n),$$

and for the lognormal number of thermal fatigue cycles *pdf*

$$N_0 = e^{\mu + z_{0.99} \ \sigma},$$

or

$$N_0 = e^{\mu + 2.326 \ \sigma}. \tag{9.8}$$

The inverse transforms of Eqs. (9.6) and (9.7) are

$$N_{RV} = M_0 \ X, \tag{9.9}$$

and

$$N_{TC} = N_0 \ Y. \tag{9.10}$$

Then, the *pdf's* and *cdf's* of N_{TC} and N_{RV} in terms of the transformed variables X and Y, $f_{N_{TC}}(y)$, $F_{N_{TC}}(y)$, $f_{N_{RV}}(x)$, and $F_{N_{RV}}(x)$, are given by

$$f_{N_{TC}}(y) = f_{N_{TC}}(n = N_0 y) \left| \frac{\partial N_{TC}}{\partial Y} \right|,$$

$$= \frac{1}{\sqrt{2\pi} \ \sigma \ (N_0 y)} exp \left\{ -\frac{1}{2} \left[\frac{\log_e(N_0 y) - \mu}{\sigma} \right]^2 \right\} \left| N_0 \right|,$$

where

$$\left| \frac{\partial N_{TC}}{\partial Y} \right| = \left| N_0 \right|,$$

$$= \text{Jacobian of the transform given by Eq. (9.10).}$$

Note that for any one-to-one variable transform, say, from V_{old} to V_{new},

$$V_{\text{new}} = g(V_{\text{old}}),$$

or

$$V_{\text{old}} = g^{-1}(V_{\text{new}}).$$

The Jacobian of this transform, J, is given by

$$J = \frac{\partial V_{\text{old}}}{\partial V_{\text{new}}},$$

or

$$J = \frac{\partial g^{-1}(V_{\text{new}})}{\partial V_{\text{new}}}.$$

Consequently,

$$f_{N_{TC}}(y) = \frac{1}{\sqrt{2\pi}\,\sigma\,y} exp\left\{-\frac{1}{2}\left[\frac{\log_e y - (\mu - \log_e N_0)}{\sigma}\right]^2\right\}, \qquad (9.11)$$

which is another lognormal distribution with log mean of $(\mu - \log_e N_0)$ and log standard deviation of σ, and

$$F_{N_{TC}}(y) = \Phi\left[\frac{\log_e y - (\mu - \log_e N_0)}{\sigma}\right]. \qquad (9.12)$$

Also

$$f_{N_{RV}}(x) = f_{N_{RV}}(n = M_0 x)\left|\frac{\partial N_{RV}}{\partial X}\right|,$$

$$= \frac{\beta}{M_0}\left(\frac{M_0\,x}{M_0}\right)^{\beta-1}\left|M_0\right|,$$

or

$$f_{N_{RV}}(x) = \begin{cases} \beta x^{\beta-1}, & \text{for } 0 \leq x \leq 1, \\ 0, & \text{for } x > 1, \end{cases} \qquad (9.13)$$

which is another asymptotically truncated Weibull distribution with the same shape parameter β but a right-truncation point at $x = 1$, and

$$F_{N_{RV}}(x) = \begin{cases} x^\beta, & \text{for } 0 \leq x \leq 1, \\ 1, & \text{for } x > 1. \end{cases} \qquad (9.14)$$

For the coupled thermal-vibration stress environment, the probability of failure is defined in the domain D covered by the following surface [1]:

$$Z = X\,Y, \qquad (9.15)$$

where

$$0 \leq X \leq 1,$$
$$0 \leq Y \leq 1,$$

and

$$0 \leq Z \leq 1.$$

Applying the transformation $Y = \frac{Z}{X}$ to Eq. (9.11) yields

$$f_{N_{TC}}\left(\frac{z}{x}\right) = \frac{1}{\sqrt{2\pi}\,\sigma\left(\frac{z}{x}\right)} exp\left\{-\frac{1}{2}\left[\frac{\log_e(\frac{z}{x}) - (\mu - \log_e N_0)}{\sigma}\right]^2\right\}\left|\frac{\partial Y}{\partial Z}\right|_{X=x},$$

$$= \frac{x}{\sqrt{2\pi}\,\sigma\,z}exp\left\{-\frac{1}{2}\left[\frac{\log_e z - \log_e x - (\mu - \log_e N_0)}{\sigma}\right]^2\right\}\left|\frac{1}{x}\right|,$$

or

$$f_{N_{TC}}\left(\frac{z}{x}\right) = \frac{1}{\sqrt{2\pi}\,\sigma\,z}exp\left\{-\frac{1}{2}\left[\frac{\log_e x - (\log_e z - \mu + \log_e N_0)}{\sigma}\right]^2\right\}.$$

$$(9.16)$$

Therefore, the probability of failure at any value of $Z = z$ is given by [2]

$$F_Z(z) = \int_{x=0}^{a} F_{N_{RV}}(x)\, f_{N_{TC}}\left(\frac{z}{x}\right)\, dx, \qquad (9.17)$$

where

> $a =$ value of x such that $n_{RV} = M_0\, x = (M_0\, a)$ corresponds to the number of vibration cycles at which 99% of the failures have occurred,

or from Eq. (9.14)

$$a = x\Big|_{F_{N_{RV}}(x)=0.99} = (0.99)^{\frac{1}{\beta}}. \qquad (9.18)$$

Substituting Eqs. (9.14) and (9.16) into Eq. (9.17) yields

$$F_Z(z) = \int_{x=0}^{a} x^{\beta}\frac{1}{\sqrt{2\pi}\,\sigma\,z}exp\left\{-\frac{1}{2}\left[\frac{\log_e x - (\log_e z - \mu + \log_e N_0)}{\sigma}\right]^2\right\}dx,$$

$$= \frac{1}{z}\int_{x=0}^{a}\frac{x^{\beta+1}}{\sqrt{2\pi}\,\sigma\,x}exp\left\{-\frac{1}{2}\left[\frac{\log_e x - (\log_e z - \mu + \log_e N_0)}{\sigma}\right]^2\right\}dx,$$

or

$$F_Z(z) = \frac{1}{z}\int_{x=0}^{a}\frac{1}{\sqrt{2\pi}\,\sigma\,x}exp\left\{(\beta+1)\log_e x - \frac{1}{2}\left[\frac{\log_e x - (\log_e z - \mu + \log_e N_0)}{\sigma}\right]^2\right\}dx.$$

Now the exponent in the integrand is

$$(\beta + 1) \log_e x - \frac{1}{2} \left[\frac{\log_e x - (\log_e z - \mu + \log_e N_0)}{\sigma} \right]^2,$$

$$= -\frac{(\log_e x)^2 - 2(\log_e N_0 z - \mu) \log_e x - 2 \sigma^2 (\beta + 1) \log_e x + (\log_e N_0 z - \mu)^2}{2 \sigma^2},$$

$$= -\frac{(\log_e x)^2 - 2[\log_e N_0 z - \mu + \sigma^2(\beta + 1)] \log_e x + [\log_e N_0 z - \mu + \sigma^2(\beta + 1)]^2}{2 \sigma^2}$$

$$- \frac{(\log_e N_0 z - \mu)^2 - [\log_e N_0 z - \mu + \sigma^2(\beta + 1)]^2}{2 \sigma^2},$$

$$= -\frac{\{\log_e x - [\log_e N_0 z - \mu + \sigma^2(\beta + 1)]\}^2}{2 \sigma^2}$$

$$+ \frac{[2(\log_e N_0 z - \mu) + \sigma^2(\beta + 1)][\sigma^2(\beta + 1)]}{2 \sigma^2},$$

$$= -\frac{1}{2} \left\{ \frac{\log_e x - [\log_e z - \mu + \log_e N_0 + \sigma^2(\beta + 1)]}{\sigma} \right\}^2$$

$$+ \left[\frac{(\beta + 1)^2 \sigma^2}{2} + (\beta + 1)(\log_e z - \mu + \log_e N_0) \right].$$

Then, the combined failure probability is given by

$$F_Z(z) = \frac{1}{z} exp \left[\frac{(\beta + 1)^2 \sigma^2}{2} + (\beta + 1)(\log_e z - \mu + \log_e N_0) \right]$$

$$\cdot \int_{x=0}^{a} \frac{1}{\sqrt{2\pi} \ \sigma \ x} \ exp \left\{ -\frac{1}{2} \left\{ \frac{\log_e x - [\log_e z - \mu + \log_e N_0 + \sigma^2(\beta + 1)]}{\sigma} \right\}^2 \right\}$$

or

$$F_Z(z) = N_0^{\beta+1} \ exp \left[\frac{(\beta + 1)^2 \sigma^2}{2} - (\beta + 1)\mu \right] \ z^\beta$$

$$\cdot \Phi \left\{ \frac{\log_e \left(\frac{a}{N_0} \right) - \log_e z + \mu - \sigma^2(\beta + 1)}{\sigma} \right\}.$$

$$(9.19)$$

9.5   OPTIMUM SCREENING STRATEGY

Assume that we are dealing with a defective population's response to screening-induced stresses far below the wear-out resisting stresses of the assemblies; i.e., the fatigue life distributions under thermal and random vibration stresses, given by Eqs. (9.1) and (9.3), are based on the defective population with latent defects. Then, the failure probability under the combined thermal-vibration stresses, given by Eq. (9.19), corresponds to the *screening probability*. Therefore, the optimum screening strategy can be obtained by maximizing the screening probability $F_Z(z)$ and solving for the optimum value of z, z^*. The value of z^* corresponds to various combinations of the optimal number of thermal cycles, N_{TC}^*, and the optimum number of random vibration cycles, N_{RV}^*, which satisfy the constraint

$$z^* = x^* \, y^* = \left(\frac{n_{RV}^*}{M_0}\right)\left(\frac{n_{TC}^*}{N_0}\right) = \frac{n_{RV}^* \, n_{TC}^*}{M_0 \, N_0}. \tag{9.20}$$

This procedure is illustrated next.

Taking the first derivative of Eq. (9.19) with respect to z, and setting it equal to zero, yields

$$\frac{\partial F_Z(z)}{\partial z} = C \, \beta \, z^{\beta-1} \, \Phi\left\{\frac{\log_e\left(\frac{a}{N_0}\right) - \log_e z + \mu - \sigma^2(\beta+1)}{\sigma}\right\}$$

$$+ C \, z^\beta \, \frac{\partial}{\partial z}\Phi\left[\frac{\log_e\left(\frac{a}{N_0}\right) - \log_e z + \mu - \sigma^2(\beta+1)}{\sigma}\right]$$

$$= 0, \tag{9.21}$$

where

$$C = N_0^{\beta+1} \, exp\left[\frac{(\beta+1)^2\sigma^2}{2} - (\beta+1)\mu\right].$$

Let

$$U = U(z) = \frac{\log_e\left(\frac{a}{N_0}\right) - \log_e z + \mu - \sigma^2(\beta+1)}{\sigma}. \tag{9.22}$$

Then, Eq. (9.21) becomes

$$\frac{\partial F_Z(z)}{\partial z} = C \, \beta \, z^{\beta-1} \, \Phi[U] + C \, z^\beta \, \frac{\partial\Phi[U]}{\partial z}$$

$$= C \, \beta \, z^{\beta-1} \, \Phi[U] + C \, z^\beta \, \phi[U] \, \frac{\partial U}{\partial z} = 0, \tag{9.23}$$

where

$$\phi(U) = pdf \text{ of the standardized normal distribution, } N(0,1),$$

$$= \frac{1}{\sqrt{2\pi}}e^{-\frac{u^2}{2}},$$

and

$$\frac{\partial U}{\partial z} = \frac{\partial U(z)}{\partial z},$$

$$= \frac{\partial}{\partial z}\left[\frac{\log_e\left(\frac{a}{N_0}\right) - \log_e z + \mu - \sigma^2(\beta+1)}{\sigma}\right],$$

or

$$\frac{\partial U}{\partial z} = -\frac{1}{\sigma\, z}.$$

Therefore, Eq. (9.23) becomes

$$C\,\beta\,z^{\beta-1}\,\Phi[U] + C\,z^\beta\,\phi[U]\left(-\frac{1}{\sigma\,z}\right) = 0,$$

or

$$C\,\beta\,z^{\beta-1}\,\Phi[U] = \left(\frac{C\,z^{\beta-1}}{\sigma}\right)\phi[U],$$

and

$$\frac{\phi(U^*)}{\Phi(U^*)} = \beta\,\sigma. \qquad (9.24)$$

Solving Eq. (9.24) numerically yields U^*. Then, z^* can be obtained by solving Eq. (9.22) for z; i.e.,

$$z^* = exp\left[\log_e\left(\frac{a}{N_0}\right) + \mu - \sigma^2(\beta+1) - \sigma\,U^*\right]. \qquad (9.25)$$

Correspondingly, the optimum screening cycle combination for the combined thermal-vibration environment can be obtained from Eq. (9.20); i.e.,

$$n^*_{RV}\,n^*_{TC} = (M_0)\,(N_0)\,z^* = (M_0)\,(N_0)\,exp\left[\log_e\left(\frac{a}{N_0}\right) + \mu - \sigma^2(\beta+1) - \sigma\,U^*\right].$$

$$(9.26)$$

EXAMPLE 9–1

A persistent and continuing problem in the electronics industry has been solder defects. Thermal cycling tests were conducted on various solders by a Japanese Company in 1975 [14]. The thermal-cycles-to-failure data for a common solder (Sn 60%, Sb 2.4%, Pb 37.6%) defect population were analyzed graphically by Lambert [8] and Chenoweth [2]. It turns out that these thermal cycles to failure follow a lognormal distribution with

$$f_{N_{TC}}(n) = \frac{1}{\sqrt{2\pi}\,(0.8225)\,n}\,exp\left[-\frac{1}{2}\left(\frac{\log_e n - 5.1614}{0.8225}\right)^2\right],$$

and

$$F_{N_{TC}}(n) = \Phi\left(\frac{\log_e n - 5.1614}{0.8225}\right),$$

where

$$\Phi(\ ) = cdf \text{ of the standardized normal distribution, } N(0,1).$$

The cycles to failure due to random vibration, for the same defect population, follow the asymptotically truncated Weibull distribution with

$$f_{N_{RV}}(n) = \begin{cases} \frac{5}{4,091}\left(\frac{n}{4,091}\right)^{5-1} = 0.001222\left(\frac{n}{4,091}\right)^4, & \text{for } n \leq M_0 = 4,091, \\ 0, & \text{for } n > M_0 = 4,091, \end{cases}$$

and

$$F_{N_{RV}}(n) = \begin{cases} \left(\frac{n}{4,091}\right)^5, & \text{for } 0 \leq n \leq 4,091, \\ 1, & \text{for } n > M_0 = 4,091. \end{cases}$$

Determine the optimum screening strategy for the combined thermal-vibration stressing environment, that is, the optimum cycles for thermal cycling and random vibration, such that the screening probability is maximized.

SOLUTION TO EXAMPLE 9–1

Given the parameters of the lognormal distribution for the number of cycles to failure of thermal cycling, or a log mean of $\mu = 5.1614$ and a log standard deviation of $\sigma = 0.8225$, the 99th percentile is given by Eq. (9.8); i.e.,

$$N_0 = e^{\mu + 2.326\,\sigma},$$
$$= e^{5.1614 + 2.326(0.8225)},$$

or

$$N_0 = 1,181 \text{ cycles.}$$

Given the parameters of the asymptotically truncated Weibull distribution for the number of cycles to failure of random vibration, or $\beta = 5$ and $M_0 = 4,091$ cycles, the value of a can be obtained using Eq. (9.18); i.e.,

$$a = (0.99)^{\frac{1}{\beta}} = (0.99)^{\frac{1}{5}} = 0.99499.$$

Then, Eq. (9.24) becomes

$$\frac{\phi(U^*)}{\Phi(U^*)} = \beta \, \sigma = 5 \, (0.8225),$$

or

$$\frac{\phi(U^*)}{\Phi(U^*)} = 4.1125.$$

This equation can be solved either numerically, or by trial-and-error using the standardized normal distribution area tables. To illustrate, let us apply the trial-and-error method. Since

$$\begin{cases} \phi(-3.88) = \frac{1}{\sqrt{2\,\pi}} \, e^{-\frac{1}{2}(-3.88)^2} = 2.1473 \times 10^{-4}, \\ \Phi(-3.88) = 5.2228 \times 10^{-5}, \\ \frac{\phi(-3.88)}{\Phi(-3.88)} = \frac{2.1473\times 10^{-4}}{5.2228\times 10^{-5}} = 4.1114, \end{cases}$$

and

$$\begin{cases} \phi(-3.89) = \frac{1}{\sqrt{2\,\pi}} \, e^{-\frac{1}{2}(-3.89)^2} = 2.0655 \times 10^{-4}, \\ \Phi(-3.89) = 5.0122 \times 10^{-5}, \\ \frac{\phi(-3.89)}{\Phi(-3.89)} = \frac{2.0655\times 10^{-4}}{5.0122\times 10^{-5}} = 4.1209. \end{cases}$$

Using linear interpolation yields

$$\frac{U^* - 3.88}{3.89 - 3.88} = \frac{4.1125 - 4.1114}{4.1209 - 4.1114},$$

or

$$U^* = -3.8812.$$

From Eq. (9.25) the optimum value of z is

$$z^* = exp\left[\log_e\left(\frac{a}{N_0}\right) + \mu - \sigma^2(\beta + 1) - \sigma \, U^*\right],$$

$$= exp\left[\log_e\left(\frac{0.99499}{1,181}\right) + 5.1614 - 0.8225^2(5 + 1) - 0.8225\,(-3.8812)\right],$$

or

$$z^* = 0.0617605.$$

Therefore, the optimum screening cycle combination can be obtained from Eq. (9.26); i.e.,

$$n^*_{RV} \, n^*_{TC} = (M_0) \, (N_0) \, z^*,$$
$$= (4,091) \, (1,181) \, (0.0617605),$$

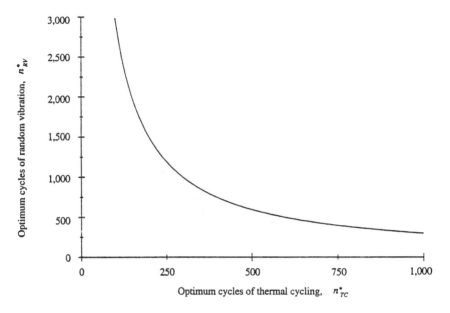

Fig. 9.2– Optimum screening cycle contour for combined thermal-vibration environment for Example 9–1.

or

$$n^*_{RV} \; n^*_{TC} = 298,394.06 \text{ cycles}^2,$$

which corresponds to a contour on the n^*_{RV} versus n^*_{TC} plane; i.e.,

$$n^*_{RV} = \frac{298,394.06}{n^*_{TC}},$$

as shown in Fig. 9.2 and tabulated in Table 9.1.

It should be noted that the optimum screening cycles, n^*_{RV} and n^*_{TC}, given in Fig. 9.2 and in Table 9.1 are valid only for those screening conditions under which the thermal fatigue, and the random vibration fatigue, life tests are conducted and the corresponding life distribution parameters are determined.

TABLE 9.1– Values of n^*_{RV} versus n^*_{TC} for Example 9–1.

No.	n^*_{TC}, cycle	n^*_{RV}, cycle
1	0	∞
2	50	5,967.88
3	100	2,983.94
4	150	1,989.29
5	200	1,491.97
6	250	1,193.58
7	300	994.65
8	350	852.55
9	400	745.99
10	450	663.10
11	500	596.79
12	550	542.53
13	600	497.32
14	650	459.07
15	700	426.28
16	750	397.86
17	800	372.99
18	850	351.05
19	900	331.55
20	950	314.10
21	1,000	298.39

REFERENCES

1. Hertzberg, R. W., *Deformation and Fracture Mechanics of Engineering Materials*, 3rd edition, John Wiley, New York, 680 pp., 1989.

2. Chenoweth, H. B., "Vibration-Thermal Screening Reliability Prediction," *Proceedings of Annual Reliability and Maintainability Symposium*, pp. 91-96, 1982.

3. Coffin, L. F. and Schenectady, N. Y., "A Study of Effects of Cyclic Thermal Stresses on a Ductile Metal," *ASME Transactions*, Vol. 76, No. 6, pp. 931-950, 1954.

4. Coffin, L. F., "The Resistance of Material to Cyclic Strains," *ASME Transactions*, Paper 57-A-286, 1957.

5. Coffin, L. F., "The Stability of Metals Under Cyclic Plastic Strain," *ASME Transactions*, Paper 59-A-100, 1959.

6. Manson, S. S., "Behavior of Materials Under Conditions of Thermal Stress," *NACA TN 2933*, July 1953.

7. Manson, S. S., "Fatigue: A Complex Subject – Some Simple Approximations," *Experimental Mechanics*, ASME, pp. 193-226, July 1965.

8. Lambert, R. G., "Mechanical Reliability for Low Cycle Fatigue," *Proceedings of Annual Reliability and Maintainability Symposium*, Los Angeles, California, pp. 179-183, 1978.

9. Mohler, J. B., "Solder Joints vs Time and Temperature," *Machine Design*, pp. 84-87, April 1971.

10. Yang, J.-N. and Heer, E., "Reliability of Randomly Excited Structures," *AIAA Journal*, Vol. 9, No. 7, pp. 1262-1268, July 1971.

11. Birnhaum, Z. W. and Saunders, S. C., "A Probabilistic Interpretation of Miner's Rule," *SIAM Journal of Applied Mathematics*, Vol. 16, No. 3, pp. 637-652, 1968.

12. Mann, N. R., Schafer, R. E. and Singpurwalla, N. D., *Methods for Statistical Analysis of Reliability and Life Data*, John Wiley, New York, 564 pp., 1974.

13. "Technical Report on Fatigue Properties," *SAEJ 1099*, February 1975.

14. Kobayaski, M., "Reliability of Soldered Connections in Single Ended Circuit Components," *Welding Research Supplement*, pp. 363-365, October 1975.

15. Miner, M. A., "Cumulative Damage in Fatigue," *Journal of Applied Mechanics*, Vol. 12, pp. A159-164, September 1945.

16. Gatts, R., R., "Applications of a Cumulative Damage Concept to Fatigue," *ASME Paper 60-WA-144*, 1960.

17. Bendat, J. et al, "Advanced Concepts of Stochastic Processes and Statistics for Flight Vehicle Vibration Estimation and Measurement," *ASD-TDR-62-973*, Aeronautical Systems Division, Wright-Patterson Air Force Base, Ohio, December 1962.

18. Curtis, A. J. and McKain, R. D., *A Quantified Method of Tailoring Input Spectra for Random Vibration Screens,* Final Report for Period March 1983 to March 1987 prepared for Office of Assistant Secretary of the Navy, 83 pp., June 1987.

19. Arnola, L, E., "Selecting Random Vibration Screening Levels," *Proceedings of the Institute of Environmental Sciences,* Orlando, Florida, pp. 312-315, 1984.

20. Lambert, R. G., "Random Vibration *ESS* Adequacy Prediction Method," *Proceedings of the Institute of Environmental Sciences,* San Diego, California, pp. 374-383, 1991.

21. Bastien, G. J., "Random Vibration of Printed Wiring Assemblies," *Proceedings of the Institute of Environmental Sciences,* Anaheim, California, pp. 204-206, 1989.

22. Freeman, M. T., "3-axis Vibration Test System Simulates Real World," *Test Engineering & Management*, pp. 10-16, December/January 1990-91.

23. Markstein, H. W., "Dealing with Vibration in Electronics," *Electronic Packaging & Production*, pp. 30-34, June 1989.

24. Kindig, W. G. and McGrath, J. D., "Vibrations, Random Required," *Proceedings of the Annual Reliability and Maintainability Symposium,* San Francisco, pp. 143-147, 1984.

Chapter 10

ESS BY APPLYING MULTIPLE STRESSES

10.1   AGING ACCELERATION UNDER MULTIPLE STRESSES — A TEST ON THE EFFECT DEPENDENCY OF COMBINED STRESSES

10.1.1   INTRODUCTION

For some types of devices, the simultaneous application of multiple stresses seems to provide extreme accelerations with typical stress levels or moderate accelerations with very modest stresses [1]. The question is how to quantify the aging acceleration that results from the simultaneous application of multiple stresses and how to determine whether there are interactions among these stresses or not.

Nachlas et al [1] proposed a general model to represent component aging, which has an additive form in the absence of any interaction and has additional polynomial terms in the presence of interactions. A factorial experimental design technique can be used to test the following two hypotheses:

H_0: The accelerations due to simultaneously applied stresses act independently,

versus

H_1: There is a synergistic effect (interaction) associated with the simultaneously applied stresses.

311

10.1.2   THE AGING ACCELERATION MODELS

The life distribution of each of the components studied is assumed to be the two-parameter Weibull with shape parameter less than one (1) or with a decreasing failure rate; i.e.,

$$f(T) = \frac{\beta}{\eta} \left(\frac{T}{\eta}\right)^{\beta-1} e^{-\left(\frac{T}{\eta}\right)^{\beta}}, \quad T \geq 0, \eta > 0, 0 < \beta < 1.$$

The *cdf* and the reliability function of this life distribution are given by

$$F(T) = 1 - e^{-\left(\frac{T}{\eta}\right)^{\beta}},$$

and

$$R(T) = e^{-\left(\frac{T}{\eta}\right)^{\beta}},$$

respectively. Let

$$\alpha = \eta^{-\beta} = \left(\frac{1}{\eta}\right)^{\beta}. \tag{10.1}$$

Then, $F(T)$ and $R(T)$ can be simplified as

$$F(T) = 1 - e^{-\alpha\, T^{\beta}},$$

and

$$R(T) = e^{-\alpha\, T^{\beta}}. \tag{10.2}$$

Denote the component's reliability function, under multiple simultaneously applied stresses, by $R_a(T)$ and assume that

$$R_a(T) = R(A\, T), \tag{10.3}$$

where

$$A = \text{acceleration factor.}$$

Obviously, this acceleration factor, A, is a function of the simultaneously applied stresses.

Assume that there are m applied stresses, $S_1, S_2, \cdots, S_m$. Let $\underline{S}$ be a vector representing these stresses; i.e.,

$$\underline{S} = (S_1, S_2, \cdots, S_m). \tag{10.4}$$

Therefore,

$$A = A(S_1, S_2, \cdots, S_m) = A(\underline{S}). \tag{10.5}$$

For the sake of easy manipulation, the logarithmic acceleration factor is often used and is defined by

$$\log_e A = g(\underline{S}). \tag{10.6}$$

Under the null hypothesis H_0, or in the absence of any synergistic effect, or if all m stresses accelerate the component independently, then $A = A(\underline{S})$ will be simply a product of m independent acceleration factors under each stress; i.e.,

$$A = A(\underline{S}) = A_1(S_1) \cdot A_2(S_2) \cdots A_m(S_m), \tag{10.7}$$

or equivalently, the logarithmic acceleration factor should have the following additive form:

$$\log_e A = g(\underline{S}) = h_1(S_1) + h_2(S_2) + \cdots + h_m(S_m), \tag{10.8}$$

where

$$h_i(S_i) = \log_e A_i(S_i), \quad i = 1, 2, \cdots, m.$$

Under the alternative hypothesis H_1, or in the presence of a synergistic effect (interaction), $g(\underline{S})$ will include additional terms representing the interactions. It is assumed that the form of $g(\underline{S})$ in this case can at least be approximated by a second-order polynomial function in $\underline{S}$; i.e.,

$$\log_e A = g(\underline{S}) = \sum_{i=1}^{m} h_i(S_i) + \sum_{i=1}^{m} \sum_{j=1}^{m} h_{ij}(S_i, S_j), \tag{10.9}$$

where the higher-order interaction terms are assumed to be negligible and therefore ignored.

Consequently, Eqs. (10.8) and (10.9) represent two candidate aging acceleration models. The remaining question is how to decide which one is appropriate for a particular device and the applied stresses. This is discussed next.

10.1.3 DEPENDENCY TEST AND EFFECTS (ACCELERATION CONTRIBUTIONS) EVALUATION

Let α and β represent the Weibull distribution parameters in Eq. (10.2) for a component under use stress conditions and let α' and β' denote the Weibull distribution parameters for a component under simultaneously applied accelerated stresses. Then,

$$R(A\,T) = e^{-\alpha(A\,T)^{\beta}}, \tag{10.10}$$

and

$$R_a(T) = e^{-\alpha' \ T^{\beta'}}. \tag{10.11}$$

According to Eq. (10.3)

$$R(A \ T) = R_a(T),$$

then,

$$e^{-\alpha(A \ T)^{\beta}} = e^{-\alpha' \ T^{\beta'}},$$

or

$$A = \left(\frac{\alpha'}{\alpha}\right)^{\frac{1}{\beta}} \ T^{\left(\frac{\beta'}{\beta}-1\right)}. \tag{10.12}$$

Consequently, the acceleration may be studied experimentally by obtaining the Weibull life distribution parameters under both use and accelerated stresses, and then analyzing the indicated acceleration factors.

The computed acceleration factors under different stress combinations can be used to determine which candidate model; i.e., Eq. (10.8) or (10.9) provides the best fit to the observed data, or equivalently to determine whether or not interactions are present. This can be realized by the following procedures:

1. Conduct a l^m complete factorial experiment (CFE) or l^{m-k} fractional factorial experiment (FFE) with n (random) replicates at each stress combination where m is the number of simultaneously applied stresses, l is the number of levels for each stress and k is an integer value such that l^{-k} is the fraction of CFE, or the FFE is conducted at; i.e., there are $(m \times l)$ total treatment (stress level) combinations for CFE or $[l \times (m - k)]$ total treatment combinations for FFE. The lowest stress level is taken as the use operating stress level for each applied stress type.

2. Under each of the $(m \times l)$ or $[l \times (m - k)]$ stress level combinations, n times-to-failure data are collected randomly. The corresponding Weibull shape and scale parameters can be estimated using the maximum likelihood estimation (MLE) method or the least-squares estimation method.

3. Calculate the acceleration factors at each stress combination using Eq. (10.12) where the use operating conditions correspond to the lowest stress levels.

4. Take the logarithms of the acceleration factors and conduct an analysis of variance ($ANOVA$) on the resulting data. This analysis will yield F statistics for each individual stress (main effect) and each possible

stress interaction. For a specified significance level, δ, the critical value for each F statistic, $F_{1-\delta}$, can be found out from *cdf* tables of the F-distribution. Then, it can be decided which stress(es) and which interaction(s) are statistically significant.

5. Verify the results in Step 4 by conducting a polynomial regression analysis using Eq. (10.9). The regression coefficient of each $h_i(S_i)$ term illustrates the significance (or contribution) of the corresponding stress S_i. The regression coefficient of each $h_{ij}(S_i, S_j)$ term illustrates the dependency (interaction) of the corresponding stresses S_i and S_j. The correlation coefficient illustrates the goodness-of-fit of the polynomial function given by Eq. (10.9). If the resultant regression equation turns out to be Eq. (10.8), then there are no interactions among the m applied stresses. Therefore, the final expression for the logarithmic acceleration factor, $g(\underline{S})$, is given by

$$g(\underline{S}) = c + d_1\ S_1 + d_2\ S_2 + \cdots + d_m\ S_m.$$

or

$$g(\underline{S}) = \sum_{i=1}^{m}[c_i + d_i\ S_i] = \sum_{i=1}^{m} g_i(S_i), \tag{10.13}$$

where

$$c = \sum_{i=1}^{m} c_i, \tag{10.14}$$

and

$$g_i(S_i) = c_i + d_i\ S_i. \tag{10.15}$$

Correspondingly, from Eq. (10.6) the natural acceleration factor, $A(\underline{s})$, is given by

$$A(\underline{S}) = e^{g(\underline{S})} = e^{\sum_{i=1}^{m} g_i(S_i)} = \prod_{i=1}^{m} e^{g_i(S_i)},$$

or

$$A(\underline{S}) = \prod_{i=1}^{m} A_i(S_i), \tag{10.16}$$

where

$$A_i(S_i) = e^{g_i(S_i)} = e^{c_i + d_i\ S_i}. \tag{10.17}$$

That is, if there are no interactions among the applied stresses, the total acceleration factor is simply the product of the acceleration factors due to all individual stresses.

10.2 AN ECONOMIC MODEL FOR OPTIMUM STRESS REGIMEN DETERMINATION FOR A MULTIPLE-STRESS ASSEMBLY-LEVEL STRESS SCREEN

10.2.1 INTRODUCTION

As described in the last section, during assembly-level or system-level stress screen, different components will experience different age accelerations under a common stress regimen. Selection of a stress screen should be based on an understanding of the different equivalent ages of the assembly components resulting from stress screen, and also on a balance between the costs of stress application and the obtained early-life field performance benefits [1]. However, most of the economic studies of stress screening in the literature focus on the thermal stress only, and most assume component level screening.

Nachlas et al [2] proposed a general cost model for selecting a multiple-stress regimen for assembly-level stress screen. The decision variables are

1. stress levels for each prechosen stress, and

2. duration of each stress application.

The proposed model is an unconstrained, non-linear program model which is non-convex. The generalized reduced gradient method is used to solve the optimum stress levels and screen duration for the minimal expected cost. The application of the model and its optimization indicate that *the cost-effective stresses should be applied at their maximum levels in order to reduce the screen duration.*

10.2.2 COST MODEL DEVELOPMENT

Assume that all components in the assembly are reliabilitywise in series; i.e., the assembly reliability is the product of the reliabilities of all of the components. Now let

N = total number of constituent components in the assembly,

m = total number of applied stresses,

T_{ss} = stress screen duration,

$\underline{S}_o = \{S_{oi}\}$ = vector of stress levels under use operating conditions, where $i = 1, 2, \cdots, m$,

$\underline{S}_a = \{S_{ai}\}$ = vector of stress levels under accelerated operating conditions, where $i = 1, 2, \cdots, m$,

A_{ji} = age acceleration experienced by component j when subjected to stress i, where $j = 1, 2, \cdots, N$ and $i = 1, 2, \cdots, m$,

$A_{j\cdot}$ = total age acceleration experienced by component j when subjected to all m stresses, where $j = 1, 2, \cdots, N$,

$R_j(T)$ = reliability of component j under use operating conditions,

$R(T)$ = assembly reliability under use operating conditions,

and

τ_j = equivalent age under use operating conditions for type j components that survive the stress screen.

Then,

$$\tau_j = A_{j\cdot} \, T_{ss}, \tag{10.18}$$

and

$$R(T_{ss}) = \prod_{j=1}^{N} R_j(\tau_j). \tag{10.19}$$

If there are no interactions among the m stresses, then according to Eqs. (10.16) and (10.17)

$$A_{j\cdot} = \prod_{i=1}^{m} A_{ji}, \tag{10.20}$$

and

$$A_{ji} = e^{c_{ji} + d_{ji} \, S_{ai}}, \tag{10.21}$$

or

$$A_{j\cdot} = e^{\sum_{i=1}^{m} (c_{ji} + d_{ji} \, S_{ai})}, \tag{10.22}$$

where

c_{ji}, d_{ji} = regression coefficients for A_{ji}.

Now Eqs. (10.18) and (10.19) become

$$\tau_j = T_{ss}\ e^{\sum_{i=1}^{m}(c_{ji}+d_{ji}\ S_{ai})}, \tag{10.23}$$

and

$$R(T_{ss}) = \prod_{j=1}^{N} R_j \left[T_{ss}\ e^{\sum_{i=1}^{m}(c_{ji}+d_{ji}\ S_{ai})} \right]. \tag{10.24}$$

Let

$\qquad t_W$ = warranty period;

then, the mission reliability of the assembly during warranty, after a stress screen, is given by

$$R(t_W | T_{ss}) = \frac{\prod_{j=1}^{N} R_j(\tau_j + t_W)}{\prod_{j=1}^{N} R_j(\tau_j)} = \prod_{j=1}^{N} \left[\frac{R_j(\tau_j + t_W)}{R_j(\tau_j)} \right] = \prod_{j=1}^{N} R_j(t_W | \tau_j).$$

$$\tag{10.25}$$

Conceptually, the total-cost-per-unit-time model is given by

Expected total cost during screen and warranty

= Cost of stress screen

+[Cost of in-plant failure during screen]

$\qquad$ ×[Probability of a unit failing during screen]

+[Cost of field failures during warranty operation]

$\qquad$ ×[Probability of a unit failing during warranty].

Therefore,

$$E[C_T(\underline{S}_a, T_{ss}, t_W)] = C_{ss}(\underline{S}_a, T_{ss})$$
$$+ C_i\ [1 - R(T_{ss})]$$
$$+ C_f\ [1 - R(t_W | T_{ss})], \tag{10.26}$$

where

$$C_T(\underline{S}_a, T_{ss}, t_W) = \text{total cost during screen and warranty},$$
$$C_{ss}(\underline{S}_a, T_{ss}) = \text{stress screen cost},$$
$$C_i = \text{unit costs of in-plant failures during screen},$$

and

C_f = unit costs of early life field failures during warranty.

Assume that the stress screen cost, $C_{ss}(\underline{S_a}, T_{ss})$, is a linear function of the screen time, T_{ss}, or

$$C_{ss}(\underline{S_a}, T_{ss}) = \sum_{i=1}^{m} \left[p_i + q_i(S_{ai} - S_{oi})T_{ss} \right], \tag{10.27}$$

where

p_i = fixed cost for stress type i,

and

q_i = linear cost coefficient of stress type i.

Assume further that each component has a Weibull life distribution with a shape parameter less than one (1) during screen and warranty period, as given in Eq. (10.2), or

$$R_j(T) = e^{-\alpha_j \, T^{\beta_j}}, \quad j = 1, 2, \cdots, N. \tag{10.28}$$

Then, from Eq. (10.24)

$$R(T_{ss}) = \prod_{j=1}^{N} R_j \left[T_{ss} \, e^{\sum_{i=1}^{m}(c_{ji}+d_{ji} \, S_{ai})} \right],$$

$$= \prod_{j=1}^{N} e^{-\alpha_j \left[T_{ss} \, e^{\sum_{i=1}^{m}(c_{ji}+d_{ji} \, S_{ai})} \right]^{\beta_j}},$$

or

$$R(T_{ss}) = e^{-\sum_{j=1}^{N} \alpha_j \left[T_{ss} \, e^{\sum_{i=1}^{m}(c_{ji}+d_{ji} \, S_{ai})} \right]^{\beta_j}}, \tag{10.29}$$

and

$$R(t_W | T_{ss}) = \prod_{j=1}^{N} \frac{R_j \left[t_W + T_{ss} \, e^{\sum_{i=1}^{m}(c_{ji}+d_{ji} \, S_{ai})} \right]}{R_j \left[T_{ss} \, e^{\sum_{i=1}^{m}(c_{ji}+d_{ji} \, S_{ai})} \right]},$$

$$= \prod_{j=1}^{N} exp\left\{-\alpha_j\left\{\left[t_W + T_{ss}\ e^{\sum_{i=1}^{m}(c_{ji}+d_{ji}\ S_{ai})}\right]^{\beta_j}\right.\right.$$

$$\left.\left.- \left[T_{ss}\ e^{\sum_{i=1}^{m}(c_{ji}+d_{ji}\ S_{ai})}\right]^{\beta_j}\right\}\right\},$$

or

$$R(t_W \big| T_{ss}) = exp\left\{-\sum_{j=1}^{N}\alpha_j\left\{\left[t_W + T_{ss}\ e^{\sum_{i=1}^{m}(c_{ji}+d_{ji}\ S_{ai})}\right]^{\beta_j}\right.\right.$$

$$\left.\left.- \left[T_{ss}\ e^{\sum_{i=1}^{m}(c_{ji}+d_{ji}\ S_{ai})}\right]^{\beta_j}\right\}\right\}. \qquad (10.30)$$

Finally, Eq. (10.26) can be written as

$$E\left[C_T(\underline{S}_a, T_{ss}, t_W)\right]$$

$$= \sum_{i=1}^{m}\left[p_i + q_i\ (S_{ai} - S_{oi})T_{ss}\right]$$

$$+ C_i\left\{1 - e^{-\sum_{j=1}^{N}\alpha_j\left[T_{AA}\ e^{\sum_{i=1}^{m}(c_{ji}+d_{ji}\ S_{ai})}\right]^{\beta_j}}\right\}$$

$$+ C_f\left\{1 - exp\left\{-\sum_{j=1}^{N}\alpha_j\left\{\left[t_W + T_{ss}\ e^{\sum_{i=1}^{m}(c_{ji}+d_{ji}\ S_{ai})}\right]^{\beta_j}\right.\right.\right.$$

$$\left.\left.\left.- \left[T_{ss}\ e^{\sum_{i=1}^{m}(c_{ji}+d_{ji}\ S_{ai})}\right]^{\beta_j}\right\}\right\}\right\}. \qquad (10.31)$$

Now, the next objective is to determine the optimum T_{ss} and $\{S_{a1}, S_{a2},$ $\cdots, S_{am}\}$ such that $E\left[C_T(\underline{S}_a, T_{ss}, t_W)\right]$ is minimized. There may be several local minima for Eq. (10.31). The best of the local minima is taken to be the optimum solution. This will be illustrated next by an example.

EXAMPLE 10-1

An assembly [2] is comprised of five ($N = 5$) components. The screening stresses are humidity, power cycling and elevated temperature, or $m = 3$. The Weibull life distribution parameters in Eq. (10.28) are given by

$$\alpha_1 = 2.25 \times 10^{-6}, \ \beta_1 = 0.40,$$
$$\alpha_2 = 4.00 \times 10^{-6}, \ \beta_2 = 0.45,$$
$$\alpha_3 = 6.25 \times 10^{-6}, \ \beta_3 = 0.50,$$
$$\alpha_4 = 8.25 \times 10^{-6}, \ \beta_4 = 0.55,$$

and

$$\alpha_5 = 9.40 \times 10^{-6}, \ \beta_5 = 0.60.$$

The corresponding acceleration factor function parameters, c_{ji} and d_{ji}, in Eq. (10.31), which correspond to the regression coefficients in Eq. (10.21), have been obtained through previous experimental design analysis using the procedures described in Section 10.1.3 and are given by

$$[C_{ji}]_{N \times m} = [C_{ji}]_{5 \times 3} = \begin{bmatrix} -0.039 & -0.304 & 38.915 \\ -0.037 & -0.276 & 38.915 \\ -0.035 & -0.280 & 38.915 \\ -0.038 & -0.310 & 31.132 \\ -0.037 & -0.310 & 31.132 \end{bmatrix},$$

and

$$[d_{ji}]_{N \times m} = [d_{ji}]_{5 \times 3} = \begin{bmatrix} 1.55 \times 10^{-3} & 0.152 & -11,596.9 \\ 1.46 \times 10^{-3} & 0.138 & -11,596.9 \\ 1.38 \times 10^{-3} & 0.140 & -11,596.9 \\ 1.52 \times 10^{-3} & 0.155 & -9,277.5 \\ 1.48 \times 10^{-3} & 0.155 & -9,277.5 \end{bmatrix}.$$

The cost parameters are given by

$$(p_1, p_2, p_3) = (0.10, 0.07, 0.05),$$
$$(q_1, q_2, q_3) = (9 \times 10^{-7}, 8 \times 10^{-4}, 2.25 \times 10^{-4}),$$
$$C_i = \$1/\text{unit},$$

and

$$C_f = \$100/\text{unit}.$$

The warranty time is assumed to be

$t_W = 2,000$ hr.

The nominal operating environment is defined by

$\underline{S}_o$ = (humidity, power cycling, temperature),

= (25%, 2 cycle/hr, 298°K(25°C)) .

Determine the optimum screen regimen; i.e.,

1. the optimum screen length T_{ss}^*, and

2. the optimum stress levels: S_{a1}^*, S_{a2}^* and S_{a3}^*,

which will minimize the total expected cost given by Eq. (10.31).

SOLUTIONS TO EXAMPLE 10–1

In applying the optimization algorithm, the following bounds are placed on the stress variables due to the components' design ability to tolerate the stresses:

$S_{o1} = 25\%$ $\leq S_{a1} \leq 75\%,$
$S_{o2} = 2$ cycles/hr $\leq S_{a2} \leq 8$ cycles/hr,
$S_{o3} = 298°K$ $\leq S_{a3} \leq 350°K,$

and

$0 \leq T_{ss} \leq 300$ hr.

Imposing bounds also simplifies the use of an optimization algorithm by restricting the search region.

The generalized reduced gradient method is used [2] to minimize the cost function given by Eq. (10.31) using an efficient software package, $GINO$ [2; 3], which is available for use on a personal computer. The best solution, identified from several local minima, is

$S_{a1}^* = 75\%,$
$S_{a2}^* = 2$ cycles/hr,
$S_{a3}^* = 350°K = 77°C$

and

$T_{ss}^* = 22.44$ hr.

The total expected cost is

$C_T(\underline{S}_a^*, T_{ss}^*, t_W) = \$0.325$ per assembly.

It may be seen that not all stresses are necessarily cost effective. In this example, for power cycling the optimum level is the lowest stress value. For humidity and temperature, the optimum levels are the upper stress bounds so that the screen duration is reduced.

10.3 COMMENTS

The procedure for aging acceleration evaluation using the design of experiments technique, and the cost model for optimum stress screen regimen determination, developed by Nachlas, et al [1; 2], are quite general in describing the effects of multiple-stresses on equipment assemblies during *ESS*. However, the implementation of the cost model requires quite a few parameters in costs and life distributions which may cause difficulties in engineering applications.

REFERENCES

1. Nachlas, J. A., Binney, B. A. and Gruber, S. S., "Aging Acceleration Under Multiple Stresses," *Proceedings of the Annual Reliability and Maintainability Symposium*, Philadelphia, Pennsylvania, pp. 438-440, 1985.

2. Nachlas, J. A., Seward, L. E. and Binney, B. A., "Multiple-Stress Assembly-Level Stress Screening," *Proceedings of the Annual Reliability and Maintainability Symposium*, Las Vegas, Nevada, pp. 40-43, 1986.

3. Lasdon, L. and Waren, A., *General Interactive Optimizer (GINO)*, Scientific Press, Palo Alto, California, 1985.

Chapter 11

SCREENING STRENGTH AND THE RELATED EMPIRICAL EQUATIONS IN *ESS*

11.1 DEFINITION OF SCREENING STRENGTH

The screening strength of a given stress screen profile, denoted by SS, is defined as the probability that the stress screen will precipitate a latent defect into a detectable failure, given that a defect is present. Mathematically it can be expressed by the ratio of the mean number of observed failures during the screen to the total number of inherent latent defects in the equipment; i.e.,

$$SS = \frac{\text{mean number of observed failures during the screen}}{\text{total number of inherent latent defects in the equipment}}.$$

(11.1)

Summarized next are some empirical equations developed by industry for screen strength and average defect failure rate estimation under different stresses.

11.2 SCREENING STRENGTH UNDER DIFFERENT SCREENING STRESSES

11.2.1 THE HUGHES MODELS

The screening strength equations presented next were formulated by Hughes based on data obtained from an industry survey conducted by Hughes and Hughes' own internal data:

1. **Screening strength for random vibration**

The screening strength equation for random vibration is given by [1]

$$SS_{RV}(T_{RV}) = D \left(1 - e^{-\sqrt{T_{RV}/B}}\right), \qquad (11.2)$$

where

> T_{RV} = duration of applied vibration excitation, in min,
> $D = 0.144 \, g - 0.0862,$
> $B = 0.266 \, G + 1.402,$

and

> $G = g_{rms}$, rms value of applied acceleration power spectral density over the frequency spectrum,

Equation (11.2) is applicable for

> $0 \leq G \leq 6 \, g,$

and

> $0 \leq T_{RV} \leq 45$ min.

2. **Screening strength for thermal cycling**

The stress screening equation for temperature cycling is given by [1]

$$SS_{TC}(N_{TC}) = 0.85 \left(1 - e^{-A}\right), \qquad (11.3)$$

where

> $A = 0.0023 \, (T_R^*)^{0.6} \sqrt{N_{TC}} \, [\log_e(e + \dot{T})]^{2.7},$
> T_R^* = temperature range = $T_{max}^* - T_{min}^*$, in $°C$,
> N_{TC} = number of temperature cycles,
> $e = 2.71828,$
> $\dot{T}^*$ = average rate of temperature change, or temperature gradient, in $°C/min,$

or

$$\dot{T}^* = \frac{1}{2}\left[\left(\frac{T_{max}^* - T_{min}^*}{t_1}\right) + \left(\frac{T_{max}^* - T_{min}^*}{t_2}\right)\right], \qquad (11.4)$$

> t_1 = transition time from T_{min} to T_{max}, for the heating segment, in min,

and

> t_2 = transition time from T_{max} to T_{min}, for the cooling segment, in min.

Equation (11.3) is applicable for

$$-55°C \le T_{min} \le 25°C,$$
$$25°C \le T_{max} \le 75°C,$$
$$1°C/min \le \dot{T} \le 20°C/min,$$

and

$$0 \le N_{TC} \le 20 \text{ cycles}.$$

3. **Screening strength for constant temperature burn-in**

The stress screening equation for constant temperature burn-in is given by [1]

$$SS_{CT}(T_b) = 0.85 \left(1 - e^{-A'}\right), \tag{11.5}$$

where

$$A' = 0.0048 \, (T_R^*)^{0.6},$$

$T_R^* = $ absolute temperature difference between
the burn-in temperature and $25°C$,

$$T_R^* = \left|T^* - 25\right|,$$

$T^* = $ burn-in temperature, in $°C$,

and

$T_b = $ burn-in time in hr.

Note that for $-1 < T_R^* < +1$, set $(T_R^*)^{0.6} = 1$. Equation (11.5) is applicable for

$$-55°C \le T^* \le 75°C.$$

11.2.2 RADC MODELS – THE REVISED HUGHES MODELS

The Rome Air Development Center (RADC) (now Rome Laboratory) [1; 2] revised the screen strength equations developed by Hughes, based on the data in [3] and [4]. The given general form of the screening strength equation, expressed as a function of screening time, T_s, is

$$SS(T_s) = 1 - e^{-\overline{\lambda}_D T_s}, \tag{11.6}$$

where

$SS(T_s) = $ screening strength as a function of screening time, T_s,

and

$\overline{\lambda}_D = $ average failure rate of defects under a given
set of stresses.

1. **Screening strength for random vibration**

   The given screening strength equation for random vibration [2] is

   $$SS_{RV}(T_{RV}) = 1 - e^{-0.0046\,(G)^{1.71}\,T_{RV}}, \qquad (11.7)$$

   where

   > $G = g_{rms}$, rms value of applied acceleration power spectral density over the frequency spectrum,

   and

   > T_{RV} = duration of applied vibration excitation, in min.

   Equation (11.7) is tabulated in Table 11.1 for various combinations of G and T_{RV}.

2. **Screening strength for swept-sine vibration**

   The given stress screening equation for swept-sine vibration [2] is

   $$SS_{SSV}(T_{SSV}) = 1 - e^{-0.000727\,(G)^{0.863}\,T_{SSV}}, \qquad (11.8)$$

   where

   > G = g-level, the constant acceleration applied to the equipment being screened throughout the frequency range above 40 Hz. The g-level below 40 Hz may be less,

   and

   > T_{SSV} = duration of swept-sine vibration, in min.

3. **Screening strength for single-(fixed)-frequency (sine) vibration**

   The given screening strength equation for single-frequency vibration [2] is

   $$SS_{SFV}(T_{SFV}) = 1 - e^{-0.00047\,(G)^{0.49}\,T_{SFV}}, \qquad (11.9)$$

   where

   > G = g-level of the applied constant acceleration,

   and

   > T_{SFV} = duration of single-frequency vibration, in min.

4. **Screening strength for temperature cycling**

   The given stress screening equation for temperature cycle [2] is

   $$SS_{TC}(N_{TC}) = 1 - e^{-0.0017(T_R^* + 0.6)^{0.6}\,[\log_e(e + \dot{T}^*)]^3\,N_{TC}}, \qquad (11.10)$$

TABLE 11.1– **Screening strengths for random vibration screens evaluated using the RADC model.**

Duration, T_{RV}, min	rms value of applied acceleration spectral density, G, in $g's$						
	0.5	1.0	1.5	2.0	2.5	3.0	3.5
5	0.007	0.023	0.045	0.072	0.104	0.140	0.178
10	0.014	0.045	0.088	0.140	0.198	0.260	0.324
15	0.021	0.067	0.129	0.202	0.282	0.363	0.444
20	0.028	0.088	0.168	0.260	0.356	0.452	0.543
25	0.035	0.109	0.206	0.314	0.424	0.529	0.625
30	0.041	0.129	0.241	0.363	0.484	0.595	0.691
35	0.048	0.149	0.275	0.409	0.538	0.651	0.746
40	0.055	0.168	0.308	0.452	0.586	0.700	0.791
45	0.061	0.187	0.339	0.492	0.629	0.742	0.829
50	0.068	0.205	0.369	0.529	0.668	0.778	0.859
55	0.074	0.224	0.397	0.563	0.702	0.809	0.884
60	0.081	0.241	0.424	0.595	0.734	0.836	0.905
Duration, T_{RV}, min	rms value of applied acceleration spectral density, G, in $g's$						
	4.0	4.5	5.0	5.5	6.0	6.5	7.0
5	0.218	0.260	0.303	0.346	0.389	0.431	0.473
10	0.389	0.452	0.514	0.572	0.627	0.677	0.723
15	0.522	0.595	0.661	0.720	0.772	0.816	0.854
20	0.626	0.700	0.764	0.817	0.861	0.896	0.923
25	0.708	0.778	0.835	0.880	0.915	0.941	0.959
30	0.772	0.836	0.885	0.922	0.948	0.966	0.979
35	0.822	0.878	0.920	0.949	0.968	0.981	0.989
40	0.860	0.910	0.944	0.966	0.981	0.989	0.994
45	0.891	0.933	0.961	0.978	0.988	0.994	0.997
50	0.915	0.951	0.973	0.986	0.993	0.996	0.998
55	0.933	0.964	0.981	0.991	0.996	0.998	0.999
60	0.948	0.973	0.987	0.994	0.997	0.999	1.000

where

$$T_R^* = \text{temperature range} = T_{max}^* - T_{min}^*, \text{ in } °C,$$
$$N_{TC} = \text{number of temperature cycles},$$
$$e = 2.71828,$$
$$\dot{T}^* = \text{temperature rate of change},$$

or

$$\dot{T}^* = \frac{1}{2}\left[\left(\frac{T_{max}^* - T_{min}^*}{t_1}\right) + \left(\frac{T_{max}^* - T_{min}^*}{t_2}\right)\right],$$

$t_1 = $ transition time from T_{min} to T_{max}, for the heating
segment, in min,

and

$t_2 = $ transition time from T_{max} to T_{min}, for the cooling
segment, in min.

Equation (11.10) is tabulated in Table 11.2 for various combinations of T_R^*, $\dot{T}^*$ and N_{TC}.

5. **Screening strength for constant temperature burn-in**

The given stress screening equation for constant temperature burn-in
[2] is

$$SS_{CT}(T_b) = 1 - e^{-0.0017\,(T_R^* + 0.6)^{0.6}\,T_b}, \tag{11.11}$$

where

$T_R^* = $ temperature range defined as the absolute value
of the difference between the screening
temperature and 25°C,

and

$T_b = $ burn-in time in hr.

11.3 AVERAGE DEFECT FAILURE RATE ESTIMATION UNDER DIFFERENT SCREENING STRESSES

According to Eq. (11.6), which gives the general form of the screening
strength as a function of the screening time, or cycles, and the average defect

TABLE 11.2– Screening strengths for thermal cycling screens evaluated using the RADC model.

Number of cycles, N_{TC}	Rate of change, $\dot{T}^*$, °C/min	Temperature range, T_R^*, °C								
		20	40	60	80	100	120	140	160	180
2	5	0.1633	0.2349	0.2886	0.3324	0.3697	0.4023	0.4312	0.4572	0.4809
	10	0.2907	0.4031	0.4812	0.5410	0.5891	0.6290	0.6629	0.6920	0.7173
	15	0.3911	0.5254	0.6124	0.6752	0.7232	0.7612	0.7920	0.8175	0.8388
	20	0.4707	0.6155	0.7034	0.7636	0.8075	0.8407	0.8665	0.8871	0.9037
4	5	0.2998	0.4147	0.4939	0.5543	0.6027	0.6427	0.6765	0.7054	0.7305
	10	0.4969	0.6437	0.7308	0.7893	0.8312	0.8624	0.8863	0.9051	0.9201
	15	0.6292	0.7748	0.8498	0.8945	0.9234	0.9430	0.9567	0.9667	0.9740
	20	0.7198	0.8522	0.9120	0.9441	0.9629	0.9746	0.9822	0.9873	0.9907
6	5	0.4141	0.5522	0.6400	0.7025	0.7496	0.7864	0.8160	0.8401	0.8601
	10	0.6431	0.7873	0.8603	0.9033	0.9306	0.9489	0.9617	0.9708	0.9774
	15	0.7742	0.8931	0.9418	0.9657	0.9788	0.9864	0.9910	0.0.9939	0.9958
	20	0.8517	0.9432	0.9739	0.9868	0.9929	0.9960	0.9976	0.9986	0.9991
8	5	0.5098	0.6574	0.7439	0.8014	0.8422	0.8723	0.8953	0.9132	0.9274
	10	0.7469	0.8731	0.9275	0.9556	0.9715	0.9811	0.9871	0.9910	0.9936
	15	0.8625	0.9493	0.9774	0.9889	0.9941	0.9967	0.9981	0.9989	0.9993
	20	0.9215	0.9781	0.9923	0.9969	0.9986	0.9994	0.9997	0.9998	0.9999
10	5	0.5898	0.7379	0.8178	0.8674	0.9005	0.9237	0.9405	0.9529	0.9623
	10	0.8204	0.9242	0.9624	0.9796	0.9883	0.9930	0.9956	0.9972	0.9982
	15	0.9163	0.9759	0.9913	0.9964	0.9984	0.9992	0.9996	0.9998	0.9999
	20	0.9585	0.9916	0.9977	0.9993	0.9997	0.9999	0.9999	0.9999	0.9999
12	5	0.6568	0.7994	0.8704	0.9115	0.9373	0.9544	0.9661	0.9744	0.9804
	10	0.8726	0.9548	0.9805	0.9906	0.9952	0.9974	0.9985	0.9991	0.9995
	15	0.9490	0.9886	0.9966	0.9988	0.9996	0.9998	0.9999	0.9999	0.9999
	20	0.9780	0.9968	0.9993	0.9998	0.9999	0.9999	0.9999	0.9999	0.9999

failure rate, the average defect failure rate, $\overline{\lambda}$, is given by

$$\overline{\lambda} = \frac{1}{T_s} \log_e \left(\frac{1}{1 - SS} \right), \tag{11.12}$$

where

T_s = screening time or number of cycles,

and

SS = screening strength.

Substitution of the screening strength Eqs. (11.2) through (11.11) into Eq. (11.12) yields the following average defect failure rate models:

11.3.1   THE HUGHES MODELS

1. **Average defect failure rate during random vibration**

   The given average defect failure rate during random vibration is

   $$\overline{\lambda}_{RV} = \frac{1}{T_{RV}} \log_e \left[\frac{1}{1 - D \left(1 - e^{-\sqrt{T_{RV}/B}} \right)} \right], \tag{11.13}$$

   where

   T_{RV} = duration of applied vibration excitation, in min,

   $D = 0.144\, g - 0.0862$,

   $B = 0.266\, G + 1.402$,

   and

   $G = g_{rms}$, rms value of applied acceleration power spectral density over the frequency spectrum.

   Similar to Eq. (11.2), Eq. (11.13) is applicable for

   $0 \leq G \leq 6\, g$,

   and

   $0 \leq T_{RV} \leq 45$ min.

2. **Average defect failure rate during thermal cycling**

   The given average defect failure rate during temperature cycling is

   $$\overline{\lambda}_{TC} = \frac{1}{N_{TC}} \log_e \left[\frac{1}{1 - 0.85 \left(1 - e^{-A} \right)} \right], \tag{11.14}$$

where

$$A = 0.0023 \, (T_R^*)^{0.6} \sqrt{N_{TC}} \, [\log_e(e + \dot{T})]^{2.7},$$

$$T_R^* = \text{temperature range} = T_{max}^* - T_{min}^*, \text{ in } °C,$$

$$N_{TC} = \text{number of temperature cycles},$$

$$e = 2.71828,$$

$$\dot{T}^* = \text{average rate of temperature change},$$
$$\text{or temperature gradient, in } °C/\text{min},$$

or

$$\dot{T}^* = \frac{1}{2}\left[\left(\frac{T_{max}^* - T_{min}^*}{t_1}\right) + \left(\frac{T_{max}^* - T_{min}^*}{t_2}\right)\right],$$

$$t_1 = \text{transition time from } T_{min} \text{ to } T_{max}, \text{ in min},$$

and

$$t_2 = \text{transition time from } T_{max} \text{ to } T_{min}, \text{ in min}.$$

Similar to Eq. (11.3), Eq. (11.14) is applicable for

$$-55°C \leq T_{min} \leq 25°C,$$

$$25°C \leq T_{max} \leq 75°C,$$

$$1°C/\text{min} \leq \dot{T} \leq 20°C/\text{min},$$

and

$$0 \leq N_{TC} \leq 20 \text{ cycles}.$$

3. **Average defect failure rate during constant temperature burn-in**

The given average defect failure rate during constant temperature burn-in is

$$\bar{\lambda}_b = \frac{1}{T_b} \log_e \left[\frac{1}{1 - 0.85 \, (1 - e^{-A'})}\right], \qquad (11.15)$$

where

$$A' = 0.0048 \, (T_R^*)^{0.6},$$

$$T_R^* = \text{absolute temperature difference between}$$
$$\text{the burn-in temperature and } 25°C,$$

$$T_R^* = |T^* - 25|,$$

$$T^* = \text{burn-in temperature, in } °C,$$

and

$$T_b = \text{burn-in time, in hr}.$$

Note that for $-1 < T_R^* < +1$, set $(T_R^*)^{0.6} = 1$. Similar to Eq. (11.5), Eq. (11.15) is applicable for

$$-55°C \leq T^* \leq 75°C.$$

11.3.2   RADC (ROME LABORATORY) MODELS

1. **Average defect failure rate during random vibration**

   The given average defect failure rate during random vibration is

   $$\overline{\lambda}_{RV} = 0.0046 \, G^{1.71}, \tag{11.16}$$

   where

   > $G = g_{rms}$, rms value of applied acceleration power spectral density over the frequency spectrum.

2. **Average defect failure rate during swept-sine vibration**

   The given average defect failure rate during swept-sine vibration is

   $$\overline{\lambda}_{SSV} = 0.000727 \, G^{0.863}, \tag{11.17}$$

   where

   > $G =$ g-level, the constant acceleration applied to the equipment being screened throughout the frequency range above 40 Hz. The g-level below 40 Hz may be less.

3. **Average defect failure rate during single-(fixed)-frequency (sine) vibration**

   The given average defect failure rate during single-frequency vibration is

   $$\overline{\lambda}_{SFV} = 0.00047 \, G^{0.49}, \tag{11.18}$$

   where

   > $G =$ g-level of the applied constant acceleration.

4. **Average defect failure rate during temperature cycling**

   The given average defect failure rate during temperature cycling is

   $$\overline{\lambda}_{TC} = 0.0017(T_R^* + 0.6)^{0.6} \, [\log_e(e + \dot{T}^*)]^3,$$

   $$\tag{11.19}$$

where

$$T_R^* = \text{temperature range} = T_{max}^* - T_{min}^*, \text{ in } °C,$$
$$e = 2.71828,$$
$$\dot{T}^* = \text{temperature rate of change},$$

or

$$\dot{T}^* = \frac{1}{2}\left[\left(\frac{T_{max}^* - T_{min}^*}{t_1}\right) + \left(\frac{T_{max}^* - T_{min}^*}{t_2}\right)\right],$$

$$t_1 = \text{transition time from } T_{min} \text{ to } T_{max}, \text{ in min},$$

and

$$t_2 = \text{transition time from } T_{max} \text{ to } T_{min}, \text{ in min}.$$

5. **Average defect failure rate during constant temperature burn-in**

The given average defect failure rate during constant temperature burn-in is

$$\overline{\lambda}_b = 0.0017\,(T_R^* + 0.6)^{0.6}, \tag{11.20}$$

where

$$T_R^* = \text{temperature range defined as the absolute value}$$
$$\text{of the difference between the screening}$$
$$\text{temperature and } 25°C.$$

11.3.3 OTHER MODELS

In [5], other empirical equations are given for the average defect failure rate estimation under thermal cycling and random vibration screens. They are summarized as follows:

1. **Average defect failure rate under thermal cycling screening:**

$$\overline{\lambda}_{TC} = \left[\left(\frac{4,280}{MTBF_I}\right) \cdot C_1 \cdot LF \cdot (0.0208\,T_{max}^* - 0.46)\right.$$
$$\left.(0.06\,\dot{T}^* + 0.7) - C_2\right]\left(1 - e^{-0.219N_{TC}}\right)$$
$$\frac{(T_{on} \cdot LF \cdot N_{TC})}{MTBF_I} - 0.008 \cdot LF \cdot N_{sru},$$

$$\tag{11.21}$$

where

$MTBF_I$ = inherent $MTBF$ based on the MIL-HDBK-217 prediction,

C_1 = first calibration factor for the dedication of the reliability program,

C_1 = 0.2,

C_2 = second calibration factor for the dedication of the reliability program,

C_2 = 0.0745,

LF = learning factor from MIL-HDBK-217 ,

T_{max}^* = upper limit of chamber temperature during thermal cycling,

$\dot{T}^*$ = rate of change of chamber temperature, °C/min, as given in Eq. (11.4),

N_{TC} = number of thermal cycles,

T_{on} = power-on time per cycle,

and

N_{sru} = number of shop replaceable units (SRU's) in the system.

2. **Average defect failure rate under random vibration screening:**

$$\overline{\lambda}_{RV} = C_3 \left(\frac{N_{parts}}{1,900}\right) (0.0926\ G + 0.008)\ (0.025\ T_{RV} + 0.625)$$
$$\cdot(1.1\ e^{-0.03\ N_{sys}} + 0.1)\ (1 - 0.75\ I)\ N_{axis},$$

$$(11.22)$$

where

C_3 = calibration factor for degree of built-in test (BIT) fault detection,

C_3 = 0.8875,

N_{parts} = number of parts in the system,

G = g_{rms}, rms value of applied acceleration power spectral density over the frequency spectrum,

T_{RV} = total time of applied vibration excitation, in min,

N_{sys} = number of systems already built,

I = indicating number as to whether the test is monitored or not while vibrating,

$$I = \begin{cases} 0, \text{ monitored}, \\ 1, \text{ not monitored}, \end{cases}$$

and

N_{axis} = number of axes in vibration ranging from 1 to 3.

11.4 MODEL APPLICABILITY

The empirical equations for screening strength and average defect failure rate estimation, developed by industry, are summarized in this chapter. These models were based on the data available at the time they were developed for selected products. Therefore, they are dependent on the failure mechanisms of the specific product screened, the screening equipment and their setup, and the stress profiles applied. They are not universally applicable since the parameters, or the regression coefficients, in these models are obtained from the regression analysis of the specific screening results which may behave differently than your own data.

Then, the question arises as how to use these models in practical situations. The following are some recommendations:

CASE 1 – Historical data are available.

Step 1– Select an empirical model for your screen effectiveness evaluation, say, the Hughes model or the RADC model for the screening strengths of thermal cycling and random vibration.

Step 2– Check whether or not the selected models fit your observed data satisfactorily. This can be done by substituing the relevant parameters of the screen profiles and the screen durations into the model, and comparing the calculated screening strengths with the corresponding observed screening strengths.

If the selected model tracts the observed data very well; i.e., the observed data points fluctuate very closely about the model-calculated values, then the selected model can be used for current or short-term prediction but will be subject to future modification as more data are accumulated.

If the selected model does not tract the observed data satisfactorily, then either another model may be selected and its goodness-of-fit checked or a modification may be made to the presently selected model which is described in Step 3.

Step 3– Modify, if desired according to Step 2, the selected model. If Step 2 indicates an unsatisfactory goodness-of-fit, then it may be modified assuming that one or more of the constants in the model are unknown and need to be determined from the data.

For example, the RADC model for random vibration screen strength given by Eq. (11.7) can be rewritten as

$$SS_{RV}(T_{RV}) = 1 - e^{-a_1\ G^{b_1}\ T_{RV}^{c_1}}, \tag{11.23}$$

where a_1, b_1 and c_1 are unknown constants to be determined from the actual observed screen strength data.

Similarly, the RADC screen strength models for swept-sine vibration, for single-frequency vibration, for temperature cycling and for constant temperature burn-in, given by Eqs. (11.8) through (11.11), respectively, can be rewritten as

$$SS_{SSV}(T_{SSV}) = 1 - e^{-a_2\ G^{b_2}\ T_{SSV}^{c_2}}, \tag{11.24}$$

$$SS_{SFV}(T_{SFV}) = 1 - e^{-a_3\ G^{b_3}\ T_{SFV}^{c_3}}, \tag{11.25}$$

$$SS_{TC}(N_{TC}) = 1 - e^{-a_4\ (T_R^*+0.6)^{b_4}\ [\log_e(e+\dot{T}^*)]^{c_4}\ N_{TC}^{d_4}}, \tag{11.26}$$

and

$$SS_{CT}(T_b) = 1 - e^{-a_5\ (T_R^*+0.6)^{b_5}\ T_b^{c_5}}, \tag{11.27}$$

respectively, where (a_2, b_2, c_2), (a_3, b_3, c_3), (a_4, b_4, c_4, d_4) and (a_5, b_5, c_5) are unknown constants to be estimated from the corresponding observed screening strength data.

The least-squares estimates can be obtained for (a_1, b_1, c_1), (a_2, b_2, c_2), (a_3, b_3, c_3), (a_4, b_4, c_4, d_4) and (a_5, b_5, c_5) by linearizing Eqs. (11.23) through (11.27) first and then conducting a multiple linear regression analysis; i.e., Eq. (11.23) may be rewritten as

$$Y_{RV} = \log_e \log_e \frac{1}{1 - SS_{RV}(T_{RV})},$$
$$= \log_e a_1 + b_1\ \log_e G + c_1\ \log_e T_{RV},$$

or

$$Y_{RV} = a_1' + b_1\ G' + c_1\ T_{RV}'. \tag{11.28}$$

Similarly, Eq. (11.24) becomes

$$Y_{SSV} = \log_e \log_e \frac{1}{1 - SS_{SSV}(T_{SSV})},$$
$$= \log_e a_2 + b_2\ \log_e G + c_2\ \log_e T_{SSV},$$

or

$$Y_{SSV} = a'_2 + b_2\, G' + c_2\, T'_{SSV},$$ (11.29)

Eq. (11.25) becomes

$$Y_{SFV} = \log_e \log_e \frac{1}{1 - SS_{SFV}(T_{SFV})},$$
$$= \log_e a_3 + b_3\, \log_e G + c_3\, \log_e T_{SFV},$$

or

$$Y_{SFV} = a'_3 + b_3\, G' + c_3\, T'_{SFV},$$ (11.30)

Eq. (11.26) becomes

$$Y_{TC} = \log_e \log_e \frac{1}{1 - SS_{TC}(N_{TC})},$$
$$= \log_e a_4 + b_4\, \log_e(T_R^* + 0.6)$$
$$+ c_4\, \log_e[\log_e(e + \dot{T}^*)] + d_4\, \log_e N_{TC},$$

or

$$Y_{TC} = a'_4 + b_4\, (T_R^*)' + c_4\, (\dot{T}^*)' + d_4\, N'_{TC},$$ (11.31)

and Eq. (11.27) becomes

$$Y_{CT} = \log_e \log_e \frac{1}{1 - SS_{CT}(T_b)},$$
$$= \log_e a_5 + b_5\, \log_e(T_R^* + 0.6) + c_5\, \log_e T_b,$$

or

$$Y_{CT} = a'_5 + b_5\, (T_R^*)' + c_5\, T'_b,$$ (11.32)

where

$$Y_{RV} = \log_e \log_e \frac{1}{1 - SS_{RV}(T_{RV})},$$
$$Y_{SSV} = \log_e \log_e \frac{1}{1 - SS_{SSV}(T_{SSV})},$$
$$Y_{SFV} = \log_e \log_e \frac{1}{1 - SS_{SFV}(T_{SFV})},$$
$$Y_{TC} = \log_e \log_e \frac{1}{1 - SS_{TC}(N_{TC})},$$
$$Y_{CT} = \log_e \log_e \frac{1}{1 - SS_{CT}(T_b)},$$

$$a_i' = \log_e a_i, \text{ for } i = 1, 2, 3, 4, 5,$$
$$G' = \log_e G,$$
$$T_{RV}' = \log_e T_{RV},$$
$$T_{SSV}' = \log_e T_{SSV},$$
$$T_{SFV}' = \log_e T_{SFV},$$
$$(T_R^*)' = \log_e (T_R^* + 0.6),$$
$$(\dot{T}^*)' = \log_e (e + \dot{T}^*),$$
$$N_{TC}' = \log_e N_{TC},$$

and

$$T_b' = \log_e T_b.$$

Equations (11.28) through (11.32) can be represented by the following general linear equation:

$$Y = A_0 + A_1\,X_1 + A_2\,X_2 + A_3\,X_3 + \cdots + A_k\,X_k, \qquad (11.33)$$

where

Y = dependent variable,
k = number of independent variables,
A_i = ith regression coefficient, $i = 0, 1, 2, 3, \cdots, k$,

and

X_i = ith independent variable or ith regressor variable,
     $i = 1, 2, \cdots, k.$

The least-squares estimate of A_i, $\hat{A}_i$, for $i = 0, 1, 2, \cdots, k$, for a data set of n observations $(X_{11}, X_{21}, \cdots, X_{k1}; Y_1)$, $(X_{12}, X_{22}, \cdots, X_{k2}; Y_2)$, $\cdots$, $(X_{1n}, X_{2n}, \cdots, X_{kn}; Y_n)$, can be obtained using the following equation [6, p. 501]:

$$\hat{\mathbf{A}} = (\mathbf{X}^T\,\mathbf{X})^{-1}\,(\mathbf{X}^T\,\mathbf{Y}), \qquad (11.34)$$

where

$\hat{\mathbf{A}}$ = $(k+1) \times 1$ matrix for least-squares estimates of
    correlation coefficients, $\hat{A}_0, \hat{A}_1, \cdots, \hat{A}_k$, or

$$\hat{\mathbf{A}} = \begin{bmatrix} \hat{A}_0' \\ \hat{A}_1 \\ \cdot \\ \cdot \\ \cdot \\ \hat{A}_k \end{bmatrix}_{(k+1) \times 1},$$

$$\hat{A}'_0 = \hat{A}_0 + \sum_{i=1}^{k} \hat{A}_i \overline{X}_i,$$

$\mathbf{X} = n \times (k+1)$ matrix for observed values of the independent variables,

$$\mathbf{X} = \begin{bmatrix} 1 & (X_{11} - \overline{X_1}) & (X_{21} - \overline{X_2}) & \cdots & (X_{k1} - \overline{X_k}) \\ 1 & (X_{12} - \overline{X_1}) & (X_{22} - \overline{X_2}) & \cdots & (X_{k2} - \overline{X_k}) \\ \cdot & \cdot & \cdot & & \cdot \\ \cdot & \cdot & \cdot & & \cdot \\ \cdot & \cdot & \cdot & & \cdot \\ 1 & (X_{1n} - \overline{X_1}) & (X_{2n} - \overline{X_2}) & \cdots & (X_{kn} - \overline{X_k}) \end{bmatrix}_{n \times (k+1)},$$

$\mathbf{X}^T = $ transpose of matrix $\mathbf{X}$, a $(k+1) \times n$ matrix,

$$\mathbf{X}^T = \begin{bmatrix} 1 & 1 & \cdots & 1 \\ (X_{11} - \overline{X_1}) & (X_{12} - \overline{X_1}) & \cdots & (X_{1n} - \overline{X_1}) \\ (X_{21} - \overline{X_2}) & (X_{22} - \overline{X_2}) & \cdots & (X_{2n} - \overline{X_2}) \\ \cdot & \cdot & & \cdot \\ \cdot & \cdot & & \cdot \\ \cdot & \cdot & & \cdot \\ (X_{k1} - \overline{X_k}) & (X_{k2} - \overline{X_k}) & \cdots & (X_{kn} - \overline{X_k}) \end{bmatrix}_{(k+1) \times n},$$

$\mathbf{Y} = n \times 1$ matrix for observed values of the dependent variable,

$$\mathbf{Y} = \begin{bmatrix} Y_1 \\ Y_2 \\ \cdot \\ \cdot \\ \cdot \\ Y_n \end{bmatrix}_{n \times 1},$$

$X_{ij} = j$th observation of X_i, $i = 1, 2, \cdots, k$, $j = 1, 2, \cdots, n$,

$Y_j = j$th observation of Y, $j = 1, 2, \cdots, n$,

$\overline{X_i} = $ mean value of n observations of X_i, $i = 1, 2, \cdots, k$,

$$\overline{X_i} = \frac{1}{n} \sum_{j=1}^{n} X_{ij}, i = 1, 2, \cdots, k,$$

and

$(\mathbf{X}^T \mathbf{X})^{-1} = $ inverse matrix of $(\mathbf{X}^T \mathbf{X})_{(k+1) \times (k+1)}$.

Manual evaluation of $(\mathbf{X}^T \mathbf{X})$, $(\mathbf{X}^T \mathbf{X})^{-1}$, $(\mathbf{X}^T \mathbf{Y})$ and Eq. (11.34) can be very tedious even for small-sized matrices $\mathbf{X}$ and $\mathbf{Y}$. For example, given are the following two matrices,

$$\mathbf{X} = \begin{bmatrix} 1 & (X_{11} - \overline{X_1}) \\ 1 & (X_{12} - \overline{X_1}) \\ 1 & (X_{13} - \overline{X_1}) \end{bmatrix}_{3 \times (1+1)} = \begin{bmatrix} 1 & 0.45 \\ 1 & -0.66 \\ 1 & 0.21 \end{bmatrix}_{3 \times 2},$$

and

$$\mathbf{Y} = \begin{bmatrix} Y_1 \\ Y_2 \\ Y_3 \end{bmatrix}_{3\times1} = \begin{bmatrix} 0.36 \\ 0.51 \\ 0.77 \end{bmatrix}_{3\times1},$$

where the number of observations is only three, or $n = 3$, and the number of independent variables is only one, or $k = 1$. Then,

$$\mathbf{X}^T = \begin{bmatrix} 1 & 1 & 1 \\ 0.45 & -0.66 & 0.21 \end{bmatrix}_{2\times3},$$

$$\mathbf{X}^T\,\mathbf{X} = \begin{bmatrix} 1 & 1 & 1 \\ 0.45 & -0.66 & 0.21 \end{bmatrix}_{2\times3} \cdot \begin{bmatrix} 1 & 0.45 \\ 1 & -0.66 \\ 1 & 0.21 \end{bmatrix}_{3\times2},$$

or

$$\mathbf{X}^T\,\mathbf{X} = \begin{bmatrix} 1\times1+1\times1+1\times1 & 1\times0.45+1\times(-0.66)+1\times0.21 \\ 0.45\times1+(-0.66)\times1+0.21\times1 & 0.45^2+(-0.66)^2+0.21^2 \end{bmatrix}_{2\times2},$$

or

$$\mathbf{X}^T\,\mathbf{X} = \begin{bmatrix} 3 & 0 \\ 0 & 0.6822 \end{bmatrix}_{2\times2},$$

and

$$(\mathbf{X}^T\,\mathbf{X})^{-1} = \frac{1}{det(\mathbf{X}^T\,\mathbf{X})} \begin{bmatrix} (-1)^{1+1}\times0.6822 & (-1)^{1+2}\times0 \\ (-1)^{2+1}\times0 & (-1)^{2+2}\times3 \end{bmatrix}^T_{2\times2},$$

where

$$det(\mathbf{X}^T\,\mathbf{X}) = \text{determinant of } (\mathbf{X}^T\,\mathbf{X}),$$
$$= \begin{vmatrix} 3 & 0 \\ 0 & 0.6822 \end{vmatrix}$$
$$= 3\times0.6822 - 0\times0,$$
$$= 2.0466.$$

Then,

$$(\mathbf{X}^T\,\mathbf{X})^{-1} = \frac{1}{2.0466} \begin{bmatrix} 0.6822 & 0 \\ 0 & 3 \end{bmatrix}^T_{2\times2},$$
$$= \frac{1}{2.0466} \begin{bmatrix} 0.6822 & 0 \\ 0 & 3 \end{bmatrix}_{2\times2},$$
$$= \begin{bmatrix} 0.3333 & 0 \\ 0 & 1.4658 \end{bmatrix}_{2\times2}.$$

Also

$$\mathbf{X}^T \, \mathbf{Y} = \begin{bmatrix} 1 & 1 & 1 \\ 0.45 & -0.66 & 0.21 \end{bmatrix}_{2 \times 3}$$

$$\cdot \begin{bmatrix} 0.36 \\ 0.51 \\ 0.77 \end{bmatrix}_{3 \times 1},$$

$$= \begin{bmatrix} 1 \times 0.36 + 1 \times 0.51 + 1 \times 0.77 \\ 0.45 \times 0.36 + (-0.66) \times 0.51 + 0.21 \times 0.77 \end{bmatrix}_{2 \times 1},$$

$$= \begin{bmatrix} 1.64 \\ -0.0129 \end{bmatrix}_{2 \times 1}.$$

Finally,

$$(\mathbf{X}^T \, \mathbf{Y})^{-1} \, (\mathbf{X}^T \, \mathbf{Y}) = \begin{bmatrix} 0.3333 & 0 \\ 0 & 1.4658 \end{bmatrix}_{2 \times 2}$$

$$\cdot \begin{bmatrix} 1.64 \\ -0.0129 \end{bmatrix}_{2 \times 1},$$

$$= \begin{bmatrix} 0.3333 \times 1.64 + 0 \times (-0.0129) \\ 0 \times 1.64 + 1.4658 \times (-0.0129) \end{bmatrix}_{2 \times 1},$$

or

$$(\mathbf{X}^T \, \mathbf{Y})^{-1} \, (\mathbf{X}^T \, \mathbf{Y}) = \begin{bmatrix} 0.5467 \\ -0.0189 \end{bmatrix}_{2 \times 1}.$$

It may be seen that even for such a simple case when $(n = 3, k = 1)$ evaluating Eq. (11.34) manually is still quite time consuming. If the number of observations, n, and the number of independent variables, k, become large, this work may become impossible. In this case, a computer software, such as MATHCAD, may be needed to evaluate Eq. (11.34) quickly and correctly.

As a special case, when there is only one independent variable in Eq. (11.33), or $k = 1$, Eq. (11.33) becomes

$$Y = A_0 + A_1 \, X_1. \tag{11.35}$$

The least-squares estimates for A_0 and A_1 for a data set of size n, (X_{11}, Y_1), (X_{12}, Y_2), $\cdots$, (X_{1n}, Y_n), can be obtained directly from the following two equations:

$$\begin{cases} \hat{A}_0 = \overline{Y} - \hat{A}_1 \, \overline{X_1}, \\ \hat{A}_1 = \frac{L_{xy}}{L_{xx}}, \end{cases} \tag{11.36}$$

where

$$\overline{X_1} = \frac{1}{n} \sum_{j=1}^{n} X_{1j},$$

$$\overline{Y} = \frac{1}{n}\sum_{j=1}^{n} Y_j,$$

$$L_{XY} = \sum_{j=1}^{n}(X_{1j} - \overline{X_1})(Y_j - \overline{Y}),$$

and

$$L_{XX} = \sum_{j=1}^{n}(X_{1j} - \overline{X_1})^2.$$

Presented next are five examples illustrating the applications of the techniques discussed earlier. All the matrix operations are accomplished using the MATHCAD software.

EXAMPLE 11–1

An aircraft company is conducting a screening strength evaluation on its random vibration stress screens. The root-mean-square (rms) value of the applied acceleration power spectral density is 6 $g's$. The screening strengths for the various vibration durations are found to be as follows:

Duration, T_{RV}, min	10	20	30	40	50	60
Screening strength, SS_{RV}	0.605	0.810	0.906	0.970	0.986	0.990

Do the following:

1. Check whether or not the RADC model for random vibration screening strength fits the data.

2. Determine the optimum screening strength equation for this screen using the least-squares method.

SOLUTIONS TO EXAMPLE 11–1

1. The RADC model of screening strength for random vibration screens is given by Eq. (11.7), or

$$SS_{RV}(T_{RV}) = 1 - e^{-0.0046\,(G)^{1.71}\,T_{RV}}, \tag{11.37}$$

where $G = 6g$ in this case. Then,

$$SS_{RV}(T_{RV}) = 1 - e^{-0.0046\,(6)^{1.71}\,T_{RV}},$$

TABLE 11.3– The observed and calculated screening strengths for a random vibration screen with $G = 6g$ for Example 11–1.

T_{RV}, min	10	20	30	40	50	60
Observed SS_{RV} values	0.605	0.810	0.906	0.970	0.986	0.990
Calculated SS_{RV} values using the RADC model	0.627	0.861	0.948	0.981	0.993	0.997
Calculated SS_{RV} values using the fitted model	0.594	0.822	0.919	0.963	0.983	0.992

or

$$SS_{RV}(T_{RV}) = 1 - e^{-0.0985\, T_{RV}}. \tag{11.38}$$

The calculated SS_{RV} values for $T_{RV} = 10, 20, \cdots, 60$ min using this equation are tabulated, together with the actual observed SS_{RV} values, in Table 11.3.

It may be seen that the RADC model for random vibration is overestimating the screening strength for this particular case. Therefore, it needs to be modified accordingly.

2. Since the *rms* value of the applied acceleration spectral density function is fixed at $G = 6g$ and only the vibration duration, T_{RV}, is varying, or $k = 1$, then Eq. (11.38) can be modified into the following generalized form:

$$SS_{RV}(T_{RV}) = 1 - e^{-a\, T_{RV}^b}, \tag{11.39}$$

where a and b are two unknown constants to be determined from the observed data using the least squares method.

Linearizing Eq. (11.39) yields

$$Y = \log_e[-\log_e(1 - SS_{RV})] = \log_e a + b\, \log_e T_{RV},$$

or

$$Y = a' + b\, X,$$

where

$$Y = \log_e[-\log_e(1 - SS_{RV})],$$
$$a' = \log_e a, \tag{11.40}$$

ESS EMPIRICAL EQUATIONS

TABLE 11.4– The calculation summaries for the least-squares estimation for Example 11–1.

j	1	2	3	4	5	6	$\sum$
$(T_{RV})_j$	10	20	30	40	50	60	
X_j [†,*]	2.303	2.996	3.401	3.689	3.912	4.094	20.395
$X_j - \overline{X}$	-1.096	-0.403	0.002	0.290	0.513	0.695	
$(X_j - \overline{X})^2$	1.201	0.162	4×10^{-6}	0.084	0.263	0.483	$L_{XX} = 2.193$
$(SS_{RV})_j$	0.605	0.810	0.906	0.970	0.986	0.990	
Y_j [‡,**]	-0.074	0.507	0.861	1.255	1.451	1.527	5.527
$Y_j - \overline{Y}$	-0.995	-0.414	-0.006	0.334	0.530	0.606	
$(X_j - \overline{X})(Y_j - \overline{Y})$	1.091	0.167	0.00012	0.097	0.272	0.421	$L_{XY} = 2.048$

† $X_j = \log_e(T_{RV})_j$;   * Mean $\overline{X} = 3.399$.
‡ $Y_j = \log_e\{-\log_e[1 - (SS_{RV})_j]\}$;   ** Mean $\overline{Y} = 0.921$.

and

$$X = \log_e T_{RV}.$$

The least-squares estimates of a' and b are obtained using Eq. (11.36) as follows:

$$\overline{X} = 3.399,$$
$$\overline{Y} = 0.921,$$
$$L_{XX} = 2.193,$$
$$L_{XY} = 2.048,$$
$$\hat{b} = \frac{L_{XY}}{L_{XX}} = \frac{2.048}{2.193} = 0.934,$$

and

$$\hat{a}' = \overline{Y} - \hat{b}\,\overline{X} = 0.921 - 0.934 \times 3.399 = -2.254.$$

Then, according to Eq. (11.40),

$$\hat{a} = e^{\hat{a}'} = e^{-2.254} = 0.105.$$

The associated calculations are summarized in Table 11.4. Therefore, the optimum screening strength equation using the least-squares method is given by

$$SS_{RV}(T_{RV}) = 1 - e^{-\hat{a}\,T_{RV}^{\hat{b}}} = 1 - e^{-0.105\,T_{RV}^{0.934}}. \tag{11.41}$$

The screening strength values for $T_{RV} = 10, 20, \cdots, 60$ min are caculated using this model and tabulated in the last row of Table 11.3. Both Eq. (11.38) and Eq. (11.41) are plotted together in Fig. 11.1 against the observed raw data. It may be seen that Eq. (11.41) fits the observed data much better than the RADC model.

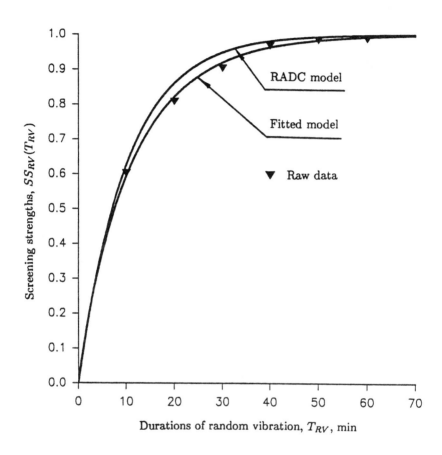

Fig. 11.1– Plots of screening strengths versus the vibration times using the observed raw data, using the RADC model, or Eq. (11.38), and using the fitted model, or Eq. (11.41), respectively.

EXAMPLE 11–2

An aircraft company is conducting a screening strength evaluation on its random vibration stress screens. The observed screening strength values for various combinations of acceleration $grms$ level, G, and vibration duration, T_{RV}, are found to be as follows:

G, $g's$	1.0	2.0	3.0	4.0	5.0	6.0
T_{RV}, min	60	50	40	30	20	10
Observed SS_{RV} values	0.260	0.550	0.761	0.804	0.798	0.663

Do the following:

1. Check whether or not the RADC model for random vibration screening strength fits the data.

2. Determine the optimum screening strength equation for this screen using the least-squares method.

SOLUTIONS TO EXAMPLE 11–2

1. The RADC model of screening strength for random vibration screens is given by Eq. (11.7), or

$$SS_{RV}(T_{RV}) = 1 - e^{-0.0046\,(G)^{1.71}\,T_{RV}}. \qquad (11.42)$$

The calculated SS_{RV} for the given combinations of (G, T_{RV}) using this equation are tabulated, together with the actual observed SS_{RV}, in Table 11.5.

It may be seen that the RADC model for random vibration is under-estimating the screening strength for this particular case. Therefore, it needs to be modified accordingly.

2. Since both G and T_{RV} are varying in this case, or $k = 2$, then Eq. (11.42) can be rewritten in the following generalized form:

$$SS_{RV}(T_{RV}) = 1 - e^{-a_1\,G^{b_1}\,T_{RV}^{c_1}}, \qquad (11.43)$$

as also given by Eq. (11.23), where a_1, b_1 and c_1 are three unknown constants to be determined from the observed data using the least-squares method.

Linearizing Eq. (11.43) yields

$$Y = A_0 + A_1\,X_1 + A_2\,X_2,$$

TABLE 11.5– The observed and calculated screening strengths for a random vibration screen for Example 11–2.

G, $g's$	1.0	2.0	3.0	4.0	5.0	6.0
T_{RV}, min	60	50	40	30	20	10
Observed SS_{RV} values	0.260	0.550	0.761	0.804	0.798	0.663
Calculated SS_{RV} values using the RADC model	0.241	0.529	0.700	0.772	0.764	0.627
Calculated SS_{RV} values using the fitted model	0.256	0.563	0.737	0.807	0.799	0.662

where

$$Y = \log_e[-\log_e(1 - SS_{RV})],$$
$$A_0 = \log_e a_1, \tag{11.44}$$
$$A_1 = b_1, \tag{11.45}$$
$$A_2 = c_1, \tag{11.46}$$
$$X_1 = \log_e G,$$

and

$$X_2 = \log_e T_{RV}.$$

Then, the least-squares estimates of A_0, A_1 and A_2 are given by Eq. (11.34), or

$$\hat{\mathbf{A}} = (\mathbf{X}^T \mathbf{X})^{-1} (\mathbf{X}^T \mathbf{Y}),$$

where

$$\hat{\mathbf{A}} = \begin{bmatrix} \hat{A}'_0 \\ \hat{A}_1 \\ \hat{A}_2 \end{bmatrix}_{3 \times 1},$$

$$\hat{A}'_0 = \hat{A}_0 + (\hat{A}_1 \overline{X}_1 + \hat{A}_2 \overline{X}_2), \tag{11.47}$$

$$\overline{X_1} = \frac{1}{n} \sum_{j=1}^{n} X_{1j} = \frac{1}{6} \sum_{j=1}^{6} X_{1j} = 1.097,$$

$$\overline{X_2} = \frac{1}{n} \sum_{j=1}^{n} X_{2j} = \frac{1}{6} \sum_{j=1}^{6} X_{2j} = 3.399,$$

n = number of (G, T_{RV}) combinations,

= 6 in this case,

$$\mathbf{X} = \begin{bmatrix} 1 & (X_{11} - \overline{X_1}) & (X_{21} - \overline{X_2}) \\ 1 & (X_{12} - \overline{X_1}) & (X_{22} - \overline{X_2}) \\ \cdot & \cdot & \cdot \\ \cdot & \cdot & \cdot \\ \cdot & \cdot & \cdot \\ 1 & (X_{1n} - \overline{X_1}) & (X_{2n} - \overline{X_2}) \end{bmatrix}_{n \times 3} ,$$

$$= \begin{bmatrix} 1 & -1.097 & 0.695 \\ 1 & -0.404 & 0.513 \\ 1 & 0.002 & 0.290 \\ 1 & 0.289 & 0.002 \\ 1 & 0.512 & -0.403 \\ 1 & 0.695 & -1.096 \end{bmatrix}_{6 \times 3} ,$$

$$\mathbf{X}^T = \begin{bmatrix} 1 & 1 & \cdots & 1 \\ (X_{11} - \overline{X_1}) & (X_{12} - \overline{X_1}) & \cdots & (X_{1n} - \overline{X_1}) \\ (X_{21} - \overline{X_2}) & (X_{22} - \overline{X_2}) & \cdots & (X_{2n} - \overline{X_2}) \\ \cdot & \cdot & \cdot & \cdot \\ \cdot & \cdot & \cdot & \cdot \\ \cdot & \cdot & \cdot & \cdot \\ (X_{k1} - \overline{X_k}) & (X_{k2} - \overline{X_k}) & \cdots & (X_{kn} - \overline{X_k}) \end{bmatrix}_{(k+1) \times n} ,$$

$$= \begin{bmatrix} 1 & 1 & 1 & 1 & 1 & 1 \\ -1.097 & -0.404 & 0.002 & 0.289 & 0.512 & 0.695 \\ 0.695 & 0.513 & 0.290 & 0.002 & -0.403 & -1.096 \end{bmatrix}_{3 \times 6} ,$$

$$\mathbf{Y} = \begin{bmatrix} Y_1 \\ Y_2 \\ \cdot \\ \cdot \\ \cdot \\ Y_n \end{bmatrix}_{n \times 1} = \begin{bmatrix} -1.200 \\ -0.225 \\ 0.359 \\ 0.488 \\ 0.470 \\ 0.084 \end{bmatrix}_{6 \times 1} .$$

The associated calculations and results are listed in Table 11.6.

Then,

$$\mathbf{X}^T \mathbf{X} = \begin{bmatrix} 1 & 1 & 1 & 1 & 1 & 1 \\ -1.097 & -0.404 & 0.002 & 0.289 & 0.512 & 0.695 \\ 0.695 & 0.513 & 0.290 & 0.002 & -0.403 & -1.096 \end{bmatrix}_{3 \times 6}$$

TABLE 11.6– The calculation summaries for the least-squares estimation for Example 11–2.

j	1	2	3	4	5	6	Mean
G_j	1	2	3	4	5	6	
X_{1j} †	0.000	0.693	1.099	1.386	1.609	1.792	$\overline{X_1} = 1.097$
$X_{1j} - \overline{X_1}$	-1.097	-0.404	0.002	0.289	0.512	0.695	
$(T_{RV})_j$	60	50	40	30	20	10	
X_{2j} ‡	4.094	3.912	3.689	3.401	2.996	2.303	$\overline{X_2} = 3.399$
$X_{2j} - \overline{X_2}$	0.695	0.513	0.290	0.002	-0.403	-1.096	
$(SS_{RV})_j$	0.260	0.550	0.761	0.804	0.798	0.663	
Y_j ‡†	-1.200	-0.225	0.359	0.488	0.470	0.084	

† $X_{1j} = \log_e G_j$.
‡ $X_{2j} = \log_e (T_{RV})_j$.
‡† $Y_j = \log_e\{-\log_e[1 - (SS_{RV})_j]\}$.

$$
\cdot \begin{bmatrix} 1 & -1.097 & 0.695 \\ 1 & -0.404 & 0.513 \\ 1 & 0.002 & 0.290 \\ 1 & 0.289 & 0.002 \\ 1 & 0.512 & -0.403 \\ 1 & 0.695 & -1.096 \end{bmatrix}_{6\times3} ,
$$

$$
= \begin{bmatrix} 6 & -0.009 & 0.001 \\ -0.009 & 2.187 & -1.930 \\ 0.001 & -1.930 & 2.194 \end{bmatrix}_{3\times3} ,
$$

$$
(\mathbf{X}^T\,\mathbf{X})^{-1} = \begin{bmatrix} 0.167 & 0.003 & 0.002 \\ 0.003 & 2.044 & 1.798 \\ 0.002 & 1.798 & 2.308 \end{bmatrix}_{3\times3} ,
$$

$$
\mathbf{X}^T\,\mathbf{Y} = \begin{bmatrix} 1 & 1 & 1 & 1 & 1 & 1 \\ -1.097 & -0.404 & 0.002 & 0.289 & 0.512 & 0.695 \\ 0.695 & 0.513 & 0.290 & 0.002 & -0.403 & -1.096 \end{bmatrix}_{3\times6}
$$

$$
\cdot \begin{bmatrix} -1.200 \\ -0.225 \\ 0.359 \\ 0.488 \\ 0.470 \\ 0.084 \end{bmatrix}_{6\times1} ,
$$

$$= \begin{bmatrix} -0.024 \\ 1.848 \\ -1.126 \end{bmatrix}_{3 \times 1} .$$

Finally,

$$\hat{\mathbf{A}} = (\mathbf{X}^T \mathbf{X})^{-1} (\mathbf{X}^T \mathbf{Y}),$$

$$= \begin{bmatrix} 0.167 & 0.003 & 0.002 \\ 0.003 & 2.044 & 1.798 \\ 0.002 & 1.798 & 2.308 \end{bmatrix}_{3 \times 3}$$

$$\cdot \begin{bmatrix} -0.024 \\ 1.848 \\ -1.126 \end{bmatrix}_{3 \times 1} ,$$

or

$$\hat{\mathbf{A}} = \begin{bmatrix} \hat{A}_0' \\ \hat{A}_1 \\ \hat{A}_2 \end{bmatrix}_{3 \times 1} = \begin{bmatrix} -0.002 \\ 1.752 \\ 1.028 \end{bmatrix}_{3 \times 1} .$$

Therefore,

$$\hat{A}_0' = -0.002,$$
$$\hat{A}_1 = 1.752,$$

and

$$\hat{A}_2 = 1.028.$$

According to Eq. (11.47),

$$\hat{A}_0 = \hat{A}_0' - (\hat{A}_1 \overline{X_1} + \hat{A}_2 \overline{X_2}),$$
$$= -0.002 - (1.752 \times 1.097 + 1.028 \times 3.399),$$

or

$$\hat{A}_0 = -5.417.$$

According to Eqs. (11.44), (11.45) and (11.46), we have

$$\begin{cases} \hat{a}_1 = e^{\hat{A}_0} = e^{-5.417} = 0.0044, \\ \hat{b}_1 = \hat{A}_1 = 1.752, \\ \hat{c}_1 = \hat{A}_2 = 1.028. \end{cases}$$

Consequently, the optimum screening strength equation using the least-squares method is given by

$$SS_{RV}(T_{RV}) = 1 - e^{-\hat{a}_1 \, G^{\hat{b}_1} \, T_{RV}^{\hat{c}_1}},$$

or

$$SS_{RV}(T_{RV}) = 1 - e^{-0.0044 \, G^{1.752} \, T_{RV}^{1.028}} . \tag{11.48}$$

The screening strength values for the given combinations of (G, T_{RV}) are calculated using this model and tabulated in the last row of Table 11.5. It may be seen that Eq. (11.48) fits the observed data much better than the RADC model.

EXAMPLE 11–3

A company is conducting a screening strength evaluation on its thermal cycling stress screens. The temperature range is

$$T_R^* = T_{max}^* - T_{min}^* = 80 - (-40) = 120°C,$$

and the rate of temperature change is $\dot{T}^* = 10\ °C/min$. The observed screening strength values for the various numbers of thermal cycles are found to be as follows:

N_{TC}	2	4	6	8	10	12
Observed SS_{TC} values	0.6000	0.8315	0.9126	0.9734	0.9800	0.9908

Do the following:

1. Check whether or not the RADC model for thermal cycling screening strength fits the data.

2. Determine the optimum screening strength equation for this screen using the least-squares method.

SOLUTIONS TO EXAMPLE 11–3

1. The RADC model of screening strength for thermal cycling screens is given by Eq. (11.10), or

$$SS_{TC}(N_{TC}) = 1 - e^{-0.0017(T_R^*+0.6)^{0.6}\,[\log_e(e+\dot{T}^*)]^3\,N_{TC}},$$

$$(11.49)$$

where

$$T_R^* = 120\ °C,$$

and

$$\dot{T}^* = 10\ °C/min.$$

Then,

$$SS_{TC}(N_{TC}) = 1 - e^{-0.0017(120+0.6)^{0.6}\,[\log_e(e+10)]^3\,N_{TC}},$$

TABLE 11.7– The observed and calculated screening strengths for a thermal cycling screen with $T_R^* = 120\ ^\circ C$ and $\dot{T}^* = 10\ ^\circ C/min$ for Example 11–3.

N_{TC}	2	4	6	8	10	12
Observed SS_{TC} values	0.6000	0.8315	0.9126	0.9734	0.9800	0.9908
Calculated SS_{TC} values using the RADC model	0.6290	0.8624	0.9489	0.9811	0.9930	0.9974
Calculated SS_{TC} values using the fitted model	0.6048	0.8265	0.9211	0.9633	0.9827	0.9917

or

$$SS_{TC}(N_{TC}) = 1 - e^{-0.4958\ N_{TC}}. \tag{11.50}$$

The calculated SS_{TC} values for $N_{TC} = 2, 4, \cdots, 12$ using this equation are tabulated, together with the actual observed SS_{TC} values, in Table 11.7.

It may be seen that the RADC model for thermal cycling screen is overestimating the screening strength for this particular case. Therefore, it needs to be modified accordingly.

2. Since both T_R^* and $\dot{T}^*$ are fixed in this case, and the number of temperature cycles is the only varying variable, or $k = 1$, then Eq. (11.50) can be rewritten in the following generalized form:

$$SS_{TC}(N_{TC}) = 1 - e^{-a\ (N_{TC})^b}, \tag{11.51}$$

where a and b are two unknown constants to be determined from the observed data using the least-squares method.

Linearizing Eq. (11.51) yields

$$Y = A_0 + A_1\ X_1, \tag{11.52}$$

where

$$Y = \log_e[-\log_e(1 - SS_{TC})],$$
$$A_0 = \log_e a, \tag{11.53}$$
$$A_1 = b, \tag{11.54}$$

and

$$X_1 = \log_e N_{TC}.$$

Then, the least-squares estimates of A_0 and A_1 in Eq. (11.52) can be obtained using Eq. (11.34), or

$$\hat{\mathbf{A}} = (\mathbf{X}^T \mathbf{X})^{-1} (\mathbf{X}^T \mathbf{Y}),$$

where

$$\hat{\mathbf{A}} = \begin{bmatrix} \hat{A}'_0 \\ \hat{A}_1 \end{bmatrix}_{2\times 1},$$

$$\hat{A}'_0 = \hat{A}_0 + \hat{A}_1 \overline{X}_1, \tag{11.55}$$

$$\overline{X}_1 = \frac{1}{n} \sum_{j=1}^{n} X_{1j} = \frac{1}{6} \sum_{j=1}^{6} X_{1j} = 1.790,$$

n = number of observations,

= 6 in this case,

$$\mathbf{X} = \begin{bmatrix} 1 & (X_{11} - \overline{X}_1) \\ 1 & (X_{12} - \overline{X}_1) \\ \cdot & \cdot \\ \cdot & \cdot \\ \cdot & \cdot \\ 1 & (X_{16} - \overline{X}_1) \end{bmatrix}_{6\times 2},$$

$$= \begin{bmatrix} 1 & -1.097 \\ 1 & -0.404 \\ 1 & 0.002 \\ 1 & 0.289 \\ 1 & 0.513 \\ 1 & 0.695 \end{bmatrix}_{6\times 2},$$

$$\mathbf{X}^T = \begin{bmatrix} 1 & 1 & 1 & 1 & 1 & 1 \\ -1.097 & -0.404 & 0.002 & 0.289 & 0.513 & 0.695 \end{bmatrix}_{2\times 6},$$

$$\mathbf{Y} = \begin{bmatrix} Y_1 \\ Y_2 \\ \cdot \\ \cdot \\ \cdot \\ Y_6 \end{bmatrix}_{6\times 1} = \begin{bmatrix} -0.087 \\ 0.577 \\ 0.891 \\ 1.288 \\ 1.364 \\ 1.545 \end{bmatrix}_{6\times 1}.$$

The associated calculations and results are listed in Table 11.8.

Then,

$$\mathbf{X}^T \mathbf{X} = \begin{bmatrix} 6 & -0.002 \\ -0.002 & 2.196 \end{bmatrix}_{2\times 2},$$

$$(\mathbf{X}^T \mathbf{X})^{-1} = \begin{bmatrix} 0.167 & 1.518 \times 10^{-4} \\ 1.518 \times 10^{-4} & 0.455 \end{bmatrix}_{2\times 2},$$

TABLE 11.8– The calculation summaries for the least-squares estimation for Example 11–3.

j	1	2	3	4	5	6
$(N_{TC})_j$	2	4	6	8	10	12
X_{1j} †,*	0.693	1.386	1.792	2.079	2.303	2.485
$X_{1j} - \overline{X_1}$	-1.097	-0.404	0.002	0.289	0.513	0.695
$(SS_{TC})_j$	0.6000	0.8315	0.9126	0.9734	0.9800	0.9908
Y_j ‡	-0.087	0.577	0.891	1.288	1.364	1.545

† $X_{1j} = \log_e (N_{TC})_j$;   * Mean $\overline{X_1} = 1.790$.

‡ $Y_j = \log_e\{-\log_e[1 - (SS_{TC})_j]\}$.

and

$$\mathbf{X}^T\,\mathbf{Y} = \begin{bmatrix} 5.578 \\ 2.010 \end{bmatrix}_{2\times 1}.$$

Finally,

$$\hat{\mathbf{A}} = (\mathbf{X}^T\,\mathbf{X})^{-1}\,(\mathbf{X}^T\,\mathbf{Y}),$$

$$= \begin{bmatrix} 0.167 & 1.518 \times 10^{-4} \\ 1.518 \times 10^{-4} & 0.455 \end{bmatrix}_{2\times 2}$$

$$\cdot \begin{bmatrix} 5.578 \\ 2.010 \end{bmatrix}_{2\times 1},$$

or

$$\hat{\mathbf{A}} = \begin{bmatrix} \hat{A}_0' \\ \hat{A}_1 \end{bmatrix}_{2\times 1} = \begin{bmatrix} 0.930 \\ 0.916 \end{bmatrix}_{3\times 1}.$$

Therefore,

$$\hat{A}_0' = 0.930,$$

and

$$\hat{A}_1 = 0.916.$$

According to Eqs. (11.55), (11.53) and (11.54),

$$\hat{A}_0 = \hat{A}_0' - \hat{A}_1\overline{X_1},$$

$$= 0.930 - 0.916 \times 1.790,$$

or

$$\hat{A}_0 = -0.710,$$

$$\hat{a} = e^{\hat{A}_0} = e^{-0.710} = 0.492,$$

and

$$\hat{b} = \hat{A}_1 = 0.916.$$

Consequently, the optimum screening strength equation using the least-squares method is given by

$$SS_{TC}(N_{TC}) = 1 - e^{-\hat{a} \ N_{TC}^{\hat{b}}},$$

or

$$SS_{TC}(N_{TC}) = 1 - e^{-0.492 \ N_{TC}^{0.916}}. \tag{11.56}$$

The screening strength values for the given number of temperature cycles are calculated using this model and tabulated in the last row of Table 11.7. It may be seen that Eq. (11.56) fits the observed data much better than the RADC model.

EXAMPLE 11-4

A company is conducting a screening strength evaluation on its thermal cycling stress screens. The temperature range was fixed at

$$T_R^* = T_{max}^* - T_{min}^* = 80 - (-40) = 120 \ °C,$$

while the rate of temperature change, $\dot{T}^*$, and the number of temperature cycles, N_{TC}, are varying. The observed screening strength values for the various combinations of $(\dot{T}^*, N_{TC})$ are found to be as follows:

$\dot{T}^*$, °C	20	15	15	10	10	5
N_{TC}	2	4	6	8	10	12
Observed SS_{TC} values	0.8660	0.9671	0.9955	0.9938	0.9987	0.9700

Do the following:

1. Check whether or not the RADC model for thermal cycling screening strength fits the data.

2. Determine the optimum screening strength equation for this screen using the least-squares method.

TABLE 11.9– The observed and calculated screening strengths for a thermal cycling screen with $T_R^* = 120$ °C for Example 11–4.

$\dot{T}^*$, °C/min	20	15	15	10	10	5
N_{TC}	2	4	6	8	10	12
Observed SS_{TC} values	0.8660	0.9671	0.9955	0.9938	0.9987	0.9700
Calculated SS_{TC} values using the RADC model	0.8407	0.9430	0.9864	0.9811	0.9930	0.9544
Calculated SS_{TC} values using the fitted model	0.8638	0.9663	0.9962	0.9930	0.9985	0.9707

SOLUTIONS TO EXAMPLE 11–4

1. The RADC model of screening strength for thermal cycling screens is given by Eq. (11.10), or

$$SS_{TC}(N_{TC}) = 1 - e^{-0.0017(T_R^*+0.6)^{0.6} \, [\log_e(e+\dot{T}^*)]^3 \, N_{TC}},$$

(11.57)

where

$T_R^* = 120°C$ in this case.

Then,

$$SS_{TC}(N_{TC}) = 1 - e^{-0.0017(120+0.6)^{0.6} \, [\log_e(e+\dot{T}^*)]^3 \, N_{TC}},$$

or

$$SS_{TC}(N_{TC}) = 1 - e^{-0.0301 \, [\log_e(e+\dot{T}^*)]^3 \, N_{TC}}.$$

(11.58)

The calculated SS_{TC} values for the given combinations of $(\dot{T}^*, N_{TC})$ using this equation are tabulated, together with the actual observed SS_{TC} values, in Table 11.9.

It may be seen that the RADC model for thermal cycling screen is underestimating the screening strength for this particular case. Therefore, it needs to be modified accordingly.

2. Since both $\dot{T}^*$ and N_{TC} varying in this case, or $k = 2$, then Eq. (11.58) can be rewritten in the following generalized form:

$$SS_{TC}(N_{TC}) = 1 - e^{-a \, [\log_e(e+\dot{T}^*)]^b \, (N_{TC})^c},$$

(11.59)

where a, b and c are three unknown constants to be determined from the observed data using the least-squares method.

Linearizing Eq. (11.59) yields

$$Y = A_0 + A_1 X_1 + A_2 X_2, \tag{11.60}$$

where

$$
\begin{aligned}
Y &= \log_e[-\log_e(1 - SS_{TC})], \\
A_0 &= \log_e a, && \text{(11.61)} \\
A_1 &= b, && \text{(11.62)} \\
A_2 &= c, && \text{(11.63)} \\
X_1 &= \log_e \log_e(e + \dot{T}^*),
\end{aligned}
$$

and

$$X_2 = \log_e N_{TC}.$$

Then, the least-squares estimates of A_0, A_1 and A_2 are given by Eq. (11.34), or

$$\hat{\mathbf{A}} = (\mathbf{X}^T \mathbf{X})^{-1} (\mathbf{X}^T \mathbf{Y}),$$

where

$$
\hat{\mathbf{A}} = \begin{bmatrix} \hat{A}_0' \\ \hat{A}_1 \\ \hat{A}_2 \end{bmatrix}_{3 \times 1},
$$

$$\hat{A}_0' = \hat{A}_0 + (\hat{A}_1 \overline{X}_1 + \hat{A}_2 \overline{X}_2), \tag{11.64}$$

$$\overline{X_1} = \frac{1}{n} \sum_{j=1}^{n} X_{1j} = \frac{1}{6} \sum_{j=1}^{6} X_{1j} = 0.9720,$$

$$\overline{X_2} = \frac{1}{n} \sum_{j=1}^{n} X_{2j} = \frac{1}{6} \sum_{j=1}^{6} X_{2j} = 1.7897,$$

n = number of $(\dot{T}^*, N_{TC})$ combinations,

= 6 in this case,

$$
\mathbf{X} = \begin{bmatrix}
1 & (X_{11} - \overline{X_1}) & (X_{21} - \overline{X_2}) \\
1 & (X_{12} - \overline{X_1}) & (X_{22} - \overline{X_2}) \\
\cdot & \cdot & \cdot \\
\cdot & \cdot & \cdot \\
\cdot & \cdot & \cdot \\
1 & (X_{16} - \overline{X_1}) & (X_{26} - \overline{X_2})
\end{bmatrix}_{6 \times 3},
$$

TABLE 11.10– The calculation summaries for the least-squares estimation for Example 11–4.

j	1	2	3	4	5	6
$(T^*)_j$, °C/min	20	15	15	10	10	5
X_{1j} †,*	1.1388	1.0559	1.0559	0.9334	0.9334	0.7147
$X_{1j} - \overline{X_1}$	0.1668	0.0839	0.0839	-0.0386	-0.0386	-0.2573
$(N_{TC})_j$	2	4	6	8	10	12
X_{2j} †,**	0.6931	1.3863	1.7918	2.0794	2.3026	2.4849
$X_{2j} - \overline{X_2}$	-1.0966	-0.4034	0.0021	0.2897	0.5129	0.6952
$(SS_{TC})_j$	0.8600	0.9671	0.9955	0.9938	0.9987	0.9700
Y_j ‡†	0.6981	1.2280	1.6871	1.6259	1.8939	1.2546

† $X_{1j} = \log_e \log_e[e + (\dot{T}^*)_j]$;   * Mean $\overline{X_1} = 0.9720$.

‡ $X_{2j} = \log_e(N_{TC})_j$;   ** Mean $\overline{X_2} = 1.7897$.

‡† $Y_j = \log_e\{-\log_e[1 - (SS_{TC})_j]\}$.

$$= \begin{bmatrix} 1 & 0.1668 & -1.0966 \\ 1 & 0.0839 & -0.4034 \\ 1 & 0.0839 & 0.0021 \\ 1 & -0.0386 & 0.2897 \\ 1 & -0.0386 & 0.5129 \\ 1 & -0.2573 & 0.6952 \end{bmatrix}_{6\times 3},$$

$$\mathbf{X}^T = \begin{bmatrix} 1 & 1 & 1 & 1 & 1 & 1 \\ 0.1668 & 0.0839 & 0.0839 & -0.0386 & -0.0386 & -0.2573 \\ -1.0966 & -0.4034 & 0.0021 & 0.2897 & 0.5129 & 0.6952 \end{bmatrix}_{3\times 6},$$

$$\mathbf{Y} = \begin{bmatrix} Y_1 \\ Y_2 \\ . \\ . \\ . \\ Y_6 \end{bmatrix}_{6\times 1} = \begin{bmatrix} 0.6981 \\ 1.2280 \\ 1.6871 \\ 1.6259 \\ 1.8939 \\ 1.2546 \end{bmatrix}_{6\times 1}.$$

The associated calculations and results are listed in Table 11.10.

Then,

$$\mathbf{X}^T\mathbf{X} = \begin{bmatrix} 6 & 10^{-4} & -10^{-4} \\ 10^{-4} & 0.1111 & -0.4264 \\ -10^{-4} & -0.4264 & 2.1956 \end{bmatrix}_{3\times 3},$$

$$(\mathbf{X}^T\,\mathbf{X})^{-1} = \begin{bmatrix} 0.1667 & -4.7524 \times 10^{-4} & -8.4714 \times 10^{-5} \\ -4.7524 \times 10^{-4} & 35.3878 & 6.8733 \\ -8.4714 \times 10^{-5} & 6.8733 & 1.7904 \end{bmatrix}_{3\times3},$$

and

$$\mathbf{X}^T\,\mathbf{Y} = \begin{bmatrix} 8.3876 \\ -0.0977 \\ 1.0572 \end{bmatrix}_{3\times1}.$$

Finally,

$$\hat{\mathbf{A}} = (\mathbf{X}^T\,\mathbf{X})^{-1}\,(\mathbf{X}^T\,\mathbf{Y}),$$

$$= \begin{bmatrix} 0.1667 & -4.7524 \times 10^{-4} & -8.4714 \times 10^{-5} \\ -4.7524 \times 10^{-4} & 35.3878 & 6.8733 \\ -8.4714 \times 10^{-5} & 6.8733 & 1.7904 \end{bmatrix}_{3\times3}$$

$$\cdot \begin{bmatrix} 8.3876 \\ -0.0977 \\ 1.0572 \end{bmatrix}_{3\times1},$$

or

$$\hat{\mathbf{A}} = \begin{bmatrix} \hat{A}_0' \\ \hat{A}_1 \\ \hat{A}_2 \end{bmatrix}_{3\times1} = \begin{bmatrix} 1.3979 \\ 3.8069 \\ 1.2210 \end{bmatrix}_{3\times1}.$$

Therefore,

$$\hat{A}_0' = 1.3979,$$
$$\hat{A}_1 = 3.8069,$$

and

$$\hat{A}_2 = 1.2210.$$

According to Eqs. (11.64), (11.61), (11.62) and (11.63),

$$\hat{A}_0 = \hat{A}_0' - (\hat{A}_1\overline{X_1} + \hat{A}_2\overline{X_2}),$$
$$= 1.3979 - (3.8069 \times 0.9720 + 1.2210 \times 1.7897),$$

or

$$\hat{A}_0 = -4.4876,$$
$$\hat{a} = e^{\hat{A}_0} = e^{-4.4876} = 0.0112,$$
$$\hat{b} = \hat{A}_1 = 3.8069,$$

and

$$\hat{c} = \hat{A}_2 = 1.2210.$$

Consequently, the optimum screening strength equation using the least-squares method is given by

$$SS_{TC}(N_{TC}) = 1 - e^{-\hat{a} \, [\log_e(e+\dot{T}^*)]^{\hat{b}} \, N_{TC}^{\hat{c}}},$$

or

$$SS_{TC}(N_{TC}) = 1 - e^{-0.0112 \, [\log_e(e+\dot{T}^*)]^{3.8069} \, N_{TC}^{1.221}}.$$

(11.65)

The screening strength values for the given combinations of $(\dot{T}^*, N_{TC})$ are calculated using this model and tabulated in the last row of Table 11.9. It may be seen that Eq. (11.65) fits the observed data much better than the RADC model.

EXAMPLE 11-5

A company is conducting a screening strength evaluation on its thermal cycling stress screens. The observed screening strength values for various combinations of $(T_R^*, \dot{T}^*, N_{TC})$ are found to be as follows:

T_R^*, °C	180	160	140	120	100	80
$\dot{T}^*$, °C/min	20	15	15	10	10	5
N_{TC}	2	4	6	8	10	12
Observed SS_{TC} values	0.8842	0.9500	0.9741	0.9683	0.9705	0.8964

Do the following:

1. Check whether or not the RADC model for thermal cycling screening strength fits the data.

2. Determine the optimum screening strength equation for this screen using the least-squares method.

SOLUTIONS TO EXAMPLE 11-5

1. The RADC model of screening strength for thermal cycling screens is given by Eq. (11.10), or

$$SS_{TC}(N_{TC}) = 1 - e^{-0.0017(T_R^*+0.6)^{0.6} \, [\log_e(e+\dot{T}^*)]^3 \, N_{TC}}. \qquad (11.66)$$

The calculated SS_{TC} values for the given combinations of $(T_R^*, \dot{T}^*, N_{TC})$ using this equation are tabulated, together with the actual observed SS_{TC} values, in Table 11.11.

It may be seen that the RADC model for thermal cycling screen is over-estimating the screening strength for this particular case. Therefore, it needs to be modified accordingly.

TABLE 11.11– The observed and calculated screening strengths for a thermal cycling screen for Example 11–5.

T_R^*, °C	180	160	140	120	100	80
$\dot{T}^*$, °C/min	20	15	15	10	10	5
N_{TC}	2	4	6	8	10	12
Observed SS_{TC} values	0.8842	0.9500	0.9741	0.9683	0.9705	0.8964
Calculated SS_{TC} values using the RADC model	0.9037	0.9667	0.9910	0.9811	0.9883	0.9115
Calculated SS_{TC} values using the fitted model	0.8834	0.9501	0.9765	0.9659	0.9694	0.8991

2. Since all three variables, T_R^*, $\dot{T}^*$ and N_{TC} are varying in this case, or $k = 3$, then Eq. (11.66) can be rewritten in the following generalized form:

$$SS_{TC}(N_{TC}) = 1 - e^{-a\ (T_R^*+0.6)^b\ [\log_e(e+\dot{T}^*)]^c\ (N_{TC})^d}, \qquad (11.67)$$

where a, b, c and d are four unknown constants to be determined from the observed data using the least-squares method.

Linearizing Eq. (11.67) yields

$$Y = A_0 + A_1\ X_1 + A_2\ X_2 + A_3\ X_3, \qquad (11.68)$$

where

$$Y = \log_e[-\log_e(1 - SS_{TC})],$$
$$A_0 = \log_e a, \qquad (11.69)$$
$$A_1 = b, \qquad (11.70)$$
$$A_2 = c, \qquad (11.71)$$
$$A_3 = d, \qquad (11.72)$$
$$X_1 = \log_e(T_R^* + 0.6),$$
$$X_2 = \log_e\log_e(e + \dot{T}^*),$$

and

$$X_3 = \log_e N_{TC}.$$

Then, the least-squares estimates of A_0, A_1, A_2 and A_3 are given by Eq. (11.34), or

$$\hat{A} = (X^T\ X)^{-1}\ (X^T\ Y),$$

where

$$\hat{\mathbf{A}} = \begin{bmatrix} \hat{A}'_0 \\ \hat{A}_1 \\ \hat{A}_2 \\ \hat{A}_3 \end{bmatrix}_{4 \times 1},$$

$$\hat{A}'_0 = \hat{A}_0 + (\hat{A}_1 \overline{X}_1 + \hat{A}_2 \overline{X}_2 + \hat{A}_3 \overline{X}_3), \tag{11.73}$$

$$\overline{X_1} = \frac{1}{n} \sum_{j=1}^{n} X_{1j} = \frac{1}{6} \sum_{j=1}^{6} X_{1j} = 4.8357,$$

$$\overline{X_2} = \frac{1}{n} \sum_{j=1}^{n} X_{2j} = \frac{1}{6} \sum_{j=1}^{6} X_{2j} = 0.9720,$$

$$\overline{X_3} = \frac{1}{n} \sum_{j=1}^{n} X_{3j} = \frac{1}{6} \sum_{j=1}^{6} X_{3j} = 1.7897,$$

n = number of $(T_R^*, \dot{T}^*, N_{TC})$ combinations,

= 6 in this case,

$$\mathbf{X} = \begin{bmatrix} 1 & (X_{11} - \overline{X_1}) & (X_{21} - \overline{X_2}) & (X_{31} - \overline{X_3}) \\ 1 & (X_{12} - \overline{X_1}) & (X_{22} - \overline{X_2}) & (X_{32} - \overline{X_3}) \\ \cdot & \cdot & \cdot & \cdot \\ \cdot & \cdot & \cdot & \cdot \\ \cdot & \cdot & \cdot & \cdot \\ 1 & (X_{16} - \overline{X_1}) & (X_{26} - \overline{X_2}) & (X_{36} - \overline{X_3}) \end{bmatrix}_{6 \times 4},$$

$$= \begin{bmatrix} 1 & 0.3606 & 0.1668 & -1.0966 \\ 1 & 0.2432 & 0.0839 & -0.4034 \\ 1 & 0.1102 & 0.0839 & 0.0021 \\ 1 & -0.0432 & -0.0386 & 0.2897 \\ 1 & -0.2245 & -0.0386 & 0.5129 \\ 1 & -0.4462 & -0.2573 & 0.6952 \end{bmatrix}_{6 \times 4},$$

$$\mathbf{X}^T = \begin{bmatrix} 1 & 1 & 1 & 1 & 1 & 1 \\ 0.3606 & 0.2432 & 0.1102 & -0.0432 & -0.2245 & -0.4462 \\ 0.1668 & 0.0839 & 0.0839 & -0.0386 & -0.0386 & -0.2573 \\ -1.0966 & -0.4034 & 0.0021 & 0.2897 & 0.5129 & 0.6952 \end{bmatrix}_{4 \times 6},$$

$$\mathbf{Y} = \begin{bmatrix} Y_1 \\ Y_2 \\ \cdot \\ \cdot \\ \cdot \\ Y_6 \end{bmatrix}_{6 \times 1} = \begin{bmatrix} 0.7682 \\ 1.0972 \\ 1.2957 \\ 1.2388 \\ 1.2594 \\ 0.8186 \end{bmatrix}_{6 \times 1}.$$

The associated calculations and results are listed in Table 11.12.

TABLE 11.12– The calculation summaries for the least-squares estimation for Example 11–5.

j	1	2	3	4	5	6
$(T_R^*)_j$, °C	180	160	140	120	100	80
X_{1j} [†,*]	5.1963	5.0789	4.9459	4.7925	4.6112	4.3895
$X_{1j} - \overline{X_1}$	0.3606	0.2432	0.1102	-0.0432	-0.2245	-0.4462
$(\dot{T}^*)_j$, °C/min	20	15	15	10	10	5
X_{2j} [†,**]	1.1388	1.0559	1.0559	0.9334	0.9334	0.7147
$X_{2j} - \overline{X_2}$	0.1668	0.0839	0.0839	-0.0386	-0.0386	-0.2573
$(N_{TC})_j$	2	4	6	8	10	12
X_{3j} [††,***]	0.6931	1.3863	1.7918	2.0794	2.3026	2.4849
$X_{3j} - \overline{X_3}$	-1.0966	-0.4034	0.0021	0.2897	0.5129	0.6952
$(SS_{TC})_j$	0.8842	0.9500	0.9741	0.9683	0.9705	0.8964
Y_j [‡‡]	0.7682	1.0972	1.2957	1.2388	1.2594	0.8186

† $X_{1j} = \log_e[(T_R^*)_j + 0.6]$;

‡ $X_{2j} = \log_e \log_e[e + (\dot{T}^*)_j]$;

‡† $X_{3j} = \log_e(N_{TC})_j$;

‡‡ $Y_j = \log_e\{-\log_e[1 - (SS_{TC})_j]\}$.

* Mean $\overline{X_1} = 4.8357$.

** Mean $\overline{X_2} = 0.9720$.

*** Mean $\overline{X_3} = 1.7897$.

Then,

$$\mathbf{X}^T\mathbf{X} = \begin{bmatrix} 6 & 10^{-4} & 10^{-4} & -10^{-4} \\ 10^{-4} & 0.4527 & 0.2149 & -0.9312 \\ 10^{-4} & 0.2149 & 0.1111 & -0.4264 \\ -10^{-4} & -0.9312 & -0.4264 & 2.1956 \end{bmatrix}_{4\times4},$$

$$(\mathbf{X}^T\mathbf{X})^{-1} = \begin{bmatrix} 0.1667 & 3.9610 \times 10^{-4} & -9.5296 \times 10^{-4} & -9.5074 \times 10^{-6} \\ 3.9610 \times 10^{-4} & 60.0281 & -72.3968 & 11.3974 \\ -9.5296 \times 10^{-4} & -72.3968 & 122.7018 & -6.8725 \\ -9.5074 \times 10^{-6} & 11.3974 & -6.8725 & 3.9544 \end{bmatrix}_{4\times4},$$

$$\mathbf{X}^T\mathbf{Y} = \begin{bmatrix} 6.4779 \\ -0.0149 \\ 0.0218 \\ 0.2916 \end{bmatrix}_{4\times1}.$$

Finally,

$$\hat{\mathbf{A}} = (\mathbf{X}^T \mathbf{X})^{-1} (\mathbf{X}^T \mathbf{Y}),$$

$$= \begin{bmatrix} 0.1667 & 3.9610 \times 10^{-4} & -9.5296 \times 10^{-4} & -9.5074 \times 10^{-6} \\ 3.9610 \times 10^{-4} & 60.0281 & -72.3968 & 11.3974 \\ -9.5296 \times 10^{-4} & -72.3968 & 122.7018 & -6.8725 \\ -9.5074 \times 10^{-6} & 11.3974 & -6.8725 & 3.9544 \end{bmatrix}_{4 \times}$$

$$\cdot \begin{bmatrix} 6.4779 \\ -0.0149 \\ 0.0218 \\ 0.2916 \end{bmatrix}_{4 \times 1},$$

or

$$\hat{\mathbf{A}} = \begin{bmatrix} \hat{A}'_0 \\ \hat{A}_1 \\ \hat{A}_2 \\ \hat{A}_3 \end{bmatrix}_{4 \times 1} = \begin{bmatrix} 1.0796 \\ 0.8521 \\ 1.7467 \\ 0.8335 \end{bmatrix}_{4 \times 1}.$$

Therefore,

$$\hat{A}'_0 = 1.0796,$$
$$\hat{A}_1 = 0.8521,$$
$$\hat{A}_2 = 1.7467,$$

and

$$\hat{A}_3 = 0.8335.$$

According to Eqs. (11.73), (11.69), (11.70), (11.71) and (11.72),

$$\hat{A}_0 = \hat{A}'_0 - (\hat{A}_1 \overline{X_1} + \hat{A}_2 \overline{X_2} + \hat{A}_3 \overline{X_3}),$$
$$= 1.0796 - (0.8521 \times 4.8357 + 1.7467 \times 0.9720$$
$$+ 0.8335 \times 1.7897),$$

or

$$\hat{A}_0 = -6.2304,$$
$$\hat{a} = e^{\hat{A}_0} = e^{-6.2304} = 0.00197,$$
$$\hat{b} = \hat{A}_1 = 0.8521,$$
$$\hat{c} = \hat{A}_2 = 1.7467,$$

and

$$\hat{d} = \hat{A}_3 = 0.8335.$$

Consequently, the optimum screening strength equation using the least-squares method is given by

$$SS_{TC}(N_{TC}) = 1 - e^{-\hat{a} \ (T_R^*+0.6)^{\hat{b}} \ [\log_c(e+\dot{T}^*)]^{\hat{c}} \ N_{TC}^{\hat{d}}},$$

or

$$SS_{TC}(N_{TC}) = 1 - e^{-0.00197 \ (T_R^*+0.6)^{0.8521} \ [\log_c(e+\dot{T}^*)]^{1.7467} \ N_{TC}^{0.8335}}.$$

(11.74)

The screening strength values for the given combinations of $(T_R^*, \dot{T}^*, N_{TC})$ are calculated using this model and tabulated in the last row of Table 11.11. It may be seen that Eq. (11.74) fits the observed data much better than the RADC model.

CASE 2 – Historical data are not available.

If there are no historical data available, the empirical models summarized in this chapter can be utilized as a starting point for your screening strength prediction and *ESS* planning. As the data accumulate, you may substitute the available information into your selected models to check the goodness-of-fit and make any necessary modifications using the techniques discussed and illustrated in Case 1. The more data you accumulate, the more improvements you are going to make on your models and the more confidence you are going to gain upon the prediction and planning.

Note that the techniques presented and illustrated in Case 1 apply not only to RADC screening strength models but to Hughes screening strength models as well, and to the failure rate models, and all other models as well.

REFERENCES

1. Victor, S. D., "A Practical Method for Tailoring Environmental Stress Screens," *Proceedings of the Institute of Environmental Sciences*, Anaheim, California, pp. 198-203, 1989.

2. Saari, A. E., et al, *Environmental Stress Screening, RADC TR-86-149*, September 1986.

3. Anderson, J. R., *Environmental Burn-In Effectiveness*, McDonnell Aircraft Company, August 1980.

4. Kube, F., *An Investigation to Determine Effective Equipment Environmental Acceptance Test Methods*, Grumman Aerospace Corporation, April 1973.

5. Hovis, J. B., "Reliability Math Model to Plan and Evaluate an *ESS* or PRAT Program," *Proceedings of the Institute of Environmental Sciences*, San Diego, California, pp. 368-373, 1991.

6. Montgomery, D. C., *Design of Experiments*, 3rd edition, John Wiley, New York, 649 pp., 1991.

Chapter 12

ESS PLANNING

12.1 INTRODUCTION

There have been two basic approaches to the application of environmental stress screening. In one approach, the government explicitly specifies the screens and screening parameters to be used at various assembly levels [1]. Another approach is to have the contractor propose a screening program which is tailored to the specific product and is subject to the approval of the procuring activity. Although both approaches suggest that the screen parameters and levels of each screening program should be tailorable to the equipment to be screened to suit existing requirements, design and factory capabilities, there are some unknowns in the implementation of stress screening. Some of these are:

1. How effective is the stress screen?

2. Has the reliability goal been achieved after a stress screen?

3. What is considered acceptable or unacceptable fallout from a screen?

4. Is an *ESS* inadequate or too excessive?

To answer these questions quantitatively, some key factors should be known prior to the planning of an *ESS* program. These key factors are:

1. Part initial latent defect level.

2. Defect level introduced during production.

3. Effectiveness of the screen.

4. Acceptable level of the number of latent defects which escape the screen.

The *ESS* programs developed in [2] and [3] provide quantitative approaches for the following:

1. Assessing the number of part defects; i.e., incoming defect density.

2. Evaluating the effectiveness of screening for various stress screening parameters; e.g., random vibration, temperature cycling, etc.

3. Designing an *ESS* procedure to remove and thereby reduce the defects to an acceptable level thus ensuring that the customer receives the equipment that meets its reliability requirement.

The application of quantitative approaches to stress screening enables the establishment of explicit quantitative objectives and evaluation for effectiveness of screening, and provides a basis for planning, monitoring and controlling the screening process to meet those objectives.

These methods were developed empirically from scant data and are not considered generally applicable to all product types. In applying these procedures, the actual fallout should be compared with the expected fallout, and adjustments should be made to the initial defect and screening effectiveness estimates to obtain a quantitative model tailored to each manufacturer's products.

12.2   ESTABLISHMENT OF THE REMAINING DEFECT DENSITY OBJECTIVES

The fundamental purpose of *ESS* is to precipitate the latent defects and remove them from the product prior to field delivery. Not all latent defects, however, are capable of being eliminated from the equipment in the factory by use of stress screening. The quantitative approach focuses on the defects which remain in the product at delivery and their impact on field reliability. The remaining defect density is a quantitative objective which should be reached after the stress screening and be consistent with the reliability requirements.

Presented next are two methods for the estimation of the remaining defect density in accordance with the required reliability. One is called the chance defect exponential (*CDE*) model which makes use of the failure rate function. The other uses a lower bound on part fraction defectives of 50 PPM (parts per million). Both methods lead to reasonably consistent estimates of the remaining defect density.

12.2.1 *CDE* MODEL FOR REMAINING DEFECT DENSITY ESTIMATION

In the *CDE* model, the system's failure rate in the field environment is given by

$$\lambda_S(T) = \lambda_0 + D_R \bar{\lambda}_D \, e^{-\bar{\lambda}_D \, T}, \tag{12.1}$$

where

$$\lambda_S(T) = \text{system's failure rate at time } T,$$
$$\lambda_0 = \text{predicted failure rate of the system,}$$
$$D_R = \text{average number of latent defects remaining per}$$
$$\text{system at delivery,}$$

and

$$\overline{\lambda}_D = \text{average failure rate of latent defects in}$$
$$\text{the field environment.}$$

For $T = 0$, Eq. (12.1) becomes

$$\lambda_S(0) = \lambda_0 + D_R \overline{\lambda}_D. \tag{12.2}$$

Equation (12.2) means that at the beginning of equipment life, the system's failure rate can be expressed as the sum of λ_0, the predicted failure rate, and $D_R \overline{\lambda}_D$, the product of remaining defect density and the field failure rate of a defect.

Dividing both sides of Eq. (12.2) by λ_0 yields

$$\frac{\lambda_S(0)}{\lambda_0} = 1 + \frac{D_R \overline{\lambda}_D}{\lambda_0}. \tag{12.3}$$

Let

$$\frac{\lambda_S(0)}{\lambda_0} = \frac{\text{Failure rate of system at T=0}}{\text{Specified failure rate of the system}} = K_1,$$

and

$$\frac{\overline{\lambda}_D}{\lambda_0} = \frac{\text{Failure rate of a defect}}{\text{Specified failure rate of the system}} = K_2.$$

Substituting K_1 and K_2 into Eq. (12.3) and rearranging yields

$$D_R = \frac{K_1 - 1}{K_2}. \tag{12.4}$$

According to MIL-STD-217, the average failure rate of a "good" part is in the range of approximately one (1) failure per 10^6 to 10^7 operating hours. It would therefore be reasonable to assume that the average failure rate of a defect in the field, $\overline{\lambda}_D$, might be in the range of 1 failure per 10^3 to 10^5 hours, or $\overline{\lambda}_D \leq 10^{-3}$ fr/hr. It would be also acceptable to assume that the failure rate of the system is 10% greater than the specified failure rate at the start of equipment life; i.e., $K_1 = 1.1$. Then,

$$D_R = \frac{K_1 - 1}{K_2},$$
$$= \frac{1.1 - 1}{\overline{\lambda}_D/\lambda_0},$$

TABLE 12.1– Remaining defect density, D_R, versus failure rate for the CDE model.

Failure rate, λ_0, fr/hr	MTBF, hr	Remaining defect density, D_R			
		$\overline{\lambda}_D = 10^{-1}$, fr/hr	$\overline{\lambda}_D = 10^{-2}$, fr/hr	$\overline{\lambda}_D = 10^{-3}$, fr/hr	$\overline{\lambda}_D = 10^{-4}$, fr/hr
0.1	10	0.1	1	10	100
0.01	100	0.01	0.1	1	10
0.005	200	0.005	0.05	0.5	5
0.002	500	0.002	0.02	0.2	2
0.001	1,000	0.001	0.01	0.1	1
0.0005	2,000	0.0005	0.005	0.05	0.5
0.0002	5,000	0.0002	0.002	0.02	0.2
0.0001	10,000	0.0001	0.001	0.01	0.1
0.00001	100,000	0.00001	0.0001	0.001	0.01

or

$$D_R = \frac{0.1\lambda_0}{\overline{\lambda}_D}. \tag{12.5}$$

Table 12.1 gives the values for the remaining defect density, D_R, using Eq. (12.5) and values of $\overline{\lambda}_D$ ranging from 10^{-1} fr/hr to 10^{-4} fr/hr.

If the upper bound of $\overline{\lambda}_D = 10^{-3}$ fr/hr is used, then Eq. (12.5) becomes a simpler relation for the estimate of D_R from λ_0; i.e.,

$$D_R = \frac{0.1\lambda_0}{10^{-3}},$$

or

$$D_R = 100\lambda_0. \tag{12.6}$$

It will be shown in the following section that the CDE model using $K_1 = 1.1$ and $\overline{\lambda}_D = 10^{-3}$, or Eq. (12.6), is equivalent to the PPM model with a remaining defect density goal of 50 PPM.

12.2.2 PPM MODEL FOR REMAINING DEFECT DENSITY ESTIMATION

The estimate of the remaining defect density for an equipment can also be made by using remaining part fraction defectives goals. The following method uses a goal of 50 PPM for the part fraction defectives to estimate the remaining defect density. Given the average failure rate of the good parts and good interconnections in the equipment, the system's predicted failure rate, λ_0, is given by

$$\lambda_0 = N_P \overline{\lambda}_P + N_C \overline{\lambda}_C, \tag{12.7}$$

where

N_P = number of electronic parts,

N_C = number of interconnections,

$\overline{\lambda}_P$ = average failure rate of the parts,

and

$\overline{\lambda}_C$ = average failure rate of interconnections.

From MIL-HDBK-217, reasonable values for $\overline{\lambda}_P$ and $\overline{\lambda}_C$ are 0.5×10^{-6} fr/hr and 0.0003×10^{-6} fr/hr or 0.3 BITS, respectively. It has been shown [3] that the average number of interconnections per part is about 3. Substituting these values into Eq. (12.7) yields

$$\lambda_0 = 0.5\,N_P \times 10^{-6} + 3 \times 0.0003\,N_P \times 10^{-6}. \tag{12.8}$$

Equation (12.8) shows that the interconnection contribution to the system failure rate can be neglected. Thus, the system's predicted failure rate can be approximated by

$$\lambda_0 \cong 0.5\,N_P \times 10^{-6}. \tag{12.9}$$

Therefore, using Eq. (12.6) yields

$$\begin{aligned} D_R &= 100\lambda_0, \\ &= 100\left(0.5\,N_P \times 10^{-6}\right), \\ &= 50\,N_P \times 10^{-6}, \end{aligned}$$

or

$$D_R^* = 50\,N_P \ (PPM), \text{ or } N_P \cdot (50\,PPM). \tag{12.10}$$

That is, Eq. (12.6) corresponds to a goal of 50 PPM for the part fraction defectives.

The estimate of the system's predicted failure rate as a function of the number of parts used in the system is shown in Table 12.2 using Eq. (12.9). By comparing Table 12.1 with Table 12.2, it may be seen that the remaining defect density, D_R^*, obtained by the PPM method, is consistent with Eq. (12.4) in the CDE model, for $K_1 = 1.1$ and $\overline{\lambda}_D = 10^{-3}$ fr/hr, or equivalently with Eq. (12.6).

The estimate using the PPM method is consistent with the results of the CDE method. In practice, the CDE method, and particularly Eq. (12.6), is used in estimating the remaining defect density.

TABLE 12.2 – Remaining defect density versus predicted failure rate for the *PPM* method.

N_P, K	$\lambda_S = N_P(0.5 \times 10^{-6})$ fr/hr	MTBF, hr	$D_R^* = N_P \cdot 50 \ PPM$
20	0.01	100	1
10	0.05	200	0.5
5	0.0025	400	0.25
2	0.001	1,000	0.1
1	0.0005	2,000	0.05
0.2	0.0001	10,000	0.01
0.1	0.00005	20,000	0.005

12.3 INCOMING DEFECT DENSITY ESTIMATION

The latent defects are introduced into equipment through defective parts and manufacturing assembly process deficiencies. The number and type of part defects present depend on part physical construction, materials and processes used in fabrication and testing, the manufacturer's internal screening, and the degree of design protection against the damages due to handling, transportation and storage.

It is expected that there is a correlation between random failure rates and the initial fraction defective for each part type.

12.3.1 DATA SOURCES

There are not enough published data from which the initial fraction defectives can be estimated.

One data source [4] which provides limited data on factory defect rates and field failure rates for parts of various quality grades was used in research conducted by Hughes Aircraft Company [2] to derive the initial part fraction defectives for several part types. The result were adopted in [3].

This data set, given in Table 12.3, includes four types of parts: Microelectronic devices, transistors, diodes and capacitors with different quality grades. The data were collected from factory and ground base operation.

12.3.2 DETERMINATION OF INITIAL FRACTION DEFECTIVES

It is assumed that the Ground Base Operation and Factory Operation correspond to the Ground, Fixed Environment of MIL-HDBK-217. The fraction

TABLE 12.3– Factory and field failure rates for various part types.

Part type	Quality grade	Part type quantity, PTQ	Factory defect rate, FDR *PPM*	Field part hours, FPH $(\times 10^6)$	Field failure rate, FFR, fr/hr, $(\times 10^{-6})$	Field defect rate[†], FIDR, *PPM*
Microelec-tronic devices	C-1	624,087	160	8,580	0.025	343
Transistors	JAN	107,398	60	1,536	0.020	286
Diodes	JAN	206,133	50	1,861	0.004	36
Capacitors	ER-M	1,292,967	32	1,735	0.022	30

† Ground, Fixed Environment.

defective for transistors, diodes and capacitors is computed as follows:

$$FD = FDR + \frac{FFR \cdot FPH}{PTQ} = FDR + FFD, \tag{12.11}$$

where

FD = fraction defective,

FDR = factory defect rate, as given in Table 12.3,

FFR = field failure rate, as given in Table 12.3,

FPH = field part hours, as given in Table 12.3,

PTQ = part type quantity, as given in Table 12.3,

$FIDR$ = field defect rate, as given in Table 12.3,

FFD = field fraction defectives,

and

$$FFD = \frac{FFR \cdot FPH}{PTQ}. \tag{12.12}$$

For example, for transistors in Table 12.3,

$FDR = 60$ PPM,

$FFR = 0.020$ fr/10^6hr,

$FPH = 1,536 \times 10^6$ hr,

and

$PTQ = 107,398.$

Substituting these into Eq. (12.11) yields

$$FD = 60 + \frac{(0.020)(1,536 \times 10^6)}{107,398},$$
$$= 60 + 286.0389,$$

or

$$FD \cong 346 \text{ PPM}.$$

Similarly, for diodes,

$$FD = 50 + \frac{(0.004)(1,861 \times 10^6)}{206,133},$$
$$= 50 + 36.1126,$$

or

$$FD \cong 86 \text{ PPM}.$$

For capacitors,

$$FD = 32 + \frac{(0.022)(1,735 \times 10^6)}{1,292,967},$$
$$= 32 + 29.5212,$$

or

$$FD \cong 62 \text{ PPM}.$$

12.3.3 INITIAL FRACTION DEFECTIVE OF DIFFERENT QUALITY GRADES

With the quality grades of the part types in the data set, the fraction defectives for different quality grades can be calculated using the quality factor, π_Q, of MIL-HDBK-217; i.e.,

$$FD = \frac{\pi_{Q1}}{\pi_{Q2}} \cdot FD_2, \qquad\qquad (12.13)$$

where

π_{Q1} = quality factor for which the fraction defectives is being calculated,

π_{Q2} = quality factor given in Table 12.3, or C-1 for microelectronic devices, JAN for transistors and diodes, and ER-M for capacitors,

and

$$FD_2 = \text{fraction defectives with quality factor } \pi_{Q2}.$$

For example, if the fraction defectives of transistors with JANTX quality level is to be determined, then

$$\pi_{Q1} = \pi_Q(JANTX).$$

From Table 12.3,

$$\pi_{Q2} = \pi_Q(JAN),$$

and

$$FD_2 = FD(JAN).$$

From MIL-HDBK-217,

$$\pi_Q(JANTX) = 0.24,$$

and

$$\pi_Q(JAN) = 1.2.$$

From Section 12.3.2, $FD(JAN)$ for transistors is

$$FD(JAN) = 346 \text{ PPM}.$$

Then,

$$FD = \frac{0.24}{1.2}(346) = 69.2 \text{ PPM}.$$

12.3.4 INITIAL FRACTION DEFECTIVES OF OTHER OPERATION ENVIRONMENTS

The fraction defectives in other operation environments can be estimated by adjusting the field fraction defectives portion of total fraction defectives with the environment factor, π_E, of MIL-HDBK-217, while the factory fraction defectives portion remains as in a fixed ground environment, GF; i.e.,

$$FD = FDR(GF) + FFD(GF)\frac{\pi_{E1}}{\pi_E(GF)}, \tag{12.14}$$

where

$$\pi_{E1} = \text{operation environment factor for which the fraction}$$
$$\text{defectives is being calculated,}$$

and

> $FFD(GF)$ = field fraction defectives under fixed ground
> environment (GF) as given in the last
> column of Table 12.3.

For example, the fraction defectives for a JAN quality level transistor in a G_M environment is

$$FD = FDR(GF) + FFD(GF)\frac{\pi_E(G_M)}{\pi_E(GF)},$$

$$= 60 + (286)\left(\frac{18}{5.8}\right),$$

or

$$FD = 948 \text{ PPM}.$$

12.3.5 INITIAL FRACTION DEFECTIVES OF ANY QUALITY GRADE IN ANY OPERATION ENVIRONMENT

The general calculation for the fraction defectives of any part type (except of microelectronic devices) of any quality factor used in any operation environment is given by combining Eqs. (12.13) and (12.14), or

$$FD = \frac{\pi_{Q1}}{\pi_{Q2}} \cdot \left[(FFD)\frac{\pi_{E1}}{\pi_E(GF)} + FDR \right], \qquad (12.15)$$

where

> π_{Q1} = quality factor for which the fraction defectives
> is being calculated,
>
> π_{Q2} = quality factor in the data set,
>
> FFD = field fraction defectives,
>
> π_{E1} = operation environment factor for which the fraction
> defectives is being calculated,

and

> FDR = factory defects rate given in Table 12.3.

The fraction defectives of transistors, diodes and capacitors for any quality grade and any operation environment are given in Tables 12.4 through 12.6, respectively.

TABLE 12.4– Part fraction defectives for transistors, PPM.

Environment	Quality level				
	JANTXV	JANTX	JAN	Lower	Plastic
GB	10.9	21.9	109.3	546.6	1,093.2
GF	34.6	69.2	346.0	1,730.2	3,460.4
GM	98.8	189.5	947.7	4,738.5	9,477.0
MP	65.2	130.4	651.8	3,259.0	6,518.0
MSB	54.3	108.7	543.3	2,716.5	5,433.1
NS	54.3	108.7	543.3	2,716.5	5,433.1
NU	109.6	219.1	1,095.7	5,478.3	1,0956.6
NH	99.7	199.4	997.0	4,985.1	9,970.2
NUU	104.6	209.3	1,046.3	5,231.7	10,463.4
ARW	139.2	278.3	1,391.6	6,957.8	13,915.6
AIC	52.9	105.7	528.5	2,642.6	5,285.1
AIT	80.0	160.0	799.8	3,998.8	7,997.5
AIB	178.6	357.2	1,786.1	8,930.5	17,860.9
AIA	104.6	209.3	1,046.3	5,231.7	10,463.4
AIF	203.3	406.5	2,032.7	10,163.4	20,326.8
AUC	80.0	160.0	799.8	3,998.8	7,997.5
AUT	129.3	258.6	1,292.9	6,464.6	12,929.2
AUB	301.9	603.8	3,019.0	15,095.1	30,190.1
AUA	178.6	357.2	1,786.1	8,930.5	17,860.9
AUF	326.6	653.1	3,265.6	16,328.0	32,656.0
SF	8.0	15.9	79.7	398.6	797.3
MFF	65.2	130.4	651.8	3,259.0	6,518.0
MFA	89.8	179.7	898.4	4,491.9	8,983.9
USL	183.5	376.1	1,835.4	9,177.0	18,354.1
ML	208.2	416.4	2,082.0	10,410.0	20,819.9
CL	3,408.9	6,817.7	34,088.7	170,443.3	340,886.7

TABLE 12.5– Part fraction defectives for diodes, PPM.

Environ-	Quality level					
ment	JANS	JANTXV	JANTX	JAN	Lower	Plastic
GB	1.2	5.9	11.8	59.2	296.2	592.3
GF	1.7	8.6	17.2	86.0	430.0	860.0
GM	4.3	21.6	43.2	216.2	1,080.8	2,161.5
MP	3.2	16.1	32.2	160.8	803.8	1,607.7
MSB	1.9	9.4	18.9	94.2	471.5	943.1
NS	1.9	9.4	18.9	94.3	471.5	943.1
NU	4.9	24.4	48.8	243.8	1,219.2	2,438.5
NH	4.5	22.5	45.1	225.4	1,126.9	2,253.8
NUU	4.7	23.5	46.9	234.6	1,173.1	2,346.2
ARW	6.0	29.9	59.6	299.2	1,496.2	2,992.3
AIC	3.8	18.8	37.7	188.5	942.3	1,884.6
AIT	4.7	23.5	46.9	234.6	1,173.1	2,346.2
AIB	6.5	32.7	65.4	326.9	1,634.6	3,269.2
AIA	5.6	28.1	56.2	280.8	1,403.8	2,807.7
AIF	7.5	37.3	74.6	373.1	1,865.4	3,730.8
AUC	5.6	28.1	56.2	280.8	1,403.8	2,807.7
AUT	6.5	32.7	65.4	326.9	1,634.6	3,269.2
AUB	10.2	51.2	102.3	511.5	2,557.7	5,115.4
AUA	8.4	41.9	83.8	419.2	2,096.2	4,192.3
AUF	10.2	51.2	102.3	511.5	2,557.7	5,115.4
SF	1.2	5.9	11.8	59.2	296.2	592.3
MFF	3.2	16.1	32.2	160.8	803.8	1,607.7
MFA	4.1	20.7	41.4	206.9	1,034.6	2,069.2
USL	7.6	38.2	76.5	382.3	1,911.5	3,823.1
ML	8.6	42.8	85.7	428.5	2,142.3	4,284.6
CL	128.4	641.9	1,283.8	6,419.2	32,096.2	64,192.3

TABLE 12.6– Part fraction defectives for capacitors, PPM.

Environ-	Quality level						
ment	S	R	P	M	L	MIL- SPEC	Lower
GB	1.2	3.8	11.5	38.4	115.3	115.3	384.4
GF	1.8	6.2	18.4	61.5	184.5	184.5	615.0
GM	9.0	30.0	89.9	299.8	899.4	899.4	2,998.1
MP	12.7	42.3	126.8	422.8	1,268.4	1,268.4	4,228.1
MSB	5.8	19.2	57.7	192.2	576.6	576.6	1,921.9
NS	6.3	21.1	63.4	211.4	634.2	634.2	2,114.1
NU	14.3	47.7	143.0	476.6	1,429.9	1,429.9	4,766.2
NH	18.4	61.5	184.5	615.0	1,845.0	1,845.0	6,150.0
NUU	20.8	69.2	207.6	691.9	2,075.6	2,075.6	6,918.7
ARW	27.7	92.2	276.7	92.25	2,767.5	2,767.5	9,225.0
AIC	3.5	11.5	34.6	115.3	345.9	345.9	1,153.1
AIT	3.5	11.5	34.6	115.3	345.9	345.9	1,153.1
AIB	5.8	19.2	57.7	192.2	576.6	576.6	1,921.9
AIA	3.5	11.5	34.6	115.3	345.9	345.9	1,153.1
AIF	6.9	23.1	69.2	230.6	691.9	691.9	2,306.2
AUC	8.6	28.8	86.5	288.3	864.8	864.8	2,882.8
AUT	9.2	30.7	92.2	307.5	922.5	922.5	3,075.0
AUB	11.5	38.4	115.3	384.4	1,153.1	1,153.1	3,843.7
AUA	9.2	30.7	92.2	307.5	922.5	922.5	3,075.0
AUF	17.3	57.7	173.0	576.6	1,729.7	1,729.7	5,765.6
SF	0.9	3.1	9.2	30.7	92.2	92.2	307.5
MFF	12.7	42.3	126.8	422.8	1,268.4	1,268.4	4,228.1
MFA	17.3	57.7	173.0	576.6	1,729.7	1,729.7	5,765.6
USL	36.9	123.0	369.0	1,230.0	3,690.0	3,690.0	12,300.0
ML	41.5	138.4	415.1	1,383.7	4,151.2	4,151.2	13,837.5
CL	703.4	2,344.7	7,034.1	23,446.9	70,340.6	70,340.6	234,468.6

12.3.6 INITIAL FRACTION DEFECTIVES FOR MICROELECTRONIC DEVICES

The microelectronic device failure rate model, from MIL-HDBK-217, is given by

$$\lambda_P = \pi_Q \left[C_1 \, \pi_T \, \pi_V + (C_2 + C_3) \, \pi_E \right] \pi_L, \qquad (12.16)$$

where

λ_P = device failure rate in failures/10^6 hours, or 10^{-6} fr/hr,

π_Q = quality factor,

π_T = temperature acceleration factor,

π_V = voltage derating stress factor,

π_E = operation environment factor,

C_1 and C_2 = circuit complexity failure rates,

C_3 = package complexity failure rate,

and

π_L = learning factor.

In Eq. (12.9) the operation environment factor, π_E, is not used as a direct multiplier. The determination of the fraction defectives for microelectronic devices is slightly different than those of the other parts.

The following parameter values apply to the microelectronic devices in Table 12.3:

$\pi_E = 4.0$,

$\pi_L = 1.0$,

$\pi_Q = 13.0$,

$\pi_T = 0.032$,

$\pi_V = 1.0$,

and

$C_1 = 0.0053$.

Substituting the foregoing parameter values into Eq. (12.16) yields

$$\lambda_P = 13.0[(0.0053)(0.032)(1.0) + (C_1 + C_2)(4.0)](1.0),$$

or

$$\lambda_P = 13.0[0.00017 + (C_2 + C_3)(4.0)]. \qquad (12.17)$$

From Table 12.3, the observed failure rate for microelectronic devices is 0.025 fr/10^6hr. Then, the observed value for $(C_2 + C_3)$ can be obtained by substituting $\lambda_P = 0.025 \times 10^{-6}$ fr/hr into Eq. (12.17) and solving for $(C_2 + C_3)$; i.e.,

$$C_2 + C_3 = \frac{0.025/13.0 - 0.00017}{4.0},$$

or

$$C_2 + C_3 = 0.00044.$$

Therefore, the field failure rate estimate for any desired quality factor, π_{Q1}, and application environment factor, π_{E1}, is given by substituting $(C_2 + C_3) = 0.00044$ into Eq. (12.17) and replacing 13.0 by π_{Q1} and 4.0 by π_{E1}; i.e.,

$$\hat{\lambda}_P = \pi_{Q1} (0.00017 + 0.00044 \, \pi_{E1}) \times 10^{-6} \, \text{fr/hr}, \tag{12.18}$$

where

$$\pi_{Q1} = \text{desired quality factor,}$$

and

$$\pi_{E1} = \text{desired operation environment factor.}$$

From Eq. (12.12),

$$FFD = \frac{FFR \cdot FPH}{PTQ},$$

where

$$FFR = \text{field failure rate} = \hat{\lambda}_P,$$
$$FPH = \text{field part hours,}$$
$$= 8,580 \times 10^6 \text{ hr for microelectronic devices as}$$
given in Table 12.3,

and

$$PTQ = \text{part type quantity,}$$
$$= 624,087 \text{ for microelectronic devices as}$$
given in Table 12.3.

Therefore,

$$FFD = \hat{\lambda}_P \left(\frac{8,580 \times 10^6}{624,087} \right) = 1.37481 \times 10^4 \, \hat{\lambda}_P. \tag{12.19}$$

Substituting Eq. (12.18) into Eq. (12.19) yields

$$FFD = (1.37481 \times 10^4)\, \pi_{Q1}\, (0.00017 + 0.00044\, \pi_{E1}) \times 10^{-6} \text{ fr/hr},$$

or

$$FFD = \pi_{Q1}\, (2.3372 + 6.0492\, \pi_E) \times 10^{-6}. \tag{12.20}$$

The factory fraction defectives can be adjusted by the quality factor. The total fraction defectives is the sum of field fraction defectives and factory fraction defectives. The results are in Table 12.7.

12.3.7 INITIAL FRACTION DEFECTIVES ESTIMATES FOR OTHER PART TYPES

With the assumption that there is a correlation between failure rate and fraction defectives, the fraction defectives for all other part types not included in Table 12.3 can be estimated based on their calculated failure rate. The baseline failure rates for each part and connection type for the highest quality level available and for a Ground, Fixed application environment are given in Table 12.8 [2].

Initial fraction defectives were calculated for each part and connection type by scaling the microelectronic device initial fraction defectives by failure rate ratios. Take the fraction defectives of resistors as an example, then

$$FD(\text{Resistor}) = \frac{\lambda(\text{Resistor}) \cdot FD(\text{Microelec. device})}{\lambda(\text{Microelec. device})},$$

where

$$\begin{aligned}
\lambda(\text{Resistor}) \quad &= \text{resistor failure rate,} \\
&= 0.00207 \times 10^{-6} \text{ fr/hr as given in Table 12.4,}
\end{aligned}$$

$$\begin{aligned}
FD(\text{Microelec. device}) &= \text{fraction defectives for C-1 quality} \\
&\quad \text{level microelectronic device in} \\
&\quad GF \text{ environment,} \\
&= 503.2 \times 10^{-6} \text{ as given in Table 12.8,}
\end{aligned}$$

and

$$\begin{aligned}
\lambda(\text{Microelec. device}) &= \text{failure rate for the microelectronic device,} \\
&= 0.05123 \times 10^{-6} \text{ fr/hr as given in} \\
&\quad \text{Table 12.4.}
\end{aligned}$$

Then,

$$FD(\text{Resistor}) = \frac{0.00207 \times 10^{-6} \times 503.2 \times 10^{-6}}{0.05123 \times 10^{-6}},$$

TABLE 12.7– Part fraction defectives for microelectronic devices, PPM.

Environ-ment	Quality level								
	S	B	B-0	B-1	B-2	C	C-1	D	D-1
GB	9.2	18.3	36.6	54.9	119.0	146.4	237.9	320.3	640.6
GF	19.4	38.7	77.4	116.1	251.6	309.6	503.2	677.3	1,354.6
GM	27.5	55.1	110.1	165.2	357.9	440.5	715.8	963.6	1,927.2
MP	25.6	51.2	102.4	153.6	332.9	409.7	665.8	896.3	1,792.5
MSB	26.6	53.1	106.3	159.4	345.4	425.1	690.8	929.9	1,859.9
NS	26.6	53.1	106.3	159.4	345.4	425.1	690.8	929.9	1,859.9
NU	34.7	69.5	139.0	208.5	451.7	556.0	903.5	1,216.2	2,432.5
NH	35.7	71.4	142.8	214.3	464.3	571.4	928.5	1,249.9	2,499.9
NUU	37.6	75.3	150.5	225.8	489.3	602.2	978.6	1,317.3	2,634.6
ARW	48.2	96.4	192.9	289.3	626.9	771.6	1,253.8	1,687.8	3,375.6
AIC	19.4	38.7	77.4	116.1	251.6	309.6	503.2	677.3	1,354.6
AIT	21.8	43.5	87.0	130.5	282.9	348.1	565.7	761.5	1,523.1
AIB	31.4	62.8	125.5	188.3	408.0	502.1	815.9	1,098.4	2,196.7
AIA	26.6	53.1	106.3	159.4	345.4	425.1	690.8	929.9	1,859.9
AIF	36.2	72.4	144.8	217.2	470.5	579.1	941.0	1,266.8	2,533.5
AUC	21.8	43.5	87.0	130.5	282.9	348.1	565.7	761.5	1,523.1
AUT	26.6	53.1	106.3	159.4	345.4	425.1	690.8	929.9	1,859.9
AUB	43.4	86.8	173.6	260.5	564.3	694.6	1,127.7	1,519.4	3,038.8
AUA	36.2	72.4	144.8	217.2	470.5	579.1	941.0	1,266.8	2,533.5
AUF	50.6	101.3	202.5	303.8	658.2	810.1	1,316.4	1,772.0	3,544.0
SF	11.7	23.3	46.6	69.9	151.5	186.4	303.0	407.9	815.7
MFF	26.1	52.2	104.4	156.5	339.2	417.4	678.3	913.1	1,826.2
MFA	33.3	66.6	133.2	199.8	433.0	532.9	866.0	1,165.7	2,331.4
USL	60.3	120.5	241.0	361.5	783.3	964.0	1,566.6	2,108.8	4,217.7
ML	69.9	139.8	279.5	419.3	908.4	1,118.0	1,816.8	2,445.7	4,891.3
CL	1,065.9	2,131.8	4,263.7	6,395.5	13,857.0	17,054.8	27,714.0	37,307.4	74,614.7

TABLE 12.8– Calculated failure rates for parts and connections.

Part or connection type	Calculated failure rate in 10^{-6} fr/hr (MIL-HDBK-217D, Notice 1)
Microelectronic devices	0.05123
Resistors	0.00207
Capacitors	0.00203
Inductive devices	0.1253
Rotating devices	1.195
Relays	0.02371
Switches	0.004506
Connectors	0.00852
Printed wiring boards	0.07073
Connections	
Hand solder	0.0026
Crimp	0.00026
Weld	0.00005
Solderless wrap	0.0000035
Wrapped and soldered	0.00014
Clip termination	0.00012
Reflow solder	0.000069

or

$$FD(\text{Resistor}) = 20.3 \times 10^{-6}, \text{ or } 20.3 \text{ PPM.}$$

This is the fraction defectives of resistors in quality level M, as given in Table 12.9. The rest of the table can be calculated using Eq. (12.11), the basic fraction defectives and the other quality factors in MIL-HDBK-217.

In extrapolating to other operation environments, the factory and field fraction defectives must be considered separately because the factory fraction defectives portion remains constant with environment changes. For example,

The total fraction defectives for microelectronic devices

= Factory fraction defectives + field fraction defectives,

= 160+343.2 $(GF, \text{C-1})$,

from Table 12.3. The fraction defectives for resistors is calculated by multiplying the terms by the failure rate ratios in Table 12.4 as follows:

Fraction defectives for resistors (GF, M)

$$= 160 \times \frac{0.00207}{0.05123} + 343.2 \times \frac{0.00207}{0.05123},$$

$$= 6.46 + 13.87.$$

Then, we can extrapolate the total fraction defectives to other operation environments by multiplying the second term by $\pi_E/2.4$ and then adding the first term which remains constant. Tables 12.9 through 12.16 are calculated using this method.

12.4   SCREENING PARAMETER SELECTION

Stress screens are not all equally effective in transforming latent defects into detectable failures. The nature of defects varies with equipment type, manufacturer and time. Different stress types have different effects in precipitating the latent defects.

The commonly used screening parameters, such as those of temperature cycling, constant temperature burn-in, random vibration and swept-sine vibration, are summarized as follows:

12.4.1   THERMAL CYCLING PARAMETERS

1. Maximum temperature, T^*_{max} – The maximum temperature to which the screened item will be exposed.

2. Minimum temperature, T^*_{min} – The minimum temperature to which the screened item will be exposed.

TABLE 12.9– Part fraction defectives for resistors, PPM.

Environment	\multicolumn{6}{c}{Quality level}

Environment	S	R	P	M	MIL-SPEC	Lower
GB	0.4	1.2	3.7	12.3	61.4	184.2
GF	0.6	2.0	6.1	20.3	101.7	305.2
GM	1.5	5.1	15.4	51.5	257.4	772.3
MP	1.7	5.7	17.2	57.2	286.2	858.7
NSB	0.9	3.1	9.2	30.7	153.6	460.9
NS	1.0	3.4	10.1	33.6	168.1	504.2
NU	2.6	8.7	26.2	87.2	436.2	1,308.5
NH	2.6	8.7	26.2	87.2	436.2	1,308.5
NUU	2.8	9.3	27.9	93.0	465.0	1,395.0
ARW	3.5	11.6	34.8	116.1	580.3	1,740.9
AIC	0.6	2.1	6.3	20.9	104.6	313.9
AIT	0.7	2.4	7.1	23.8	119.0	357.1
AIB	1.3	4.4	13.2	44.0	219.9	659.8
AIA	1.2	4.1	12.3	41.1	205.5	616.6
AIF	1.8	5.8	17.5	58.4	292.0	876.0
AUC	1.4	4.7	14.1	46.9	234.4	703.1
AUT	1.3	4.4	13.2	44.0	219.9	659.8
AUB	2.8	9.3	27.9	93.0	465.0	1,395.0
AUA	2.8	9.3	27.9	93.0	465.0	1,395.0
AUF	3.7	12.2	36.5	121.8	609.1	1,827.4
SF	0.3	0.9	2.6	8.8	44.1	132.3
MFF	1.7	5.8	17.3	57.8	289.1	867.4
MFA	2.3	7.6	22.7	75.7	378.5	1,135.5
USL	4.7	15.6	46.9	156.4	782.1	2,346.3
ML	5.4	17.9	53.8	179.5	897.4	2,692.2
CL	88.4	294.7	884.1	2,947.0	14,735.0	44,205.0

TABLE 12.10– Part fraction defectives for inductive devices, PPM.

Environment	Quality level	
	MIL-SPEC	Lower
GB	537.2	1,790.7
GF	1,222.9	4,078.4
GM	1,996.1	7,140.1
MP	2,142.0	6,653.8
NSB	1,135.4	3,784.6
NS	1,222.9	4,076.4
NU	2,433.8	8,112.7
NH	2,725.6	9,085.3
NUU	3,017.4	10,058.0
ARW	3,892.7	12,975.8
AIC	1,047.8	3,492.8
AIT	1,266.7	4,222.3
AIB	1,266.7	4,222.3
AIA	1,266.7	4,222.3
AIF	1,704.4	5,681.2
AUC	1,339.6	4,465.4
AUT	1,339.6	4,465.4
AUB	1,485.5	4,951.7
AUA	1,485.5	4,951.7
AUF	1,850.3	6,167.5
SF	537.2	1,790.7
MFF	1,996.1	6,653.8
MFA	2,579.7	8,599.0
USL	5,059.9	16,866.2
ML	5,643.4	18,811.5
CL	89,385.3	297,951.1

TABLE 12.11– Part fraction defectives for rotating devices, PPM.

Environment	Fraction defective (Defects/10^6)
GB	5,935.2
GF	11,663.1
GM	30,168.5
MP	27,965.5
NSB	14,967.6
NS	16,289.4
NU	34,574.6
NH	38,980.6
NUU	43,386.7
ARW	56,604.8
AIC	12,544.3
AIT	13,645.8
AIB	15,848.8
AIA	13,645.8
AIF	23,559.4
AUC	14,747.3
AUT	18,051.9
AUB	20,254.9
AUA	18,051.9
AUF	25,762.5
SF	5,935.2
MFF	27,965.5
USL	74,229.1
ML	83,041.2
CL	*******

TABLE 12.12– Part fraction defectives for relays, PPM.

Environment	Quality level	
	MIL-SPEC	Lower
GB	142.5	210.9
GF	231.4	388.8
GM	635.1	1,784.5
MP	1,510.8	4,384.3
NSB	621.4	1,716.0
NS	621.4	1,716.0
NU	1,031.9	2,673.9
NH	2,263.4	6,642.0
NUU	2,400.2	6,915.7
ARW	3,221.2	9,652.3
AIC	450.3	724.0
AIT	484.5	1,100.3
AIB	758.2	1,442.4
AIA	587.2	1,100.3
AIF	758.2	1,784.5
AUC	621.4	1,442.4
AUT	689.8	1,784.5
AUB	1,100.3	2,810.7
AUA	758.2	2,126.5
AUF	1,100.3	3,152.8
SF	142.5	210.9
MFF	1,510.8	4,384.3
MFA	2,058.1	5,684.2
USL	4,215.8	13,073.1
ML	4,931.6	14,441.4
CL	N/A	N/A

TABLE 12.13– Part fraction defectives for switches, PPM.

Environment	Quality level MIL-SPEC	Lower
GB	1.4	24.4
GF	2.4	44.0
GM	8.8	158.4
MP	12.8	230.6
NSB	5.3	95.5
NS	5.3	95.5
NU	12.2	220.3
NH	19.1	344.1
NUU	20.3	364.7
ARW	27.1	488.4
AIC	5.4	96.6
AIT	5.4	96.6
AIB	9.4	168.8
AIA	9.4	168.8
AIF	12.2	220.3
AUC	6.5	117.2
AUT	6.5	117.2
AUB	12.2	220.3
AUA	12.2	220.3
AUF	15.1	271.9
SF	1.4	24.4
MFF	12.8	230.6
MFA	17.4	313.1
USL	36.9	663.7
ML	41.5	746.2
CL	688.3	12,388.6

TABLE 12.14– Part fraction defectives for connections, PPM.

Environment	Connection type						Crimp			
	Hand solder	Weld	Solder-less wrap	Wrapped and soldered	Clip term	Reflow solder	Auto	Man., upper	Man., std.	Man., lower
GB	12.0	0.2	0.02	1.0	1.0	0.3	1.2	1.2	2.5	24.8
GF	26.0	0.5	0.03	1.0	1.0	0.7	2.6	2.6	5.2	52.0
GM	90.0	1.7	0.12	5.0	4.0	2.4	9.0	9.0	18.1	180.8
MP	90.0	1.7	0.12	5.0	4.0	2.4	9.0	9.0	18.1	180.8
NSB	43.0	0.8	0.06	2.0	2.0	1.1	4.3	4.3	8.7	86.7
NS	54.0	1.0	0.07	3.0	3.0	1.4	5.4	5.4	10.9	109.0
NU	123.0	2.4	0.16	7.0	6.0	3.3	12.3	12.3	24.5	245.1
NH	136.0	2.6	0.18	7.0	6.0	3.6	13.6	13.6	27.2	272.4
NUU	149.0	2.9	0.20	8.0	7.0	3.9	14.9	14.9	29.7	297.1
ARW	198.0	3.8	0.27	11.0	9.0	5.3	19.8	39.6	39.6	396.2
AIC	31.0	0.6	0.04	2.0	1.0	0.8	3.1	3.1	6.2	61.9
AIT	56.0	1.1	0.07	3.0	3.0	1.5	5.6	5.6	11.1	111.4
AIB	68.0	1.3	0.09	4.0	3.0	1.8	6.8	6.8	13.6	136.2
AIA	62.0	1.2	0.08	3.0	3.0	1.6	6.2	6.2	12.4	123.8
AIF	93.0	1.8	0.12	5.0	4.0	2.5	9.3	9.3	18.6	185.7
AUC	37.0	0.7	0.05	2.0	2.0	1.0	3.7	3.7	7.4	74.3
AUT	74.0	1.4	0.10	4.0	3.0	2.0	7.4	7.4	14.9	148.6
AUB	93.0	1.8	0.12	5.0	4.0	2.5	9.3	9.3	18.6	185.7
AUA	87.0	1.7	0.12	5.0	4.0	2.3	8.7	8.7	17.3	173.3
AUF	118.0	2.3	0.16	6.0	5.0	3.1	11.8	11.8	23.5	235.2
SF	12.0	0.2	0.02	1.0	1.0	0.3	1.2	2.5	2.5	24.8
MFF	90.0	1.7	0.12	5.0	4.0	2.4	9.0	9.0	18.1	180.8
MFA	124.0	2.4	0.17	7.0	6.0	3.3	12.4	12.4	24.8	247.6
USL	272.0	5.2	0.37	15.0	13.0	7.2	27.2	27.2	54.5	544.8
ML	310.0	6.0	0.42	11.0	14.0	8.2	31.0	31.0	61.9	619.0
CL	5200.0	100.0	7.00	280.0	240.0	138.0	520.0	520.0	1,040.0	10,400.0

TABLE 12.15– Part fraction defectives for connectors, PPM.

Environment	Quality level	
	MIL-SPEC	Lower
GB	73.7	97.3
GF	83.2	248.1
GM	417.7	1,204.6
MP	427.1	827.7
NSB	219.8	408.3
NS	276.3	544.9
NU	639.2	1,298.9
NH	639.2	1,251.8
NUU	686.3	1,346.0
ARW	921.9	1,770.1
AIC	120.9	497.8
AIT	168.0	497.8
AIB	238.7	733.4
AIA	215.1	733.4
AIF	332.9	969.0
AUC	262.2	733.4
AUT	403.6	733.4
AUB	497.8	969.0
AUA	474.3	969.0
AUF	733.4	1,440.2
SF	73.7	97.3
MFF	427.1	827.7
MFA	592.1	1,157.5
USL	1,204.6	2,382.7
ML	1,393.1	2,759.6
CL	23,115.8	45,733.8

TABLE 12.16– Part fraction defectives for printed wiring boards, PPM.

Environment	Quality level	
	MIL-SPEC	Lower
GB	425.0	4,250.0
GF	690.3	6,903.2
GM	1,792.4	17,924.3
MP	1,629.2	16,291.5
NSB	1,057.7	10,576.9
NS	1,302.6	13,026.0
NU	2,670.0	26,700.3
NH	2,874.1	28,741.2
NUU	3,078.2	30,782.2
ARW	4,098.7	40,986.9
AIC	731.1	7,311.4
AIT	1,139.3	11,393.2
AIB	1,853.7	18,536.5
AIA	1,567.9	15,679.2
AIF	2,261.8	22,618.4
AUC	1,751.6	17,516.1
AUT	3,282.3	32,823.1
AUB	5,323.3	53,232.5
AUA	4,302.8	43,027.8
AUF	7,364.2	73,641.9
SF	425.0	4,250.0
MFF	1,996.5	19,965.2
MFA	2,670.0	26,700.3
USL	5,527.3	55,273.5
ML	6,139.6	61,396.3
CL	102,267.9	*******

3. Temperature range, T_R^* – The difference between the maximum and the minimum temperature; i.e.,

$$T_R^* = T_{max}^* - T_{min}^*.$$

4. Temperature rate of change, $\dot{T}^*$ – The average rate of change of temperature for the item to be screened as it transitions between T_{max}^* and T_{min}^*; i.e.,

$$\dot{T}^* = \frac{1}{2} \left[\left(\frac{T_{max}^* - T_{min}^*}{t_1} \right) + \left(\frac{T_{max}^* - T_{min}^*}{t_2} \right) \right],$$

where

t_1 = transition time from T_{min}^* to T_{max}^*,

and

t_2 = transition time from T_{max}^* to T_{min}^*.

5. Dwell – Maintaining the temperature constant when it has reached the maximum or minimum temperature. The duration of the dwell is a function of the thermal mass of the item being screened.

6. Number of thermal cycles, N_{TC} – The number of transitions between temperature extremes divided by two.

12.4.2 CONSTANT TEMPERATURE BURN-IN PARAMETERS

1. Temperature delta, ΔT^* – The absolute value of the difference between the chamber temperature and 25°C, or

$$\Delta T^* = |\, T^* - 25°C\,|,$$

where

T^* = chamber temperature.

2. Duration – The time period over which the temperature is applied to the item, in hr.

12.4.3 RANDOM VIBRATION PARAMETERS

1. Power spectral density (PSD) function – The profile which describes the frequency band and the random vibration power spectral density values over each frequency.

2. *grms* level – The root mean square (rms) value of the applied power spectral density over the vibration frequency spectrum. It is the area under the random vibration power spectral density curve.

3. Duration – The time period over which the vibration excitation is applied to the item being screened, in min.

4. Axes – This can be a single axis or multiple axes depending on the sensitivity of defects to particular axial inputs.

12.4.4 SWEPT-SINE VIBRATION PARAMETERS

1. Frequency range – The frequency band which specifies the lower vibration frequency f_L and the upper vibration frequency f_U.

2. Frequency sweep rate – Frequency change rate against time. The mode of frequency change as a function of time usually has the form of

$$f(t) = f(0)e^{(r \log_e 2)\, t},$$ (12.21)

where

$$f(0) = \text{starting frequency when } t = 0,$$
$$t = \text{sweep time, in min,}$$

and

$$r = \text{sweep rate, in octaves/min.}$$

3. g-level – The constant acceleration applied to the equipment being screened when the frequency sweeps throughout the frequency range above 40 Hz. The g-level below 40 Hz may be less.

4. Duration – The time period over which the swept-sine vibration is applied to the equipment being screened, in min.

   The time it takes to sweep through a frequency band from f_1 to f_2 can be determined from the following expression:

$$t = \frac{\log_e (f_2/f_1)}{r \log_e 2},$$ (12.22)

where

$$f_1 = \text{starting frequency,}$$
$$f_2 = \text{ending frequency,}$$

and

$$r = \text{sweep rate in octaves/min.}$$

An alternative parameter for the duration t is the number of sweeps, N, which relates to the total sweep duration, T, as follows:

$$T = 2N \frac{\log_e (f_2/f_1)}{r \log_e 2},$$ (12.23)

where

N = number of sweeps, which is the number of times that frequency sweeps from f_1 to f_2 and then returns from f_2 to f_1,

f_1 = starting frequency,

f_2 = ending frequency,

and

r = sweep rate.

5. Axes – This can be a single axis or multiple axes depending on the sensitivity of defects to particular axial input.

12.5   SCREENING STRENGTH

Screening strength is the probability that a screen will precipitate a latent defect into a detectable failure, given that a defect is present.

Screening strength equations were developed in [5] for temperature and vibration screening. These equations were revised in [2] based on the data in [6] and [7]. The screening strength equation, expressed as a function of screening time, is given by

$$SS(T_s) = 1 - e^{-\overline{\lambda}_D\, T_s}, \qquad\qquad (12.24)$$

where

$SS(T_s)$ = screening strength as a function of screening time, T_s,

and

$\overline{\lambda}_D$ = average failure rate of defects under a given set of stress conditions.

Most screening strength equations have been summarized in the preceding chapter. For the purpose of ESS planning, the following equations are widely used in the literature.

12.5.1   SCREENING STRENGTH FOR RANDOM VIBRATION

The given screening strength equation for random vibration is [2]

$$SS_{RV}(T_{RV}) = 1 - e^{-0.0046\,(G)^{1.71}\,T_{RV}}, \qquad\qquad (12.25)$$

where

$G = g_{rms}$, rms value of applied acceleration power spectral
density over the frequency spectrum,

and

T_{RV} = duration of applied vibration excitation, in min.

12.5.2 SCREENING STRENGTH FOR SWEPT-SINE VIBRATION

The given screening strength equation for swept-sine vibration is

$$SS_{SSV}(T_{SSV}) = 1 - e^{-0.000727\,(G)^{0.863}T_{SSV}}, \tag{12.26}$$

where

G = g-level, the constant acceleration applied to the
equipment being screened throughout the frequency
range above 40 Hz. The g-level below 40 Hz may be
less,

and

T_{SSV} = duration of swept-sine vibration, in min.

12.5.3 SCREENING STRENGTH FOR SINGLE-FREQUENCY VIBRATION

The given screening strength equation for single-frequency vibration is

$$SS_{SFV}(T_{SFV}) = 1 - e^{-0.00047\,(G)^{0.49}\,T_{SFV}}, \tag{12.27}$$

where

G = g-level of the applied constant acceleration,

and

T_{SFV} = duration of single-frequency vibration, in min.

12.5.4   SCREENING STRENGTH FOR TEMPERATURE CYCLING

The given screening strength equation for temperature cycle is

$$SS_{TC}(N_{TC}) = 1 - e^{-0.0017(T_R^* + 0.6)^{0.6} [\log_e(e + \dot{T}^*)]^3 N_{TC}},$$

(12.28)

where

$$T_R^* = \text{temperature range} = T_{max}^* - T_{min}^*, \text{ in } °C,$$
$$N_{TC} = \text{number of temperature cycles},$$
$$\dot{T}^* = \text{temperature rate of change},$$

or

$$\dot{T}^* = \frac{1}{2}\left[\left(\frac{T_{max}^* - T_{min}^*}{t_1}\right) + \left(\frac{T_{max}^* - T_{min}^*}{t_2}\right)\right],$$

(12.29)

$$t_1 = \text{transition time from } T_{min} \text{ to } T_{max}, \text{ in min},$$

and

$$t_2 = \text{transition time from } T_{max} \text{ to } T_{min}, \text{ in min}.$$

12.5.5   SCREENING STRENGTH FOR CONSTANT TEMPERATURE BURN-IN

The given screening strength equation for constant temperature burn-in is

$$SS_{CT}(T_b) = 1 - e^{-0.0017(T_R^* + 0.6)^{0.6} T_b},$$

(12.30)

where

$$T_R^* = \text{temperature range defined as the absolute value}$$
$$\text{of the difference between the screening}$$
$$\text{temperature and } 25°C,$$

and

$$T_b = \text{burn-in time in hr}.$$

TABLE 12.18– Fault coverage for automatic test systems .

Circuit type	Automatic test system type			
	Loaded board shorts tester, LBST	In-circuit analyzer, ICA	In-circuit tester, ICT	Functional board tester, FBT
Digital	45% to 65 %	50% to 75 %	85% to 94 %	90% to 98 %
Analog	35% to 55 %	70% to 92 %	90% to 96 %	80% to 90 %
Hybrid	40% to 60 %	60% to 90 %	87% to 94 %	83% to 95 %

12.6 TEST DETECTION EFFICIENCY

Test detection efficiency is a measure of test thoroughness, or coverage, which is expressed as the fraction of latent defects detectable by a defined test procedure to the total possible number of latent defects which can be present.

The fault coverage (probability of detection) estimate for various automatic test systems commonly used in today's factories are given in Table 12.18 [8].

An illustration of fault coverage for a sample of 100 printed wiring assemblies (PWA) subjected to various test strategies is provided in Table 12.19 [8].

In some system procurement, the detection efficiency (DE) is a specified parameter for built-in test (BIT) and performance monitoring and fault location (PM/FL) capabilities. If BIT or PM/FL is used primarily to verify performance of an item being screened, the specified DE values should be used.

If DE is not specified, the result of a failure modes and effects analysis (FMEA) can be used to estimate the fraction of defects detectable.

If DE is not specified and FMEA has not been conducted, the estimate of test detection efficiency should be made based upon experience data. The typical values of test detection efficiency for various tests applicable to stress screens are given in Table 12.20 [3].

12.7 TEST STRENGTH

The test strength, TS, is defined as the product of screening strength and test detection efficiency. It is the joint probability that a defect will be precipitated by a screen to a detectable state and detected in a test; i.e.,

$$TS = SS \cdot DE. \qquad (12.31)$$

TABLE 12.19 – Fault detection for a 1,000 PCB (printed circuit board) lot size.

Fault classification	Actual	LBS	ICA	ICT	FBT	ICA-ICT	ICA-FBT	ICT-FBT	ICA-ICT-FBT
Shorts	216	216	216	216	216	216	216	216	216
Opens	5	5	5	5	5	5	5	5	5
Missing components	30		25	28	25	29	27	29	30
Wrong components	67		53	61	55	64	59	60	65
Reversed components	28		26	23	25	27	28	25	28
Bent leads	43		38	43	43	43	43	43	43
Analog specifications	25		13	21	18	21	21	22	23
Digital logic	27			20	27	20	27	27	27
Performance	26				26		26	26	26
Total number of faults	512	266	421	462	486	470	497	498	508
Fault coverage	100%	52%	82%	90%	95%	92%	97%	97%	99%
Fault coverage increase	-	-	-	-	-	2.2%	2.3%	2.5%	4.5%
Rejected PCBs	398	223	345	370	385	374	391	393	394
Rework yield		195	316	354	376	361	384	388	393
Undetected faulty PCB's		203	82	44	22	37	14	10	5
Rework yield		49%	79%	89%	94%	91%	96%	97%	99%
Rework yield increase	-	-	-	-	-	2%	2.1%	3.2%	4.5%
Finished units		805	918	956	978	963	986	990	995

TABLE 12.20 – Detection efficiency versus test types.

Level of assembly	Test type	Detection efficiency
Assembly	Production line "go/no-go" test	0.85
	Production line in-circuit test	0.90
	High performance automatic tester	0.95
Unit	Performance verification test (PV1)	0.90
	Factory checkout	0.95
	Final acceptance test	0.98
System	On-line performance monitoring test	0.90
	Factory checkout test	0.95
	Customer final acceptance test	0.99

For a single screening, the fallout, F, is

$$F = D_{IN} \cdot TS, \tag{12.32}$$

and the relation between remaining defect density, D_R, and incoming defect density, D_{IN}, is given by

$$D_R = D_{IN} - F = D_{IN}(1 - TS). \tag{12.33}$$

For multiple screens at different assembly levels, such as three levels: assemblies, units and systems, the combined test strength is given by

$$TS_C = 1 - (1 - TS_1)(1 - TS_2)(1 - TS_3). \tag{12.34}$$

Assuming that screening is applied on a 100% basis at these assembly levels, D_R is related to D_{IN} by

$$D_R = D_{IN1} \cdot \prod_{i=1}^{3}(1 - TS_i) + D_{IN2} \cdot \prod_{i=2}^{3}(1 - TS_i)$$
$$+ D_{IN3} \cdot (1 - TS_3), \tag{12.35}$$

and

$$D_{IN} = \sum_{i=1}^{3} D_{INi}, \quad i = \text{assembly stages.} \tag{12.36}$$

12.8 PROCEDURE FOR QUANTITATIVE SELECTION

Given the required $MTBF$, or λ_0, of the system, the quantitative approach of ESS planning for an equipment should consist of the following steps:

1. Determine the remaining defect density, D_R, for a given $MTBF$, or λ_0, based on Table 12.1.

2. Identify the equipment to be screened as to assembly level, such as units, assemblies and systems.

3. Determine the fraction defectives in parts per million (PPM) for each assembly level, using Tables 12.15 through 12.16, according to the part types, quality level or grade, quantity of each type of part and connection, and field operation environment.

4. Determine the defect density for each assembly by multiplying the quantity and the fraction defectives for each part to obtain the estimated defects and adding all estimated defects for all parts in the assembly.

Unit defects are estimated by only considering those parts and inter-connections that were not included in the assemblies.

System defects are obtained by estimating the number of defects not included in the unit and assembly levels.

Summing the defect estimates in units, assemblies, and systems is the incoming defect density for the system, D_{IN}.

5. Select screen parameter, level and duration for each assembly level under the guidance of Section 12.4. Determine screening strength, SS, for selected screens using Eqs. (12.25) through (12.30), or using the equations developed based on your own data with the techniques provided in Chapter 11.

6. Determine the Detection Efficiency (DE) of tests to be performed during and after screening for each assembly level under the guidance of Section 12.6 and Tables 12.18 through 12.20.

7. Calculate the test strength by multiplying the screening strengths by their respective detective efficiency, or by Eq. (12.31).

8. Compute D_R using Eq. (12.35). This is an estimate of the number of defects remaining after completing the selected screening sequence. This value of D_R must be equal to or less than the specified system remaining defects, D_R, in Step 1. If not, repeat Steps 5 through 8.

12.9 ECONOMIC OPTIMIZATION OF *ESS* PLANNING

The screen parameters and durations can be optimized economically. The following is a total cost model modified from [9]:

$$C_T(\mathbf{S}, \mathbf{T}_s) = CF \cdot D_R + \sum_{k=1}^{m}[CSU_k + CS_k(\mathbf{S}_k) \cdot T_{sk}], \qquad (12.37)$$

where

$\mathbf{S}$ = vector of applied stress levels, such as $g - rms$ value for random vibration, temperature change rate, lower and upper temperature extremes for thermal cycling,

$\mathbf{T}_s$ = vector of applied stressing durations, such as random vibration time and number of temperature cycles,

$C_T(\mathbf{S}, \mathbf{T}_s)$ = total expected cost of an equipment subjected to *ESS*, which is a function of the applied stress levels, $\mathbf{S}$, and their durations, $\mathbf{T}_s$,

CF = field failure cost of each defect escaping from *ESS*,

D_R = remaining defect density given by Eq.(12.33), which is a function of the applied stress levels, $\mathbf{S}$, and their durations, $\mathbf{T}_s$,

$D_R = D_{IN} \cdot TS$,

D_{IN} = incoming defect density, estimated using methods presented in Section 12.3,

TS = test strength of the given screen profiles and of defect detection equipment and procedures,

$TS = \prod_{k=1}^{m} (1 - TS_k)$, also see Eq. (12.34),

m = total number of stress types applied,

TS_k = test strength of the kth stress type, $k = 1, 2, \cdots, m$,

$TS_k = SS_k \cdot DE$, given by Eq. (12.31),

SS_k = screening strength of the kth stress type,

DE = detection efficiency,

CSU_k = set-up cost for stress type k, $k = 1, 2, \cdots, m$,

$\mathbf{S}_k$ = stress level vector for the kth stress type,

$CS_k(\mathbf{S}_k)$ = cost of screen per unit time using stress type k which is a function of the applied stress levels, $\mathbf{S}_k$,

and

T_{sk} = screen duration of stress type k.

Note that the screening strength, SS_k, can be evaluated using either a RADC model, or using a customer-developed model, obtained using the techniques presented in Chapter 11.

Therefore, the optimum stress levels, $\mathbf{S}$, and their durations, $\mathbf{T}_s$, can be obtained by solving the following minimization problem:

$$\underset{\mathbf{S}, \mathbf{T}_s}{\text{Min}} \quad C_T(\mathbf{S}, \mathbf{T}_s), \tag{12.38}$$

which may be accomplished using an appropriate multi-variate non-linear optimization routine.

REFERENCES

1. Caruso, H., "An Updated Overview of Environmental Stress Screening Standards and Documents for Electronic Assemblies," *Journal of Environmental Sciences*, Vol. 34, No. 4, pp. 49-61, March/April 1992.

2. Saari, A. E., et al, *Environmental Stress Screening, RADC TR-86-149*, September 1986.

3. DOD–HDBK–344 (USAF), *Military Handbook: Environmental Stress Screening (ESS) of Electronic Equipment*, 126 pp., 20 October 1986.

4. Institute of Environmental Sciences, *Environmental Stress Screening Guidelines*, 940 East Northwest Highway, Mount Prospect, IL 60056, 1984.

5. Saari, A. E., *Stress Screening of Electronic Hardware*, Hughes Aircraft Company, May 1982.

6. Anderson, J. R., *Environmental Burn-In Effectiveness*, McDonnell Aircraft Company, August 1980.

7. Kube, F., *An Investigation to Determine Effective Equipment Environmental Acceptance Test Methods*, Grumman Aerospace Corporation, April 1973.

8. Bateson, J. T., "Board Test Strategies – Production Testing in the Factory of the Future," *Test and Measurement World*, December 1984.

9. Xie, M., Goh, T. N. and Mok, Y. L., "Optimization in Environmental Stress Screening Planning," *Proceedings of the 2nd International Symposium on Reliability, Maintainability and Safety*, Beijing, China, pp. 584-589, 1994.

Chapter 13

TAILORING TECHNIQUES IN *ESS*

13.1 THE MECHANICS OF TAILORING

Stress screening is a manufacturing process in which the simulated environmental stresses are used to screen out those failures that would otherwise occur in the field. A basic premise of stress screening is that under specific screening stresses applied over time, the failure rates of defectives are accelerated from that which would occur under normal field operating stress conditions. All assembled hardware consists of many paths along which a stress might be transmitted. The selection of screening parameters and methods of stress application must be suited to the stress transmission characteristics of the hardware's design. The stress should be closely tailored to the equipment's design capability, to provide an effective screen without damaging good components.

A tailored screen requires that specific parameters of the equipment being screened be reviewed such that the defects are detected and removed without incurring undue damage to the equipment. Therefore, the tailoring process should be based on the following premises:

1. Good hardware should not be damaged, or have its useful service life compromised.

2. The method of stress application must be such that critical hardware components are indeed stressed as expected.

The primary emphasis in tailoring should be on the latent defect type to be precipitated, hardware design capability and program needs.

13.2   STRESS RESPONSE SURVEY

It is important to have confidence that the stress transmitted to the hardware will not exceed the design limits and the hardware will indeed behave in the expected manner after screening. The only way to obtain this confidence for assembled hardware is to perform a stress response survey in which stresses are applied and the response of critical hardware elements is measured. It should be ensured that hardware responses are large enough to generate an effective screen while not exceeding hardware design capability. The stress response survey should be performed as a part of thermal and vibration screening selection and placement process.

13.2.1   THERMAL RESPONSE SURVEY

The purpose of the thermal response survey is to determine the thermal rate of change response and the time to reach thermal stabilization, and to identify any "hot spots" in the items that could degrade the equipment's reliability. The response of critical hardware elements should be measured to determine whether maximum or minimum temperature limits are being exceeded.

An output of the survey is a time–temperature plot showing actual temperature of the high power dissipating components, chamber air temperature, chamber controller temperature commands and a time axis for the survey. This plot will yield a temperature rate of change that can be compared to the required values.

The outcome of this survey will also be used as input to the *ESS* procedure, specifically to determine the temperature levels, time durations and temperature rate of change needed to precipitate defects without damaging the equipment.

13.2.2   RANDOM VIBRATION RESPONSE SURVEY

The purpose of random vibration response survey is to determine if any resonant conditions exist within the test item or test fixture, and identify the amplification factor, Q, between input and output excitation levels. The output of this survey is a plot showing vibration input level compared to equipment response levels over the tested frequency range.

Random vibration stress levels have been mistakenly specified as input values in many programs. In all assemblies there are inherent structural characteristics which provide either amplification or attenuation of the input stress on its path to the part. Structural characteristics of the test fixture can also have an impact on the value of the stress applied at the part's location.

For these reasons *the stress levels specified for random vibration must be the response levels measured on the equipment, not the input levels.*

A vibration response survey must be performed on each assembly to be screened to determine the response of the equipment to the input forces and to develop stress levels necessary to achieve the required response.

13.3 TAILORING OF ENVIRONMENTAL STRESSES

The stress levels specified in specifications and estimated in screening strength requirements should be baseline values only. The tailoring process should be required to experiment with the equipment to determine either

1. that the baseline values are not damaging and producing an effective screen, or

2. that the different stress levels are necessary.

All baseline values are for guidance only, the final parameters should be based on the equipment response survey and analysis of the screening results.

13.3.1 TAILORING FOR THERMAL CYCLE SCREEN

The three parameters that determine the stress intensity levels for thermal cycling screen are the temperature range, the temperature rate of change and the number of thermal cycles. The parametric value must be measured on the hardware, not the chamber air.

The baseline temperature range should be established by considering the component characteristics, the equipment specifications for maximum and minimum storage, and the operating temperatures. The temperature range should be as large as component characteristics will permit.

The rate of change depends on the chamber capacity and the mass of the equipment. When measured by a thermocouple mounted on a large item, the rate of change must be less than the chamber air rate of change to overcome the item's thermal capacitance.

The number of cycles is more closely related to the temperature range and rate of change than to the equipment complexity or number of parts. At the beginning of the screening program, the number of failures per cycle should be monitored and analyzed. If the data show that a large majority of the defects have been screened out in fewer cycles, the number of cycles may be reduced in order to reduce costs.

Dwell time at maximum and minimum operating and storage temperatures should be only enough to achieve thermal stability. When the temperature of the item or typical part cannot be measured, Fig. 13.1 [1] may be used to estimate the temperature.

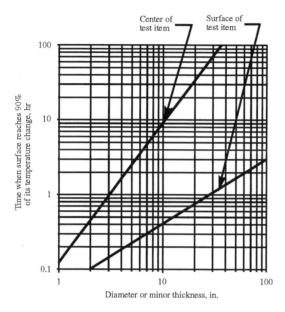

Fig. 13.1– Temperature stability curve [1].

13.3.2 TAILORING FOR RANDOM VIBRATION SCREEN

The emphasis in tailoring the random vibration stresses should be on preventing the overstress in G_{rms} and on avoiding the resonant frequencies.

13.3.2.1 NOTCHING

During the response survey of the equipment, the resonant frequency should be identified. If the equipment resonant frequencies fall within the input frequency range, excessive energy could be seen by the equipment and damage could occur. There are two ways to prevent equipment damage:

1. Modify the equipment design to achieve a more rugged item to obtain a resonance falling outside the input frequency range.

2. Make a notch on the input profile; i.e., reduce the *rms* value, and particularly reduce the power spectral density (PSD) values, near the resonant frequencies.

Notching is a process of reducing the input power spectral density level over a small frequency bandwidth so that equipment is not damaged due to overstress at the resonant frequencies. The notch should be only large enough so that the equipment's response matches the specified vibration profile.

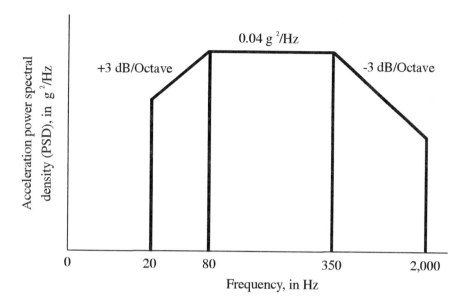

Fig. 13.2– Random vibration profile recommended by *NAVMAT*
P-9492 [2].

13.3.2.2 PROFILE

The vibration profile specified by the Naval Material Command (*NAVMAT*)
Document P-9492, Fig. 13.2 [1], has been erroneously considered the "standard" random vibration profile and has been specified on numerous *ESS*
programs.

In fact, due to the varying parameters of the electronic and electromechanical hardware, and the complexity of optimizing the vibration levels,
there is no one profile applicable to all equipment.

The random vibration profile does not have to match the *NAVMAT*
profile to be effective. Industrial studies have shown that effective screening
can be achieved by many different profiles. In one study [3] six different
militarized avionic computers were vibrated each under a different profile
and good results were obtained with all of them.

A more efficient method of determining the required profile is to use the
expected field environment as a baseline and experiment with the equipment
to determine the *PSD* levels and the profile necessary to screen out the
defects.

13.3.2.3   INPUT rms ACCELERATIONS, G_{rms}

The G_{rms} value is defined as the square root of the area under the vibration profile. In real terms, it is the time and frequency average of vibration energy produced by the profile. The energy introduced at any particular frequency is specified in terms of the PSD.

Since the random vibration profile curves are plotted on log-log paper, such as that given by Fig. 13.2, the calculation of the area under the sloped sections of the curves must use the special equation given next. The areas under the positive- or negative-slope sections can be determined using the following equations [4]:

When the slope is not -3, then

$$A = \frac{3\,P_2}{3+S}\left[f_2 - \left(\frac{f_1}{f_2}\right)^{S/3} f_1 \right], \qquad (13.1)$$

or

$$A = \frac{3\,P_1}{3+S}\left[\left(\frac{f_2}{f_1}\right)^{S/3} f_2 - f_1 \right], \qquad (13.2)$$

where

$A =$ area under the sloped section of the curve, in G^2,

$f_1 =$ frequency at the beginning of the sloped curve,

$f_2 =$ frequency at the end of the sloped curve,

$P_1 = PSD$ value at f_1,

$P_2 = PSD$ value at f_2,

and

$S =$ slope of the curve, in dB/octave.

When the slope has the value of -3, then

$$A = -f_2\,P_2\,\log_e\left(\frac{f_1}{f_2}\right), \qquad (13.3)$$

or

$$A = f_1\,P_1\,\log_e\left(\frac{f_2}{f_1}\right). \qquad (13.4)$$

The area under the flat-top section can be determined from

$$A = P\,(f_2 - f_1), \qquad (13.5)$$

where

$P = PSD$ value.

The total area, A_T, is the sum of all the sub-areas; i.e.,

$$A_T = \sum_{i=1}^{n} A_i. \tag{13.6}$$

The G_{rms} of the whole profile is given by

$$G_{rms} = (A_T)^{1/2}. \tag{13.7}$$

A very rough guide for preventing damage to the electronic enclosures that are to be subjected to *ESS*, based on the G_{rms} and the size of the electronic enclosure, is the following [4]:

$$L \cdot G_{rms} \leq 6, \tag{13.8}$$

where

L = largest box dimension in feet.

This relationship is approximate because the input *PSD* level and the resonant frequency of the enclosure will determine the response of the assembly.

13.3.2.4 DISPLACEMENT

The peak displacement of an item being screened is given by [4]

$$X = \frac{0.98 \times G_{max}}{f_n^2}, \tag{13.9}$$

where

X = peak displacement in inches,

f_n = fundamental resonant frequency, in Hz,

and

G_{max} = maximum acceleration response at resonance.

G_{max} can be determined as follows:

$$G_{max} = 3 \left(P_n \cdot BW \right)^{1/2}, \tag{13.10}$$

where

P_n = power spectral density at resonance, in G^2/HZ,

and

BW = bandwidth at resonance, 3 dB down from the peak.

Steinberg's criteria [4] give a guide to determine the displacement allowed for a printed circuit board (PCB) at the first bending mode resonance as follows:

$$X \leq 0.003\,L, \tag{13.11}$$

where

$X =$ maximum displacement of a PCB at its first
bending mode resonance,

and

$L = PCB$'s length.

13.3.2.5 FREQUENCY RANGE

The frequency range should excite a number of different modes in the equipment. The input bandwidth can be modified if the survey results show that the equipment's response is still in the specified frequency range.

The lower-frequency stress will excite the same defects that will be precipitated during transportation. All specifications require that the upper frequency not exceed 2,000 Hz.

13.3.2.6 TOLERANCE

Since the equipment response depends more on its structural characteristics than the input profile, tight control of the input profile is not necessary. It is recommended that tolerances of 20% be allowed for the overall G_{rms} level, the lower-frequency limit, and the upper-frequency limit.

REFERENCES

1. MIL-STD-810D, *Environmental Test Methods and Engineering Guidelines*, 19 July 1983.

2. *NAVMAT* P-9492, *Navy Manufacturing Screening Program*, Department of the Navy, May 1979.

3. Blake, R. K., " Random Vibration Screening of Six Militarized Avionic Computers," *The Journal of Environmental Sciences*, pp. 15-24, March/April 1982.

4. Steinberg, Save S., *Vibration Analysis for Electronic Equipment*, 2nd edition, John Wiley, A Wiley Interscience Publication, New York, 467 pp., 1988.

Chapter 14

MONITORING, CONTROL AND EVALUATION OF *ESS*

14.1 INTRODUCTION

During the production phase, the stress screening process and the screen fallout data must be monitored and controlled, and the screening program should be evaluated to assure that the program objectives are achieved. The objectives of the monitoring evaluation and control tasks are to establish assurance that the remaining defect count and the required reliability goals are being achieved by improving the capabilities of the manufacturing, screening and test processes. The manufacturing process capability can be improved by identifying defect sources and their root causes, and taking corrective actions to reduce the number of defects that are otherwise introduced into the product. The screening process capability can be improved by increasing the screening strength and ensuring that potential locations of defects in the product are being adequately stimulated by the screen. The testing process capability can be improved by increasing test detection efficiencies at all assembly levels.

14.2 DATA COLLECTION AND FAILURE CLASSIFICATION

The data elements in the list that follows should be collected in a timely fashion and accurately, during the implementation of the screening program.

1. Idenfication of the items exposed to the screen and test, by name, rating, etc.

2. Number of like items exposed to the screen and the test.

3. Number of like items that passed or failed the screen and test.

4. Description of the type of defect found, such as part, workmanship, manufacturing process, design, etc.

5. Identification of the part or interconnection site where the defect was found.

6. Identification of the assembly level, or manufacturing process, where the defect was introduced.

7. Screen conditions under which the defect was found, such as high temperature, vertical axis of vibration, etc.

8. Times to failure measured from the start of the screen.

9. Failure analysis results which identify the root cause of the defect.

10. Corrective actions taken to eliminate the cause of the defect from the process.

To establish a basis for the analysis of the screening fallout data, the failures must be properly classified so that the failure causes can be clearly established. The following classification scheme is recommended in [1]:

1. *Part defect.* A failure or malfunction that is attributable to a basic weakness or flaw in a part, such as a diode, transistor, or microcircuit.

2. *Manufacturing defect.* A failure or malfunction that is attributable to workmanship or to the manufacturing process, such as cold solder joint, cracked etch, broken wire strands, or process out of control.

3. *Design failure.* A failure or malfunction attributable to a design deficiency. If the fallout from screening indicates persistent evidence of design problems, such as electrical or thermal overstress failures due to inadequate derating, hot spots; methods of 100% stress screening should not be used, but Reliability Growth and Test-Analyze-And-Fix (TAAF) techniques are recommended to be used.

4. *Externally induced failure.* A failure attributable to external influences such as prime power disturbances, test equipment, instrumentation malfunction or test personnel.

5. *Dependent or secondary failure.* A failure that is caused by the failure of another associated item which failed independently.

6. *Software failure.* A failure attributable to an error in a computer program.

7. *Unknown cause failure.* An independent failure that requires repair and rework but that cannot be classified into any of the foregoing categories. The number of failures classified as "unknown cause" should be kept to a minimum. Every effort should be made to correlate the failure circumstance data with the other similar failure incidents, as well as to use failure analysis so as to establish the cause of the failure.

14.3   MONITORING AND CONTROL OF THE SCREENING PROCESS

The purposes of monitoring and control of a screening process are to ensure that the outgoing products meet their reliability requirements and to identify the causes and corrective actions for out-of-control conditions.

The critical parameters for monitoring and control are remaining defect density, D_R, screening strength, SS, defect density, D, fallout, F, and cost. Since the only observable statistics are fallout, procedures must be provided to determine values for the parameters of interest from the factory fallout.

14.3.1   QUALITY CONTROL CHARTS

The control charts are used for defect control based upon the observed screen fallout data and to establish whether the observed defect density falls within, or outside of, predetermined control limits.

Under the assumptions that the number of latent defects in a product are independently and identically distributed, the number of defectives in an equipment can be described approximately by the Poisson distribution with the parameter $D = N\overline{P}$, or

$$P(X = x) = \frac{e^{-D} \cdot D^x}{x!}, \tag{14.1}$$

where

$X =$ number of defects in an item,

$x =$ value of X,

$D =$ defect density, or average number of latent
      defects per item,

$\quad = N\,\overline{P}$,

$N =$ total number of parts in the item,

and

$\overline{P} =$ average part fraction defectives over all part types.

The mean of the Poisson distribution given by Eq. (14.1), or the average number of defects per item, is D, and the standard deviation is $\sqrt{D}$. When the true defect density is D, 99.73% of the time the number of defects in an item will lie between the control chart limits established by $(\mu_X - 3\sigma_X, \mu_X + 3\sigma_X) = (D - 3\sqrt{D}, D + 3\sqrt{D})$.

The primary purpose of the control chart technique is to establish baselines against which the process can be monitored and by which out-of-control conditions can be identified.

Three types of control charts and their trial values for the mean and standard deviation are the following:

1. $D_{IN} \pm 3\sqrt{D_{IN}}$, for incoming defect density, D_{IN}.

2. $D_{IN} \cdot TS \pm 3\sqrt{D_{IN} \cdot TS}$, for fallout F where TS is the test strength.

3. $D_{IN}(1 - TS) \pm 3\sqrt{D_{IN}(1 - TS)}$, for outgoing defect density.

These three types of control charts are shown in Figs. 14.1 through 14.3.

The fallout data can be used to calculate the part fraction defectives and defect density which can then be compared against the control chart baselines.

The observed part fraction defectives, $\overline{P}_0$, can be calculated using the following weighted average relationship:

$$\overline{P}_0 = \frac{n_1\,p_1 + n_2\,p_2 + \cdots + n_k\,p_k}{N}, \tag{14.2}$$

and

$$N = \sum_{i=1}^{k} n_i, \tag{14.3}$$

where

$\overline{P}_0$ = observed average part fraction defectives per item,

n_i = number of parts and interconnections of type i per item, such as diodes, transistors, hand soldered connections,

p_i = observed part fraction defectives for type i,

and

N = total number of parts and interconnections per item.

Correspondingly, the observed latent defect density, D_0, can be estimated by

$$D_0 = N\,\overline{P}_0. \tag{14.4}$$

Plotting D_0 into Figs. 14.1 through 14.3 and comparing the observed values for all batches of screened items, it can be determined whether the observed

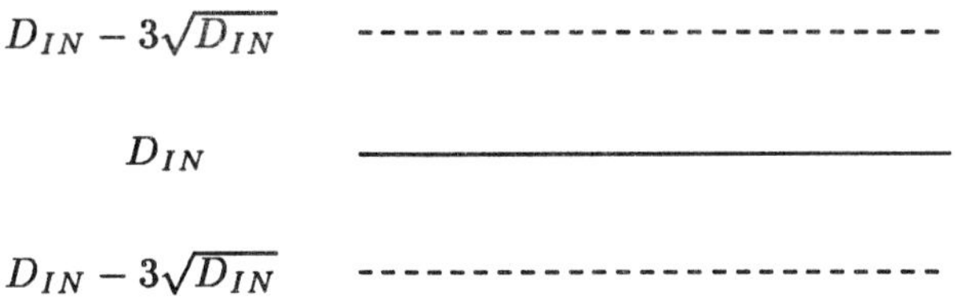

$$D_{IN} - 3\sqrt{D_{IN}} \quad \text{- -}$$

$$D_{IN} \quad \text{———————————}$$

$$D_{IN} - 3\sqrt{D_{IN}} \quad \text{- -}$$

Fig. 14.1– Control chart for incoming defect density.

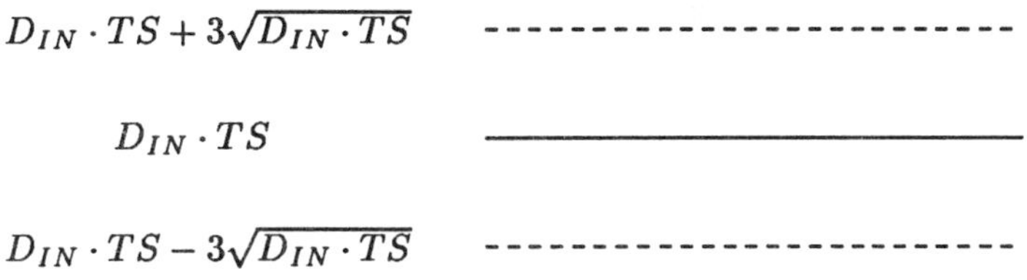

$$D_{IN} \cdot TS + 3\sqrt{D_{IN} \cdot TS} \quad \text{- -}$$

$$D_{IN} \cdot TS \quad \text{———————————}$$

$$D_{IN} \cdot TS - 3\sqrt{D_{IN} \cdot TS} \quad \text{- -}$$

Fig. 14.2– Control chart for fallout.

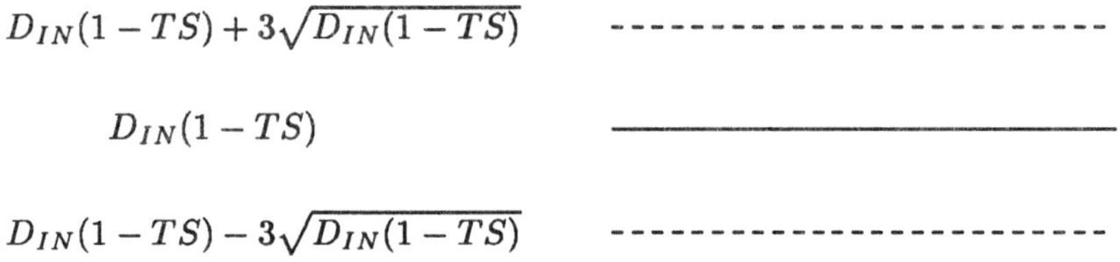

$$D_{IN}(1 - TS) + 3\sqrt{D_{IN}(1 - TS)} \quad \text{- -}$$

$$D_{IN}(1 - TS) \quad \text{———————————}$$

$$D_{IN}(1 - TS) - 3\sqrt{D_{IN}(1 - TS)} \quad \text{- -}$$

Fig. 14.3– Control chart for outgoing defect density.

TABLE 14.1– Comparison of actual versus desired defect density (D_{IN}) and test strength (TS) values.

Condition		Comparison Actual versus desired		Effect on Remaining defect density goal (D_R)	Effect on Future screening costs	Actions required D_{IN} Reduce D_{IN} by corrective actions	Actions required TS Changes to screen/test
		D_{IN}	TS				
I	a	HI	LO	Higher than expected	Increase	Essential	Increase screening strength
	b	HI	OK				
	c	OK	LO				
II	d	HI	HI	If higher Uncertain If lower ——↑—— ——↓——	——↑—— ——↓——	——↑—— ——↓——	——↑—— ——↓——
	e	LO	LO				
III	f	OK	HI	Lower than expected	Reduce	By opportunity	Reduce screening strength
	g	LO	OK				
	h	LO	HI				
IV	i	OK	OK	Likely to be achieved	Reduce	By opportunity	No change or eventually reduce

fallout is above, below or within established control limits. The results depend upon the degree to which the actual values of D_{IN} and TS differ from the desired values. The various possible conditions which may exist when the "true" values of (D_{IN}; TS) are compared against the desired values, are shown in Table 14.1.

The conditions in Table 14.1 are ranked according to severity and grouped into the following four categories, depending on whether or not the outgoing defect density or costs are affected:

1. The values of D_0 exceed the upper control limit of Fig. 14.1, the Incoming Defect Density Control Chart. It clearly indicates that the desired values for incoming defect density, D_{IN}, were too low, without having to consider TS.

2. The values of D_0 fall above the upper control limit of Fig. 14.2, the Fallout Control Chart. It indicates the following possible conditions, from Table 14.1:

Condition		D_{IN}	TS
I	b	High	OK
II	d	High	High
III	f	OK	High
I	a	High	Low
III	h	Low	High

3. The values of D_0 fall below the lower control limit of Fig. 14.2, the Fallout Control Chart. It indicates the following possible conditions from Table 14.1:

Condition		D_{IN}	TS
I	c	OK	Low
II	e	Low	Low
III	g	Low	OK
I	a	High	Low
III	h	Low	High

4. The values of D_0 fall within the control limits of Fig. 14.2, the Fallout Control Chart. It indicates the following possible conditions from Table 14.1:

Condition		D_{IN}	TS
IV	i	OK	OK
I	a	High	Low
III	h	Low	High

Regardless of the foregoing conditions which might actually exist, the questions of upmost concern are: Is the incoming defect density, D_{IN}, higher than desired and specified and is the test strength, TS, lower than expected? These questions can be addressed by using the methods of control intervals as described next and the CDE model as presented in Section 12.2.1 and further discussed in Section 14.4.

14.3.2 THE 90% CONTROL INTERVALS ON EXPECTED FALLOUT

When the user has a high degree of confidence that the test strength, TS, values are correct, a 90% control interval can be used to determine if the expected number of defects in a batch of screened items are consistent with the actual fallout. The 90% control limits are based on the binomial distribution:

$$P(\text{Fallout=K}) = \binom{M}{K} (TS)^K [1 - (TS)]^{M-K}, \tag{14.5}$$

where

TS = test strength,

M = expected number of defects entering the screen,

and

K = 0, 1, 2, ..., M.

The upper 90% control interval limit, UL, is the smallest integer satisfying the following inequality:

$$\sum_{K=UL+1}^{M} \binom{M}{K} (TS)^K [1 - (TS)]^{M-K} < 0.05. \tag{14.6}$$

The lower 90% control interval limit, LL, is the largest integer satisfying the following inequality:

$$\sum_{K=0}^{LL-1} \binom{M}{K} (TS)^K [1 - (TS)]^{M-K} < 0.05. \tag{14.7}$$

The values of $[LL, UL]$ are shown in Table 14.2 [1] as a function of the test strength, TS, and the expected number of defects, M. If the actual number of defects observed for the batch of screened items falls between LL and UL, the desired and specified incoming defect density is being met. If the actual number of defects observed falls above the upper control limits, then the corrective actions to reduce D_{IN} and/or to increase the test strength should be established. If the actual number of defects observed falls below the lower control limit, the corrective actions to reduce the screening regimen should be determined.

14.4   EVALUATION FOR THE SCREENING PROCESS

The purpose of evaluation for the screening process is to obtain estimates of D_{IN} and TS from the observed screening data using the CDE model and compare them with the results of the control chart method. The appropriate corrective action should be established to increase or reduce TS or D_{IN},

TABLE 14.2– The 90% control probability intervals of the actual fallout for various test strengths, TS, and various expected number of defects, M.

Expected number of defects, M	Test strength, TS									
	0.50	0.55	0.60	0.65	0.70	0.75	0.80	0.85	0.90	0.95
5	4	5	5	5	5	5	5	5	5	5
	1	1	1	1	2	2	2	3	3	4
6	5	5	5	6	6	6	6	6	6	6
	1	1	2	2	2	3	3	4	4	5
7	6	6	6	6	7	7	7	7	7	7
	1	2	2	2	3	3	4	4	5	6
8	6	7	7	7	8	8	8	8	8	8
	2	2	3	3	3	4	4	5	6	6
9	7	7	8	8	8	9	9	9	9	9
	2	3	3	3	4	5	5	6	6	7
10	8	8	8	9	9	10	10	10	10	10
	2	3	3	4	5	5	6	7	7	8
11	8	9	9	10	10	10	11	11	11	11
	3	3	4	4	5	6	6	7	8	9
12	9	9	10	10	11	11	12	12	12	12
	3	4	4	5	6	6	7	8	9	10
13	9	10	11	11	12	12	13	13	13	13
	4	4	5	6	6	7	8	9	10	11
14	10	11	11	12	12	13	13	14	14	14
	4	5	5	6	7	8	9	10	11	12
15	11	11	12	13	13	14	14	15	15	15
	4	5	6	7	7	8	9	10	11	13
16	11	12	13	13	14	15	15	16	16	16
	5	6	6	7	8	9	10	11	12	14
17	12	13	13	14	15	16	16	17	17	17
	5	6	7	8	9	10	11	12	13	14
18	12	13	14	15	16	16	17	18	18	18
	6	6	7	8	9	10	11	13	14	15
19	13	14	15	16	16	17	18	18	19	19
	6	7	8	9	10	11	12	13	15	16
20	14	15	16	16	17	18	19	19	20	20
	6	7	8	9	11	12	13	14	16	17

depending upon the outcome of the comparison as indicated by Table 14.1 and the results of the control chart method.

The failure rate function during screening, $\lambda(T_s)$, of the CDE model is given by

$$\lambda(T_s) = \lambda_0 + D \,\overline{\lambda}_D \, e^{-\overline{\lambda}_D \, T_s}, \tag{14.8}$$

where

$$D = \text{defect density,}$$
$$\overline{\lambda}_D = \text{failure rate of the defect,}$$
$$\lambda_0 = \text{failure rate of the main population,}$$

and

$$T_s = \text{screening time.}$$

The failure rate function of the CDE model can be fitted to the observed fallout data for a given screen so as to obtain estimates of the model's parameters. At the beginning, λ_0 can be set to zero or a prior estimate of λ_0 can be used. The estimate of the other parameters, D_{IN} and $\overline{\lambda}_D$, of the CDE model are obtained from the screening fallout over time, based upon the times-to-failure or cycles-to-failure data, by using the least-squares method or the maximum likelihood estimation (MLE) procedure.

The screening strength of the screen is calculated by substituting the estimated $\overline{\lambda}_0$ and the total duration of the screen, T_s, into

$$SS(T_s) = 1 - e^{\overline{\lambda}_D \, T_s}, \tag{14.9}$$

and the test strength is obtained from

$$TS = SS \cdot DE, \tag{14.10}$$

where

$$DE = \text{detection efficiency.}$$

The estimated D_{IN} and TS are compared with the desired and specified values of D_{IN} and TS, and the results of the control chart calculations, so as to determine which conditions of Table 14.1 exist, and what necessary action should be taken to ensure that the remaining defect density goals are achieved. This procedure should be repeated for the results from several batches of screened items to reflect the new estimates of the screening process.

REFERENCE

1. DOD–HDBK–344 (USAF), *Military Handbook: Environmental Stress Screening (ESS) of Electronic Equipment*, 126 pp., 20 October 1986.

Chapter 15

AN APPLICATION OF DOD-HDBK-344 IN DEVELOPING A CLOSED-LOOP DYNAMIC *ESS* PROGRAM [1]

15.1 INTRODUCTION

An extensive *ESS* program has been successfully conducted on an electronic countermeasures (ECM) pod system, the AN/ALQ-184 ECM pod system, using the procedures of DOD-HDBK-344 at Raytheon Company, Goleta, California. This dynamic *ESS* program includes the following tasks:

1. Prediction of the number of latent defects in the system before screen.

2. Various baseline screen setups at the subassembly and system levels with the required test strengths to precipitate the latent defects.

3. Establishment of control charts for the screen fallouts based on the latent defect and test strength estimates.

4. Periodic monitoring of the screens using control charts.

5. Implementation of corrective actions and screen modifications upon any exceedance of the control limits.

15.2 PRODUCT DESCRIPTION

The AN/ALQ-184 ECM pod system was developed to provide self-protect countermeasures to Air Force fighter aircraft. This pod resulted from a major modification and updating program for the AN/ALQ-119 pod, which was developed and fielded in the 1960's, to provide improved ECM performance and increased pod reliability. The AN/ALQ-184 pod is 13 feet long, weighs 95 pounds, and is made up of the following assemblies:

1. Eighty seven (87) printed wiring assemblies (PWA's).

2. Eighteen (18) traveling-wave tube (TWT) assemblies.

3. Major radio frequency (RF) assemblies.

4. Five (5) power supplies.

The pod generates 10 KW of heat in the transmit mode and 1 KW in the standby mode. The *ESS* program includes lower-level screens on 50 unique types of PWA's and 32 unique subcontracted assemblies and a top-level screen at the pod assembly level.

15.3   CLOSED-LOOP *ESS* PROCESS

Figure 15.1 illustrates a so-called closed-loop *ESS* process as extracted from DOD-HDBK-344 and used by Raytheon Company. This closed-loop process is a dynamic data collection, analysis, and corrective action process consisting of the following six key steps:

1. Incoming latent defect prediction.

2. Screen selection and placement.

3. Data collection.

4. Screen monitoring and data analysis.

5. Failure cause investigation.

6. Corrective actions and/or screen modifications, if required.

These six steps are summarized next.

15.4   INCOMING LATENT DEFECT PREDICTION

The DOD-HDBK-344 provides tables of incoming part fraction defective (PFD) values for piece parts, such as microelectronic devices, resistors, diodes; interconnections, such as connectors, hand solders, welds, crimps; and printed wiring boards (PWB) of different quality levels and under various environments. These tables are also given in Tables 12.4 through 12.16.

A procedure similar to MIL-HDBK-217 reliability prediction is followed for predicting the incoming latent defects for AN/ALQ-184 pod hardware. A

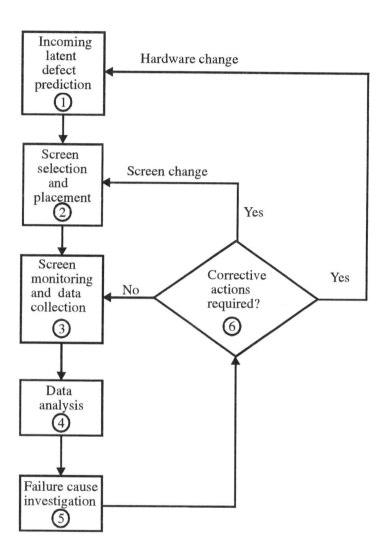

Fig. 15.1– Closed-loop *ESS* process applied in Raytheon Company [1].

sample worksheet is given in Table 15.1 for a typical subassembly. Similar worksheets were used for all subassemblies in the pod, including subcontract units, such as LVPS, HVPS, RF switches. Table 15.2 is a categorized summary of the predicted incoming latent defects.

According to Table 15.2, for planning purposes, the screening program has to be established assuming 7.014 latent defects existing in a complement of pod hardware before any screening is conducted; i.e.,

$$D_{IN} = 7.014.$$

15.5   SCREEN SELECTION AND PLACEMENT

1. **Prediction of the Maximum Allowable Remaining Defect Density, D_R, and the Required Test Strength, TS**

   The maximum required pod $MTBF$ is 150 hr while the expected mission time is 100 hr. Then, the maximum allowable remaining defect density of the pod after screening, D_R, is given by

   $$D_R = \frac{100}{MTBF} = \frac{100}{150} = 0.667.$$

   The difference between D_{IN} and D_R corresponds to the required defect density fallout, D_F; i.e.,

   $$D_F = D_{IN} - D_R = 7.014 - 0.667 = 6.347.$$

   Thus, the required test strength, TS, is given by

   $$TS = \frac{D_F}{D_{IN}} = \frac{6.347}{7.014} = 0.904.$$

2. **Screen Selection and Placement**

   A two-level screening approach has been developed for the AN/ALQ-184 to cost-effectively accomplish this test strength goal. The first-level screen is for four categories of assemblies: major shop replaceable units (SRU's), printed wiring assemblies (PWA's), minor SRU's, and other hardware in the pod. The second-level screen is for the pod system.

   (a) First-Level Screen

   The major SRU category accounts for 57% of the total D_{IN} of the system. All screens selected for these SRU's are a combination of random vibration, temperature cycling and/or thermal shock, and high temperature burn-in [1; p. 461]. Bias voltages are applied in most cases. No monitoring is applied due to the extremely high cost. Table 15.3 lists the typical screen parameters for this category. Fig. 15.2 shows a typical screen sequence with accompanying defect density fallout calculations.

TABLE 15.1– A typical worksheet for estimating incoming latent defects [1].

Program/Project AN/ALQ-184 pod production	AN/ALQ-184 ECM pod		
Unit identifier	Assembly identifier 2×8 RF switch G254058	Prepared by K. Dymoke	Date 1/9/90
Part type	Quality level/ Grade	Quantity	Fraction defective	Estimated defects*
Microelectronic devices	C	1	810.1	810.1
Transistors				
Diodes	JANTXV	38	51.2	1,945.6
Resistors	M	16	121.8	1,948.8
Capacitors	M	51	576.6	29,406.6
Inductive devices	MIL-SPEC	27	1,850.3	49,958.1
Rotating devices				
Relays				
Switches				
Connectors	MIL-SPEC	5	733.4	3,667.0
Printed wiring boards	MIL-SPEC	1	7,364.2	7,364.2
Diodes	JANTXV	8	102.3	818.4
Connections, hand solder		322	110.0	35,420.0
Connections, crimp				
Connections, weld				
Connections, solderless wrap				
Connections, wrapped and soldered				
Connections, clip termination				
Connections, reflow solder		293	3.1	908.3
Defect density/assembly				132,247.1
Total defect density (2 assemblies)				264,494.2

* Defects per 10^6.

TABLE 15.2– A categorized summary of predicted incoming latent defects [1].

Category	Defects/pod
Printed wiring assemblies	1.718
Subcontracted major SRUs	4.046
Subcontracted minor SRUs	0.355
Misc. in other pod hardware	0.895
Total	7.014

TABLE 15.3– Typical screen parameters for the major SRU category [1].

No.	Environment	Parameters	Operating	Monitoring
1	Random vibration	6 $grms$; 15 min for the most critical axis; 7.5 min for each of the critical axes if more than one axis is critical.	DC voltage applied	No
2	Temperature cycling	-54°C to 95°C; 25 cycles at 15°C/min; Dwell time = 0.5 hr.	DC voltage applied	No
3	Burn-in	85°C for 72 hr.	DC voltage applied	No

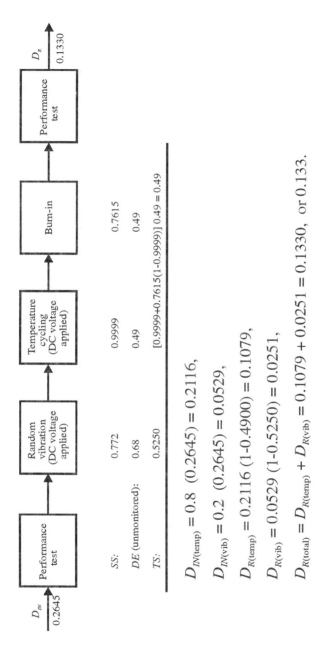

Fig. 15.2– A typical screen sequence and defect density fallout calculations at the major SRU level [1].

431

Printed wiring assemblies account for 24% of the system's D_{IN}. The screens selected are random vibration and temperature cycling preceded and followed by performance tests. The PWA's are unenergized during random vibration and operated with DC bias voltages during temperature cycling. No monitoring is applied for this category.

The other two categories account for the remaining 19% of the system's D_{IN}. For minor SRU's, such as attenuators and couplers, a thermal shock screen is selected. For other hardware, such as government furnished equipment (GFE), cables and wiring, no screens are imposed at the assembly level.

Figure 15.3 summarizes the screening results at the first level. It may be seen that the pod's D_{IN} is decreased from 7.014 to 3.137 due to the first-level screen. This in turn requires that the test strength for the screen of next level, or the pod level, be sufficient to reduce D_{IN} from 3.137 to less than the maximum allowable D_R of 0.667.

(b) Second-Level Screen

The screens selected for the AN/ALQ-184 system-level ESS program are random vibration and temperature cycling. Figure 15.4 shows the screening flow diagram with accompanying defect density fallout calculations. During both random vibration and temperature cycling, the pod is operated and monitored using built-in-test (BIT) capabilities. As may be seen from this figure, the final remaining defect density after screen is

$$D_R = 0.651,$$

which is less than the maximum allowable $D_R = 0.667$ for a required $MTBF$ of 150 hr.

15.6 DATA COLLECTION

Data for all ESS, such as dates of screens, time of each failure, type of test at failure, vendors' names, causes of failures, failed component type and location, and failure analysis document numbers, are collected and stored in the computer data base for each part by serial number. By plotting the failures across time, patterns can be identified and corrective actions can be taken in either the ESS profiles or in the manufacturing processes.

To further enhance data compatibility between vendors, a compiled data base program may be generated and supplied to each vendor. Thus, monthly data submittals can be accomplished on floppy discs rather than on hundreds of paper sheets.

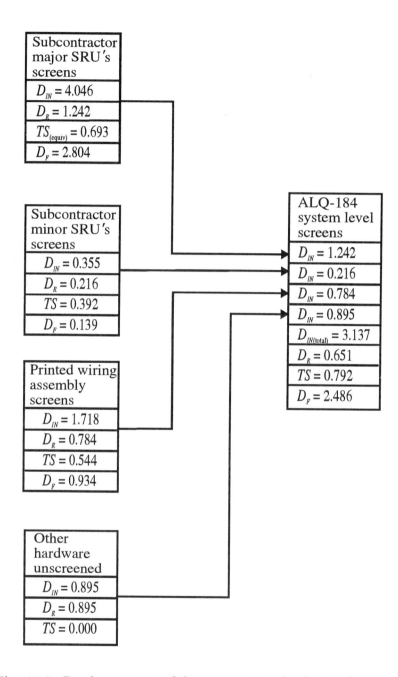

Fig. 15.3– Results summary of the screens at two levels; i.e., the
major SRU level and the pod system level [1].

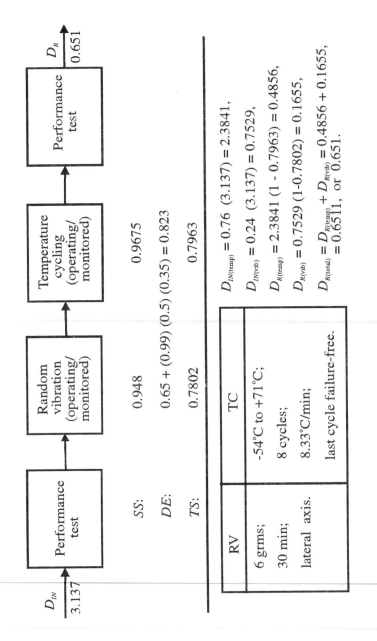

Fig. 15.4– Screening flow diagram, screening parameters, and defect density fallout calculations for the screens at the second, or pod system, level [1].

434

15.7 SCREEN MONITORING AND DATA ANALYSIS

Screen monitoring and data analysis are done at the SRU, PWA and pod-system-level screens using control charts for the screen fallout established according to the procedures of DOD-HDBK-344 or Section 14.3 of this book, and comparing the collected actual fallout data, as described in the previous section, with the lower and upper control limits, LCL and UCL, respectively.

The nominal value for the control limits, CL, is set up at the predicted fallout, $\overline{D}_F$; i.e.,

$$CL = \overline{D}_F = D_{IN} \cdot TS.$$

Under the Poisson distribution assumption for the actual fallout,

$$\sigma_{D_F} = \sqrt{\overline{D}_F} = \sqrt{D_{IN} \cdot TS}.$$

At Raytheon Company, the UCL and LCL for the fallout were set up at "$\mu \pm \sigma$"; i.e.,

$$UCL = D_{IN} + \sqrt{D_{IN} \cdot TS},$$

and

$$LCL = D_{IN} - \sqrt{D_{IN} \cdot TS}.$$

Figure 15.5 shows a typical control chart. The actual fallout is plotted each month on the control chart. If the actual fallout is higher than the UCL or lower than the LCL, then an investigation is established on the SRU, PWA or pod system.

A tabular-formated analysis is also conducted for the screen fallout. The cumulative fallout of each SRU, for both predicted and observed, is compared at the SRU-level and at the pod system screens. Table 15.4 is a typical example which offers an insight into the effectiveness of the screens for each SRU at the two screen levels based on cumulative data.

Screen monitoring and data analysis revealed failures of the following four categories which required investigation:

1. Low SRU failures/High pod failures, such as the case of SRU A in Table 15.4 where "0" fallout at the SRU level and "25" fallouts at the pod level were observed.

2. High SRU failures/High pod failures, such as the case of PWA B in Table 15.4 where "25" fallouts at the SRU level and "81" fallouts at the pod level were observed.

3. High SRU failures/Low pod failures, such as the case of RF SRU in Table 15.4, where "13" fallout at the SRU level and "3" fallouts at the pod level were observed.

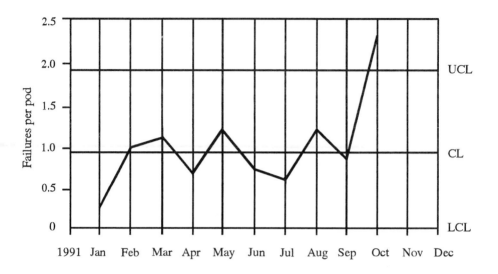

Fig. 15.5 A typical control chart used for screening fallout monitoring and data analysis [1].

TABLE 15.4–ᵀypical screen fallout data as compared with the predicted value for each major SRU category [1].

Item	PWA/SRU level		Pod level	
	Predicted	Actual	Predicted	Actual
SRU A	11	0	8	25
PWA B	7	25	5	81
RF SRU	6	13	4.5	3
SRU B	5	1	4	0

4. Low SRU failures/Low pod failures, such as the case of SRU B in Table 15.4, where "1" fallout at the SRU level and "0" fallout at pod level were observed.

15.8 FAILURE CAUSE INVESTIGATION AND CORRECTIVE ACTIONS

Each one of the previous failure categories requires unique investigation procedures as summarized below.

Category 1: Low SRU Failure/High Pod Failures

In Table 15.4, the predicted fallouts for SRU A are "11" at the SRU level and "8" at pod level. But the actual observed fallouts are "0" at the SRU level and "25" at the pod level. For failures falling into this category, attention should be focused on the test environment during SRU-level screens since the latent defects are not satisfactorily screened out at the SRU level, but detected later at the pod level. An investigation of SRU A in Table 15.4 discovered that the vendor, with whom the screens were conducted, was stacking the units, 40 at a time, during the thermal cycling screen and the SRU was not experiencing the required temperature rate of change which is the main failure-precipitating stress factor during thermal cycling. Consequently, the screening strength was much lower than required.

Had this analysis not been done, people may have been congratulating the vendor for having zero defects in the screening program. Note that the screening strength equations in DOD-HDBK-344 are based on test item temperature, not chamber air temperature. Therefore, it is also necessary to verify that the part temperature stabilizes at both extremes.

Category 2: High SRU Failure/High Pod Failures

In Table 15.4, the predicted fallouts for PWA B are "7" at the PWA level and "5" at the pod level. But the actual, observed fallouts are "25" at the PWA level and "81" at the pod level. For failures falling into this category, more comparative analysis is required due to high failure rates at both levels, since the failure causes at both levels are often not identical. An investigation on PWA B in Table 15.4 revealed that a defective lot of IC's caused the PWA failures. The pod failures were caused by a different IC that failed when the main pod power supply lost +15 V, which was a secondary fault and could not be expected to be precipitated during the PWA screen. After changing the IC type, the problem disappeared.

Category 3: High SRU Failure/Low Pod Failures

In Table 15.4, the predicted fallouts for RF SRU are "6" at the SRU level and "4.5" at the pod level. But the actual, observed fallouts are "13" at the SRU level and "3" at the pod level. For failures falling into this category, engineers who are more process oriented are assigned to the investigation, since such failures are frequently associated with the SRU manufacturing methods and inspection criteria. It turned out that the majority of the failures at the SRU level were due to broken bonds. After changing the wire forming method and introducing a new attachment technique, the problem disappeared.

Category 4: Low SRU Failure/Low Pod Failures

In Table 15.4, the predicted fallouts for SRU B are "5" at the SRU level and "4" at the pod level. But the actual, observed fallouts are "1" at the SRU level and "0" at the pod level. For failures falling into this category, investigations should focus on the evaluation of the lower-level screening conditions to make sure that the items are adequately stressed. Corresponding modifications or adjustments to the screen profiles are usually required to improve the real screen strength.

15.9   CONCLUSIONS

The significance of implementing *ESS* is not just the fact that *ESS* is being conducted, but the effectiveness of the conducted screens. Sometimes random vibration, temperature cycling, and constant temperature screens are performed in a "cookbook" fashion without regard to whether or not they are "doing any good." The example presented in this section illustrates that a scientifically planned, appropriately conducted, and periodically monitored screen will yield failure-free products, if the program is a real dynamic process with extensive data analysis, timely feedback, and corresponding corrective actions and modifications on the manufacturing process and screen profiles.

REFERENCE

1. Schmidt, R. E., Vossler, D. L., and Russell, D. B., "Making *ESS* a Dynamic Process Using the Procedures of DOD-HDBK-344," *Proceedings of the Institute of Environmental Sciences*, pp. 458-465, 1992.

Chapter 16

ESS CASE HISTORIES

16.1 *ESS* IN THE *APOLLO* SPACE PROGRAM [1]

One early application of *ESS* was by the U.S. National Aeronautics and Space Administration (NASA) during the *Apollo* Space Program for the acceptance testing of electronic equipment. Since the major sources of vibration for products used in space are random, NASA was receptive to introduce *ESS* for acceptance testing. The Grumman Corporation in Bethpage, New York, started the pioneering work in 1970 with tests on the design of the lunar module for the *Apollo* man-on-the-moon program. The company was able to precipitate and eliminate 85% to 90% of all workmanship defects from the equipment subject to random vibration and thermal cycling. Later, the *ESS* research was extended in the Grumman Corporation to Airborne Electronic Systems for military aircraft.

The *ESS* work conducted early at the Grumman Corporation was one of the main bases for the U.S. Navy's document P-9492, "Navy Manufacturing Screening Program," which was one of the first military documents to suggest effective temperature profiles and random vibration spectra for electronic equipment screening.

16.2 *ESS* ON NEW PRINTERS IN IBM [1; 2; 3]

IBM in Charlotte, North Carolina, successfully applied *ESS* in 1985 to the design and production of its new Model 4234 printers. This high-output printer is used with the company's line of System/36 and System/38 minicomputers, and is designed to print up to 410 lines a minute. Kenneth Chesney, a staff test engineering coordinator with the IBM Corporation's product development and production facility in Charlotte, North Carolina, told an exciting story of successful printer screening using a simultaneous three-axis random shaker plus a quick-ramping thermocycler. Each printer was vibrated in all three axes at 5.7 G root-mean-square (*grms*) for 5 minutes while printing a variety of selected test printing patterns. After vibration, a number of printers were placed on carts and thermally stressed in a temperature chamber. The air distribution inside the chamber ensures that all printers are uniformly and rapidly heated and cooled over a range of -25°C to 60°C. The temperature controller could raise the ambient temperature in the chamber by 40°C a minute with oversize electric resistance heaters, and it could lower it at 50°C a minute by using liquid nitrogen cooling.

Though there was an initial resistance to the idea of stress screening among IBM's test engineers at the Charlotte facility, according to Chesney, the money saved by solving the first problem detected by *ESS* more than paid for the cost of screening the equipment. The following had been achieved during the developmental phase screening:

1. Forty (40) design improvements had been made to the printer, one of which was a redesign of the structure of the center casting to reduce a ringing effect that loosened screws and produced destructive vibration in the printer itself. This problem was brought to light by random vibration screening within minutes. It might have taken weeks for such a problem to appear during qualification testing at normal operating conditions for the printers.

2. The final design, achieved after three (3) iterations, reduced the printer's self-induced vibration by 65%.

3. A significant increase of printing ability under vibration of up to 15 *grms* was achieved. The redesigned chassis withstood vibration seven (7) times that of the original screening levels.

4. The arrangement and mounting problems of power resistors and capacitors were found and corrected.

5. "Cold soldered" joints were found.

6. Capacitor lead fatigue problems were found after only three (3) minutes of shaking, whereas it might have taken hours of heavy printing before a single glitch occurred. A sister IBM plant, according to Chesney, had spent "an estimated million-plus dollars" doing warranty work on products with a similar problem which was not identified until much later.

7. Cabling problems were found which could not occur without screening until the product was shipped, requiring the installer to rectify the problem.

IBM estimates that it can save $20 in repairs for every dollar it spends on the *ESS* program for its Model 4234 printers.

16.3 *ESS* ON SWITCHING LOGIC UNITS AT LOCKHEED [1]

In 1981 the Lockheed-California Corporation screened switching logic units, which were used in U.S. Navy S-3A antisubmarine aircraft for distributing communications within and outside the aircraft. These units had been high on the "bad actor" list for 10 years since they were first used. They had, before *ESS*, resulted in incomplete missions or inoperative planes as often as the sum total of problems in the entire engine system. Due to the frequent removals of the switching logic units for maintenance, they got "banged around" a lot, causing damage to the chassis. The mean time between failures for the switching logic units was approximately 100 hr.

A 6 *grms* vibration spectrum with several notches around the natural frequencies of certain components, plus 50 hr of thermal cycling identified many workmanship defects in the unit. It took about 350 hr of screening and repairs on each unit before it could endure a 50-hr screening without failure.

The results were promising. The mean time between failures increased from 100 hr to 500 hr after screening and only a few minor design changes were required. The average number of units removed each week from nine aircraft dropped from 1.8 over a 27-week period prior to screening to 0.14 during a 7-week period following the *ESS* procedure.

16.4 AMERICAN AIRLINES' *ESS* DEMONSTRATION [4]

Presented here is a real world *ESS* demonstration conducted at the American Airlines Maintenance and Engineering Center, Tulsa, Oklahoma, during the period of December 11-15, 1989. The objective of this demonstration was to precipitate chronic intermittent failure problems (NTF) in various avionics units, using various tri-axial, quasi-random vibration and high-ramp-rate temperature cycling profiles to excite the internal components of each unit. Table 16.1 is a sample of the units which were included in the *ESS* demonstration, and of the results obtained.

16.5 POWER SUPPLY RANDOM VIBRATION STRESS SCREENING AT ZYTEC [4]

ZYTEC manufactures various kinds of high-quality power supplies. This case history describes the procedure of how the random vibration screen was designed and implemented during production, and what had been achieved.

Figure 16.1 gives the production test flow of these power supplies. Table 16.2 is a summary of the defects precipitated by the random vibration screen

TABLE 16.1– Results of the American Airlines *ESS* Demonstration [4].

No.	Screened unit	ESS profile* TC†	ESS profile* RV‡	Results
1	Air data computer	30°C → 70°C; at 15°C/min; 3 min dwells	10 grms; 12 min;	2 CCA failed: Shorted CAP/CMOS D-A NTF for 1 year.
2	Air data computer	Same as No. 1	Same as No. 1	Intermittent failure twice: Once on ramp down at 10 grms; once at-30°C + 10 grms – then hard failure 50°C + 10 grms – power supply failure.
3	Air data computer	-30°C → 70°C; at 20°C/min; 5 min dwells	15 grms; 5 min	Transducer hard failure after 0.8 min of random vibration without temperature; sent to supplier for repair.
4	ELF/FL duplex actuator	-30°C → 70°C; at 30°C/min	7.5 grms; 5 min	Resolver erratic on ramp-up from -30°C; bad wire crimps found.
5	CDC-22 pressurization control	No	7 grms; 10 min	Hard failure within first minute of random vibration (box had long history of NTF).
6	CDC-22 pressurization control	-30°C → 70°C; at 12°C/min; 5 min dwells	7 grms; 10 min	Failed in cold cycle; isolated to failed transformers.
7	Flight guidance roll computer	First sequence -30°C → 70°C; at 10°C/min; 10 min dwells Second sequence No change	15 grms; 30 min 15 grms; 10 min	No failure during first sequence, then intermittent hard failure after 5.3 min during second sequence; recovered with random vibration off. Replaced 25% of CCA's; rescreened after successful ATE; functioned normally thereafter.

* All screens performed are power-on monitored.
† TC=thermal cycling.
‡ RV=random vibration.

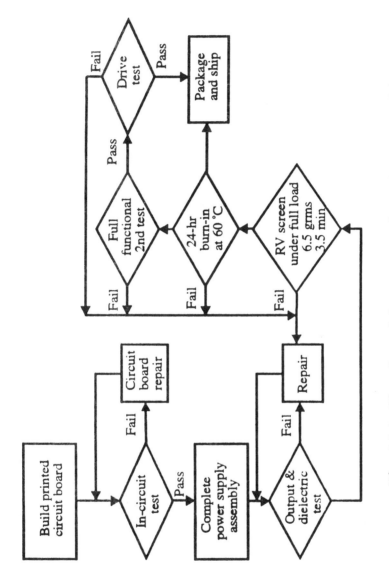

Fig. 16.1– Flow chart of the power supply production tests [4].

TABLE 16.2– Defects precipitated by the random vibration
screen on the test power supply at ZYTEC [4].

Log No.	Defects precipitated	grms level at failure	Failure time, min:sec
1	Poor plating on XFMR pins	10.0	:30
2	Intermittent open thermostat	N/A	N/A
3	Intermittent open circuit breaker	N/A	N/A
4	Fractured capacitor lead	8.0	:50
5	Loose hardware on transistors	11.0	4:00
6	Open thermal circuit breakers	10.0	:45
7	Loose plug	8.0	:20
8	Fractured solder, inductor pin	6.0	:05
9	Open EMI filter (poor solder)	8.0	1:00
10	Broken XFMR wire	10.0	1:00
11	Fractured solder joint on plug	8.0	1:24
12	Loose fuse in fuseholder	8.0	:25
13	Spread barrel in output plug	11.5	:11
14	Unseated fast on	7.0	:10
15	Cracked inductor core	9.0	:55
16	Loose bulk capacitor terminal screw	8.0	:50
17	Loose capacitor terminal screw	8.0	:50
18	Intermittent input switch	4.0	1:24
19	Lead not inserted into board	4.5	:10
20	No retention on capacitor	10.0	1:50

on the test power supply, grms levels and the times at which the failures
occurred. Figure 16.2 categorizes the failures precipitated and detected by
the random vibration screen.

According to the customer's requirement, the ESS was further tailored
for the computer power supply with 1,050 parts at the unit level. The new
ESS profile was composed of a 5-minute random vibration with 5.8 grms
of acceleration at each of three axes, followed by a 7-cycle power-on thermal
cycling with the temperature ranging from -54°C to 71°C at a change rate of
4.5°C/min.

The results achieved from the newly tailored screen profile were promising.
73% of the flaws were detected for the power supplies. The field MTBF
increased 3.3 times since the production started.

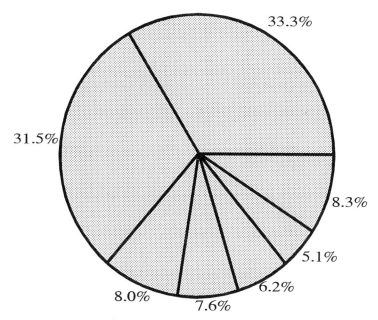

33.3%: bad solder
31.5%: bad components
8.0%: loose hardware
7.6%: reversed part
5.1%: broken wiring
6.2%: wrong component
8.3%: other (miswired, missing part, workmanship, assembly error)

Fig. 16.2– Failure categorization based on failure analysis of the random vibration screen on the power supply at ZYTEC [4].

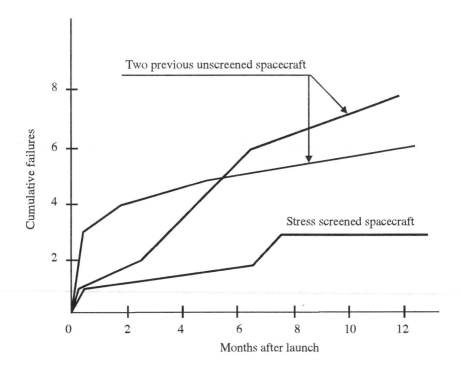

Fig. 16.3– Effectiveness of the thermal screen on spacecraft at
the Hughes Aircraft Company [5].

16.6   THERMAL CYCLING SCREEN OF A
SPACECRAFT [5]

Failures of an in-orbit spacecraft is particularly dangerous because of the inability to perform corrective maintenance. To prevent these kinds of failures
as much as possible, Hughes Aircraft Company conducted a system level thermal cycling screen on its spacecrafts. Figure 16.3 compares the in-orbit failure
histories between thermal-screened spacecrafts and two previous unscreened
spacecrafts. The results clearly demonstrate the benefits of this stress screen.

TABLE 16.3– **Dramatic improvement of field** *MTBF* **by** *ESS* [5].

Type	No. of units	No. of sites	Total hours	Total failures	MTBF, hr
Pre-screen	40	9	124,224	1,129	1,150
Post-screen					
1978	21	9	36,951	4	9,238
1981	159	31	600,631	63	9,534

16.7 *ESS* ON SHIPBOARD MILITARIZED DIGITAL COMPUTER [5]

This case history is about the stress screen on a shipboard militarized digital computer which has been shipped to almost every ship in the U.S. Navy. Previous field data revealed very low reliabilities. The primary causes have been found to be workmanship, component defects, and some design problems. A "get-well" program, which implemented the following *ESS* profile, was developed and conducted:

1. Voltage/timing: supply varied within 5V±10% at the ambient temperature of 25°C.

2. Temperature cycling: -5°C to +60°C at a change rate of 5°C/min with a 2-hr dwell at -5°C and a 4-hr dwell at 60°C; 4 cycles.

3. Swept sine vibration: 30 minutes.

4. Constant temperature soak: 125 hr at 50°C.

The results achieved after *ESS* and the corresponding corrective actions taken are summarized in Table 16.3, which shows a dramatic *MTBF* improvement.

16.8 RANDOM VIBRATION OF AN AIRBORNE DIGITAL COMPUTER [5]

This case history is about a random vibration screen on an airborne digital computer, which measured 8″ × 13″ × 16″ and weighed 40 lb with 15 shop replaceable units (*SRU*). There were 4,260 parts in the hardware of which 1,653 were integrated circuits (*IC*) and 2,607 were discrete components. Fifty (50) mature production computers were screened under the following profiles:

1. Five (5) minutes of random vibration in each of 3 axes at a power spectral density of 0.01 g^2/Hz from 20 Hz to 1,000 Hz and a 3 dB/Octave rolloff from 1,000 Hz to 2,000 Hz.

2. Vibration was conducted after a 4-hr burn-in, followed by extensive thermal cycling.

A total of 22 failures were found. They are broken down as follows:

1. 14 component failures:
   - 6 failures due to internal conductive failures.
   - 2 failures due to electrical overstress.
   - 1 failure due to internal short.
   - 1 failure due to internal binding and bit indicator.
   - 4 failures unanalysed.

2. 5 workmanship failures:
   - 2 failures due to loose/contaminated connectors.
   - 3 failures due to fractured solder joints.

3. 3 intermittent and undiagnosed failures.

The results were promising. A 6-to-1 reduction in infant mortalities were achieved.

16.9 RANDOM VIBRATION SCREEN OF AN ADVANCED SIGNAL PROCESSOR [5]

In this case history, the signal processor measured 11″ × 22″ × 50″, weighed 210 to 260 lb, and consisted of the following:

1. 2 control panels.

2. 50 to 66 plug-in assemblies.

3. 5 massive power supply assemblies.

The following stress screen profile was applied to 33 units:

1. 10 minutes of random vibration in each of the 2 axes (vertical axis and front-to-back axis) at 6 *grms* from 20 Hz to 2,000 Hz.

2. Power on.

3. Thermal cycling before and after random vibration.

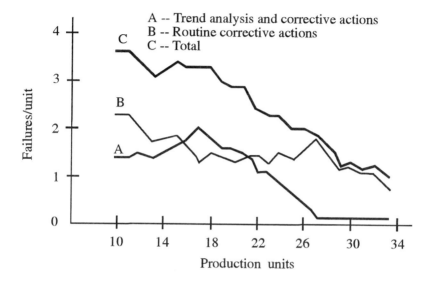

Fig. 16.4– Benefits of a stress screen plus failure analysis and corrective actions [5].

The screening of these 33 units revealed a total of 79 failures. The failure analysis identified that among these 79 failures 56% were due to workmanship, 14% due to piece parts, 13% due to design, and 17% due to unclassified causes. Units which were proven to be the most troublesome were chosen for a more extensive corrective action program. The reduction in failures per unit, as production progressed, was far more pronounced for those units receiving intensive failure analysis and corrective actions, as shown in Fig. 16.4. The conclusions from this case history are the following:

1. Workmanship flaws (56% in this case) are the primary reasons for performing a vibration screen.

2. Routine stress screening without effective failure analysis and corrective actions is at best inefficient.

TABLE 16.4– Original and modified thermal cycling screen profiles for airborne radar module screening experiment at Hughes [5].

Original profile	New profile
• Range: -60°C to 115°C	• Range: -60°C to 95°C
• Change rate: $\dot{T}^* = 6$°C/min	• Change rate: $\dot{T}^* = 15$°C/min
• Duration: 168 hr of 1-hr and 2-hr cycles	• Duration: 23 hr of 46 cycles
• No dwell during 1-hr cycle	• 5 minutes of dwell at
• 30 minutes of dwell at	temperature extremes
temperature extremes	
during a 2-hr cycle	
• Non-operating	• Non-operating

16.10   THERMAL CYCLING SCREENING OF AN AIRBORNE RADAR MODULE [5]

This case history demonstrates the advantage of a dynamic ESS program both to the manufacturer and to the customer. Hughes Aircraft Company conducted an experiment in 1970 to compare the effectiveness of the two different thermal screen profiles given in Table 16.4 for F-15 Airborne Radar Module stress screening.

It may be seen that the differences between the original and the new thermal stress profiles are temperature upper extreme, temperature change rate, and dwell duration. It was demonstrated by the experiment that the higher temperature change rate was more effective in precipitating flaws, and that the screening time could be reduced from one week to one day.

According to the conclusion of the experiment, Hughes Aircraft Company made a proposal to the customer to adopt the new screening conditions on the current production module. The customer stipulated that no degradation in delivered product reliability could be allowed, and that an improvement in module screening efficiency should be achieved. Since Hughes was confident it could meet these two conditions, based on the results of the experiment, an agreement was made to adopt the new screen profile.

As expected, the screen results using the new thermal screen profile was attractive. A savings of $800 per radar set was obtained. More importantly, six days were gained in the manufacturing cycle, which proved to be very significant.

16.11 STRIFE TEST PROGRAM IN HEWLETT-PACKARD [6]

16.11.1 INTRODUCTION

The term "STRIFE test", proposed by the Hewlett Packard (HP) Company in the early 1980's, stands for stressed life test, which subjects the product to thermal cycling in conjunction with power on-off cycling and random vibration. HP's San Diego Division has been using the product level STRIFE test since 1982. The field failure rate was reportedly decreased by a factor of two to three in the period of 1981 to 1985.

16.11.2 THERMAL CYCLING PROFILE

The temperature profile was tailored to both the product under test and the test chamber. The primary goal of this profile was to continuously keep the temperature of the components in rapid transition. The thermal profile was optimized by measuring the components versus chamber temperature change rates. Thermocouples were placed on components of various thermal masses and the board subjected to the thermal profile while recording the resulting temperatures. Plots were made to compare the proposed temperature profile of the chamber and the resulting component temperatures. The profile was then modified and this process was repeated until the components were in maximum transition for 90% of the temperature profile.

Figure 16.5 is the final temperature profile which contains a 15°C overshoot at both temperature extremes. It was found out that if the overshoots did not exist, the rate of change of the components would be greatly reduced as the component temperature asymptotically approaches the profile extremes of 75°C and −30°C. The duration of the overshoot is determined such that the component temperature achieves 90% or greater of the profile extremes.

16.11.3 POWER ON-OFF CYCLING

It has been a common practice to use power on-off in conjunction with thermal cycling to accelerate electronic failures. In Fig. 16.5, the power is applied after a suitable dwell time at the low temperature extreme, which allows the component to achieve the air temperature of the chamber. This dwell time will vary from product to product depending on the product's thermal mass. The product remains powered on until the completion of the high temperature dwell. Each time the product is powered on and off, the components experience an internal temperature cycle.

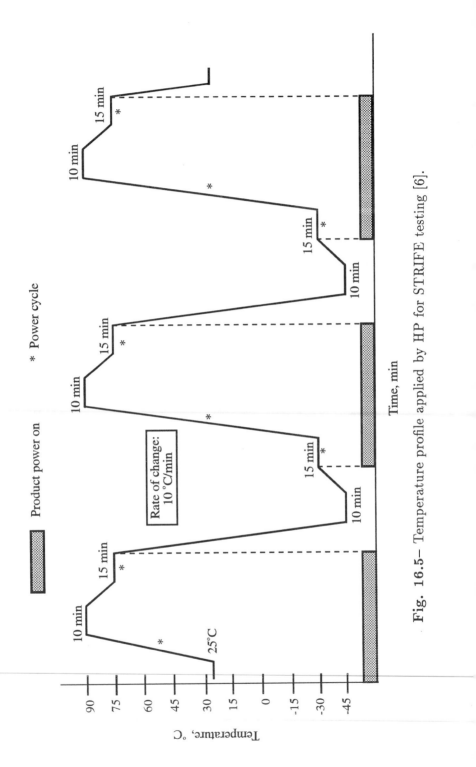

Fig. 16.5– Temperature profile applied by HP for STRIFE testing [6].

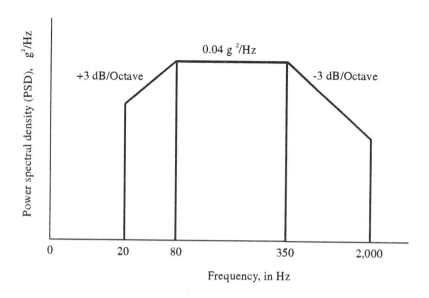

Fig. 16.6– Random vibration profile applied by HP as a starting point for STRIFE testing [6].

16.11.4 RANDOM VIBRATION PROFILE

HP's San Diego Division started its random vibration profile with the NAVMAT-9492 specification, as shown in Fig. 16.6, and tailored it to each individual product undergoing test. The set of electronics and motors were mounted in a very similar manner as they were mounted in the product. The two most susceptable axes of vibration were selected based on the product under test.

16.11.5 SCREENING RESULTS FOR PRODUCT A

Product A consists of main and head driver PC boards and motors. The main board has either an HP-IB or a Centronics interface. Twenty-four (24) sets of Product A were tested in the lab prototype phase. Using the temperature change rate of 10°C/min, the following failure modes were precipitated and detected:

1. Four (4) HP-IB buffer chips failed due to crazing.

2. Three (3) motor drivers failed due to parasitic leakage.

3. Twenty-six (26) out of forty-eight (48) power capacitors had a plastic disc on the top bulge outward due to internal out-grassing of the

capacitor sleeving.

Production prototypes were also tested in a similar manner as the lab prototype units. The following failures were precipitated using 10°C/min change rate:

1. Three (3) HP-IB buffers failed due to crazing.

2. Three (3) 64K static CMOS RAM's failed mainly due to latch up.

3. Pins of two +5V regulators broke during vibration due to a manufacturing change.

4. Three hundred (300) head trailing cable intermittents occurred.

5. 25-plus pulse width modulation cable intermittents failed due to poor crimping.

16.11.6   SCREENING RESULTS FOR PRODUCT B

Product B was a current product in production and had a similar configuration as Product A. Sixty-four (64) sets of Product B were tested using the same profile as that for Product A. A total of 28 failures were precipitated in 30 temperature cycles, 24 of which occurred within the first 10 cycles. Two predominant failure mechanisms were brought to light:

1. Manufacturing process deviation which was precipitated by the random vibration test. Misadjustment of the motor and the optical encoder resulted in a failure in a servo system.

2. Instability of the +5V power supply at low-temperature extremes due to soft failures. As much as 1.5V peak-to-peak ripple was observed. The units would shut down the servo when operated at the low-temperature extremes and would work normally upon power cycling at ambient conditions. A capacitor was replaced with one of a lower parameter drift, which reduced the power supply ripple to an acceptable level.

16.12   STRIFE TESTING AT ZYTEC HELPS THE COMPANY WIN BALDRIGE [7; 8]

In late 1991, the U.S. Commerce Department presented the Baldrige Award for manufacturing to ZYTEC of Redwood Falls and Eden Prairie, Minnesota. Doug Tersteeg, Director of Quality, Reliability, Safety and Training for ZYTEC, credited the company's design-verification test program, which is a combination of stress tests and life (STRIFE) test during the design phase, with helping it win the award.

ZYTEC is the fifth largest U.S. manufacturing company of Switching Power Supplies ranging from 150 to 1,500 watts, which go to computers,

TABLE 16.5– Results of STRIFE test on 12 power supply units [7].

Number of units	12
Test time, week	20
Component infant mortality	None
Workmanship failures Design defects	• Solder joints C24 and C25 • Loose hardware on fan • Spring clip on CR15 • C9 and C18 temperature sensitive

medical equipment, and test equipment. At ZYTEC, STRIFE testing is conducted on all new products. Testing is conducted at the prototype stage and consists of three specific test modes; i.e.,

1. random vibration testing,

2. temperature cycling using liquid nitrogen, and

3. electrical overstressing.

Figure 16.7 shows one cycle of the temperature profile, which is usually repeated 10 times during the screen. The power is on during the first half of the cycle where the temperature is above the room temperature, and is off during the second half of the cycle where the temperature is below the room temperature.

The random vibration screening is conducted at 7 *grms* for 15 minutes under full load, nominal input line and 25°C. Both temperature cycling and random vibration are conducted in the QRS-400T Environmental Chamber manufactured by the Screening Systems, Inc.

The electrical overloading test is conducted at the maximum specified operating temperature and high input line as follows:

1. 100% load for 1 hour,

2. 110% load for 1 hour,

3. 120% load for 1 hour,

4. 130% load for 2 hours.

This process is then repeated using the low input line.

Table 16.5 shows the results of a STRIFE test on 12 power supply units. It may be seen that both workmanship problems and attributable problems, such as design defects, were brought to light.

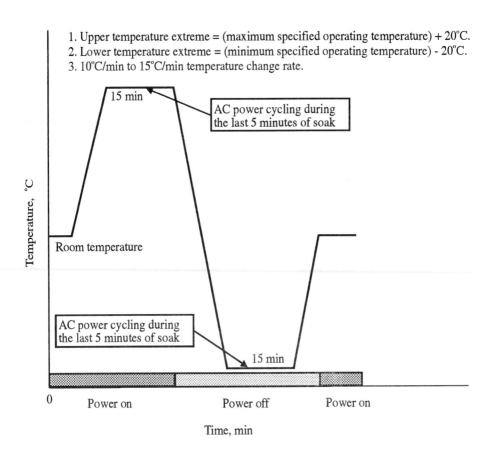

Fig. 16.7– One cycle of the temperature profile used by ZYTEC
[7].

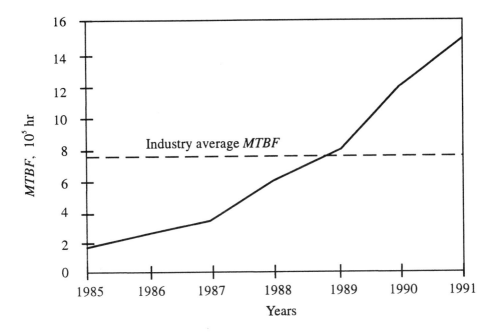

Fig. 16.8– Dramatic $MTBF$ improvement of ZYTEC's power
supplies due to STRIFE testing and corrective actions
[8].

ZYTEC's Product Qualification Process now calls for STRIFE testing
two units prior to Reliability Demonstration Testing, which has significantly
decreased attributable failures in Reliability Demonstration Testing. Reliability Demonstration Tests have been successfully conducted on 7 new products
following STRIFE Testing with *zero* attributable failures. At ZYTEC the
application of STRIFE Testing to products in the early prototype stage has
reduced development cycle time, increased field reliability and significantly
decreased qualification costs. As shown in Fig. 16.8, ZYTEC has improved
the $MTBF$ of its power supplies from less than 0.2 million hours to more
than 1.5 million hours. As a result, ZYTEC has reduced warranty costs by
50% from 1984 to 1990.

16.13   COST EFFECTIVENESS OF *ESS* DEMONSTRATED AT ROBINS AIR FORCE BASE [9]

16.13.1   BACKGROUND

In April 1988, Air Force Logistics Command (AFLC) Regulation 74-25 established the requirement for incorporating *ESS* into AFLC operations. In Section 3-2, it states that *ESS* is required for all organically manufactured electronic hardware unless a waiver is issued. For *ESS* to obtain local acceptance, a cost-effectiveness demonstration was conducted at Warner Robins Air Logistics Center in May of 1988. Twenty (20) newly manufactured Dynamic Focus Generator Circuit Board Assemblies (ALR-20 System) were exposed to an *ESS* program which resulted in a predicted cost savings of $28,000.

16.13.2   SCREEN PROGRAM

The QRS-400 portable triaxial quasirandom vibration and temperature cycling system manufactured by Screening Systems, Inc. was used, which is capable of temperature extremes of −60°C and 125°C, a maximum temperature change rate of 30°C/min, and vibration frequency range of 20 Hz to 2,000 Hz.

The screen program is composed of first exposing the circuit cards to

1. 10 minutes of random vibration of 6 *grms*, and then to

2. 5 temperature cycles with a lower extreme of −55°C and an upper extreme of 85°C, and a 3-minute dwell period.

16.13.3   SCREEN RESULTS

A total of 7 component failures due to faulty capacitors were precipitated on 5 of the 20 boards. All of these 7 failures were breaks at the base of the capacitor's leads close to the body of the component and all occurred at the same general locations on the board. Two of the 7 broken capacitor leads were still making a connection, which will possibly cause intermittent failures in the field. Failure analysis confirmed that corrosion weakened the base metal, making it susceptable to fatigue. A quality deficiency report (QDR) was proposed recommending that all capacitors under the cited contract number be screened using component screening techniques.

16.13.4   COST EFFECTIVENESS

An economic analysis was conducted which revealed the following cost savings by screening only 20 circuit cards:

Assembly level	Cost savings
Circuit card (SRU)	$3,563
Next level assembly (LRU)	$13,183
Aircraft level	$28,023

It may be seen that a great return of investment (ROI) was realized. This successful *ESS* project initiated support for an intensive *ESS* program.

16.14   RANDOM VIBRATION SCREENING OF MILITARIZED AVIONIC COMPUTERS AT IBM [10]

16.14.1   COMPUTER DESCRIPTION

From the start of utilizing random vibration at IBM in 1976 through 1982, over 600 computer sets representing more than 1,000 line replaceable units (LRU's) had been exposed to this screen. Presented here is a brief summary of the random vibration screening conducted by IBM's Federal Systems Division at Owego, New York on six types of digital computers produced for military avionic applications. These six computer types operate in five different in-service use environments and fall into two general categories:

1. Four computer types in which the computer is a singular LRU.

2. Two computer types in which the equipment sets consist of multiple (2 and 6) LRU's.

The LRU's, in general, consist of 1 or 2 major printed wire interconnection boards (backpanels) and a series of shop replaceable units (SRU's) such as digital, analog and memory circuit board assemblies and a power supply which may be either a single circuit board or an assembly containing several circuit boards.

16.14.2   RANDOM VIBRATION PROFILE

Table 16.6 summarizes the applied random vibration parameters on each computer type. The corresponding profiles are shown in Fig. 16.9.

TABLE 16.6– Random vibration parameters on each of six computer types [10].

Computer type	$grms$	Peak g^2/Hz	Minimum number of times random vibration is applied	Minimum duration per unit, min	Axes of random vibration
A	2.67	0.010	1	60	Major horizontal only
B	6.00	0.051	2	30	Major horizontal only
C	3.80	0.010	1	15	3
D	5.80	0.020	1	15	3
E	4.90	0.012	1	15	3
F	6.30	0.040	2	3	3

Table 16.7 is a summary of failure quantities and percentages, test equipment failures, intermittent failures and the number of computers tested for each computer type. Table 16.8 presents the percent contribution of various general failure types, such as components, workmanship, etc.

TABLE 16.7- Summary of random vibration precipitated failures [10].

Computer type	Computers tested	Intermittent failures	Test equipment	No. of failures (repairs)	Percent failures
A	63	10	3	45	71
B	21	2	0	8	38
C	330	20	2	85	26
D	50	0	0	3	6
E	109	1	4	11	10
F	68	9	0	30	40
Total	641	42	9	182	28

It may be seen from Table 16.7 that an average of 0.28 failures per computer were experienced during the random vibration screen. From Table 16.8, it may be seen that 55.3% of the failures were due to components. Intermittents and assembly workmanship constituted 18.7% and 15.5% of the failures, respectively.

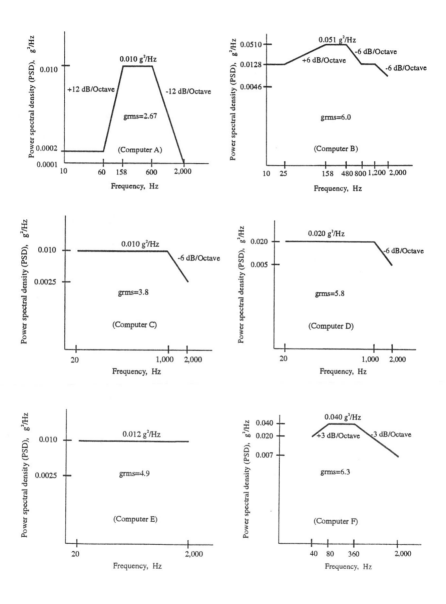

Fig. 16.9- Random vibration profiles for 6 different computer types [10].

TABLE 16.8– Percentage contribution of various failure types [10].

Failure type		Components	Intermittents	Workmanship	Unknown
Quantity	A	34	10	7	4
of	B	3	2	1	4
failure	C	59	20	10	16
of each	D	3	0	0	0
computer	E	5	1	6	0
type	F	20	9	10	0
Average percentage contribution		55.3%	18.8%	15.2%	10.7%

16.15 *ESS* OF A HIGH-DENSITY SURFACE MOUNT CIRCUIT CARD AT AT&T LITTLE ROCK [11]

This case history is about the *ESS* on a circuit card purchased from an outside supplier (hereafter referred to as Vendor A) and assembled into a computer product at the AT&T Little Rock Works.

The products were manufactured using surface mount technology. Each consisted of 6 Application Specific Integrated Circuits (ASIC's) packaged in both 25 and 50 mil pitch surface mount packages. In addition, each card contained more than 400 other Integrated Circuits (IC's) and discrete components.

In planning for integration of *ESS* into the production process, AT&T engineers summed up their past experience with similar products and recommended to conduct a monitored thermal cycling screen. At AT&T Little Rock, an important distinction is made between unmonitored and monitored *ESS*. Monitored *ESS* integrates portions of a functional test into the *ESS* process and is therefore referred to as Environmental Stress Testing (*EST*) at AT&T. Monitored *ESS* has the following two major benefits:

1. It helps optimize the duration of the process by measuring time or cycles to failure.

2. It helps identify intermittent failures plus those devices under test (DUT's) which lack sufficient temperature margin to ensure high reliability.

An *EST* thermal cycling profile was determined based on prior experience

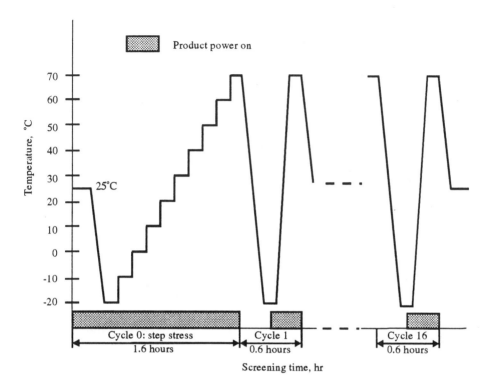

Fig. 16.10– An EST thermal cycling profile consisting of a step-stress period (Cycle 0) and 16 rapid thermal cycles [11].

with similar circuit card design. This thermal cycling profile consists of the following two parts:

1. 1.5-hour period of step-stress screening (Cycle 0) during which the product was first tested in the chamber at 25°C and then turned off and the chamber pulled down to −20°C where all cards were retested. Chamber temperature was then ramped up step-wisely by a constant increment of 10°C until the maximum temperature of 70°C was reached. This step-stress process is shown in Fig. 16.10.

2. 16 rapid thermal cycles covering an 8-hour period with a temperature change rate of

   • 9°C/min for a temperature pulldown from 70°C to 0°C,

- 6°C/min for a temperature pulldown from 70°C to −20°C, and
- 18°C/min for a temperature pullup from −20°C to 70°C.

These thermal cycles, which follow the step-stress part immediately, are also shown in Fig. 16.10.

During the thermal cycling period, the products were tested at both cold and hot extremes as well as during temperature ramp-up period. Each device under test (DUT) was tested every 4 minutes on the average during the power-on period. During one full EST run, each DUT was tested a minimum of 65 times.

The monitored step-stress portion provides the following information:

1. Initial measurement of temperature threshold of failure on DUTs which exhibit margin defects and help identify the failure mechanisms, and identify the risk of field failures for units failing beyond temperature specification limits.

2. Narrow temperature ranges of failure which may otherwise go undetected during rapid temperature ramps.

Table 16.9 is a plot of the number of failures during the corresponding screening Cycle 0 (step-stress), 1, 2, ⋯, 16 (thermal cycling).

An attempt was made to verify all EST failures at room temperature by applying a bench test and additional consecutive stresses, as shown in Fig. 16.11. This process also helped to convert intermittent failures to hard failures. Verifiable failures were then returned to the board supplier for warranty repair and failure analysis. Unverifiable failures were held for further analysis.

After verification, EST failures were divided into the following categories:

1. T1 – fails 25°C bench test after EST run.

2. T2 – passes first bench test but fails bench test at some later point in tracking.

3. T3 – passes bench test at all times but fails a four-corner test conducted at temperature and voltage specification limits of 0°C – 55°C and 4.75V – 5.25V Vcc.

4. T4 – passes bench test and four-corner test after repeated failures in EST chamber.

5. T5 – false failure indication.

Table 16.10 lists the failures by category after verification, which illustrates that 170 units, or less than half of the 413 total failures initially identified as defective by EST, excluding false failures, actually failed during the first bench test after the initial EST run. Had monitoring not been applied, all T2, T3 and T4 category defects, or 105+29+85=219 out of 170+105+29+85=389, or 56%, of all EST defects, would have been shipped

TABLE 16.9– A sample plot of failures precipitated during each cycle where Cycle 0 corresponds to the step-stress portion, and Cycles 1 through 16 correspond to the thermal cycling portion according to Fig. 16.10 [11]. Note that only the first 7 cycles are presented here.

Temperature, °C	-2†	-1	0	+1	+2	+3	+4	+5	+6	+7	+8	Cycle #	Time
26.3						P*						0	23:15
-20.8	P											0	23:30
-9.5		P										0	23:38
-9.3		P										0	23:39
0.3			P									0	23:47
10.4				P								0	23:56
10.3				P								0	23:58
20.6					P							0	00:06
30.6						P						0	00:16
40.0							P					0	00:24
40.5							P					0	00:27
50.1								F**				0	00:34
60.1									F			0	00:43
60.3									F			0	00:45
69.9										F		1	00:52
-19.9	P											1	01:14
47.7								P				1	01:18
68.8										F		1	01:24
-21.8	P											2	01:44
69.5										P		2	01:48
68.7										F		2	01:53
-19.3	P											3	02:11
-21.7	P											3	02:15
68.7										F		3	02:23
-20.0	P											4	02:40
8.4				P								4	02:45
68.6										F		4	02:52
-20.8	P											5	03:11
42.3							P					5	03:16
68.7										F		5	03:32
-21.8	P											6	03:42
69.3										P		6	03:46
68.6										F		6	03:51
-18.5	P											7	04:09

† These values in the first row are °C to be added to the temperature in the first column.

* P = Pass.

** F = Fail.

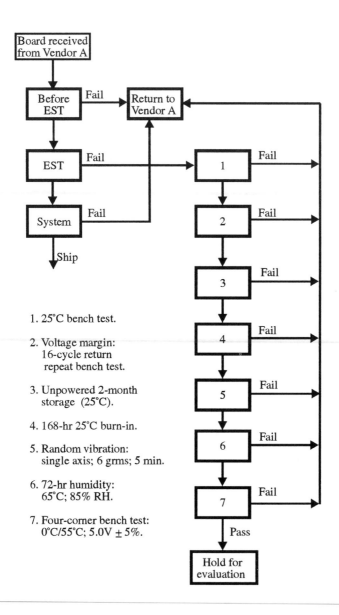

1. 25°C bench test.

2. Voltage margin:
   16-cycle return
   repeat bench test.

3. Unpowered 2-month
   storage (25°C).

4. 168-hr 25°C burn-in.

5. Random vibration:
   single axis; 6 grms; 5 min.

6. 72-hr humidity:
   65°C; 85% RH.

7. Four-corner bench test:
   0°C/55°C; 5.0V ± 5%.

Fig. 16.11– Product flow through an EST process [11].

TABLE 16.10– EST failure breakdown by category [11].

Category	Quantity	Percent
T1	170	0.99%
T2	105	0.61%
T3	29	0.17%
T4	85	0.49%
Unknown	24	0.14%
Total	413	2.40%
False	265	1.54%

to the next process step and most would have been shipped to the AT&T customers. This implies that monitored *ESS* identified 389/(389-219)=2.3 times as many potential defects as unmonitored *ESS*.

16.16 AUTOMATED *ESS* OPTIMIZATION FOR U.S. ARMY ELECTRONIC HARDWARE [12]

16.16.1 INTRODUCTION

Optimized *ESS* results in an improved quality product with significant cost savings. System analysis techniques employed by the U.S. Army provide engineering guidelines to optimized *ESS*, considering in-house manufacturing and field costs. A menu-driven and user-friendly computer program was developed by the U.S. Army to provide the quality engineers with a uniform assessment methodology by automating the calculations for the most cost-effective *ESS* profile based on all essential input information.

16.16.2 PROGRAM FUNCTIONS

The *ESS* computer program was written in FORTRAN language containing over 2,000 detailed program steps. It consists of the following three (3) independent subprograms:

1. *Subprogram 1– Optimization of Stress Screening.* This subprogram will choose an optimum stress screen based on the total cost of ownership. The computer flow chart for this subprogram is shown in Fig. 16.12.

2. *Subprogram 2– Estimation of Life-Cycle Failures.* This subprogram will allow the engineers to estimate the expected number of life-cycle-

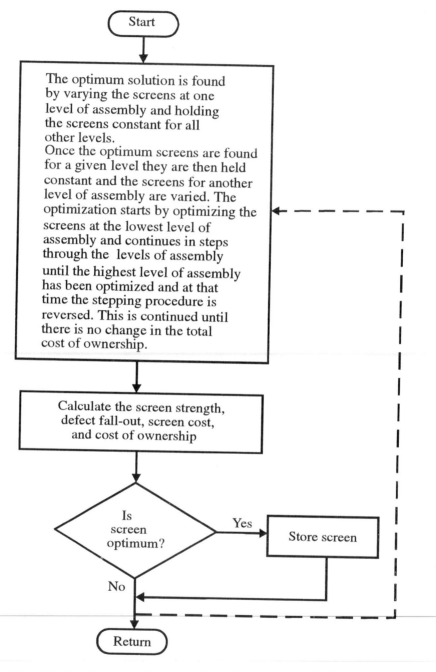

Fig. 16.12– Computer flow chart for Subprogram 1 of U.S. Army's *ESS* computer program [12].

fielded failures with and without *ESS*. The computer flow chart for this subprogram is shown in Fig. 16.13.

3. *Subprogram 3– Estimation of Owner's Cost.* This subprogram will enable the engineer to estimate the owner's cost for the preselected set of *ESS* plans. The computer flow chart for this subprogram is shown in Fig. 16.14.

16.16.3   PROGRAM PRINCIPLE

The generalized theory for Subprogram 1 follows the production flow shown in Fig. 16.15. The cost of failures due to latent defects induced either by parts or by the manufacturing process, and the cost of repairs are considered by this program as the product moves toward field use. The composition and the calculation of the owner's cost are shown in Fig. 16.16.

Subprogram 1 can model up to 5 levels of equipment as a breakdown structure with Level 1 being the first level of assembly, Level 2 being the next, etc., and the highest level being the complete system under consideration. Parts rescreening is not considered in the program. The program optimizes the screen such that the life-cycle cost is minimized.

The following equations for screen strength (*SS*) estimation are used by Subprogram 1:

1. Thermal cycling:

$$SS = D\left\{1 - e^{-0.0023\left[\log_e(e+\dot{T})\right]^{2.7} N_{TC}^{0.5} (\Delta T)^{0.6}}\right\},$$

where

$$D = 0.8,$$
$$\dot{T} = \text{temperature rate of change, } °C/\text{min},$$
$$N_{TC} = \text{number of repeated cycles},$$

and

$$\Delta T = \text{temperature range, } °C.$$

2. Random vibration:

$$SS = D\left(1 - e^{-\frac{t_{RV}}{B}}\right),$$

where

$$B = 0.266G + 1.0402,$$
$$D = 0.144G - 0.0862,$$
$$G = grms \text{ value of the input acceleration spectrum},$$

and

$$t_{RV} = \text{vibration length, min.}$$

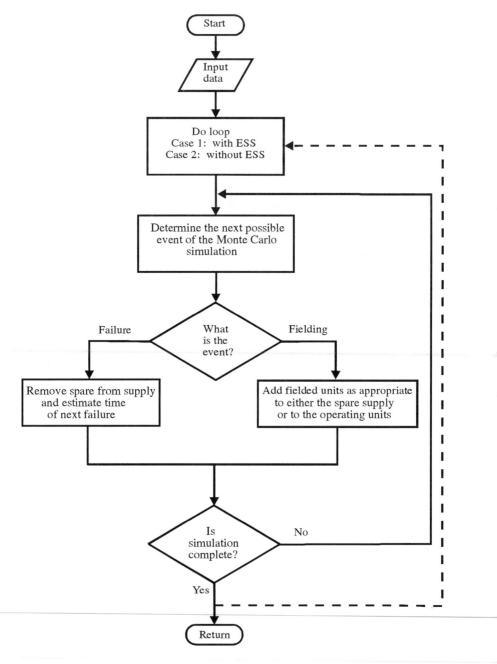

Fig. 16.13– Computer flow chart for Subprogram 2 of U.S. Army's
ESS computer program [12].

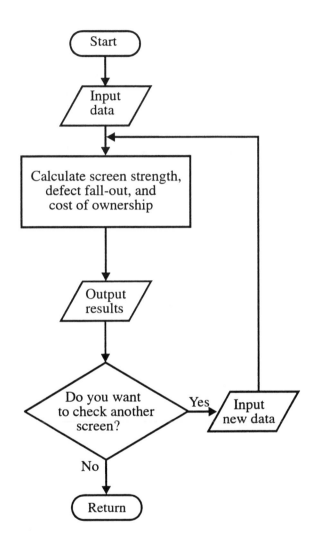

Fig. 16.14– Computer flow chart for Subprogram 3 of U.S. Army's *ESS* computer program [12].

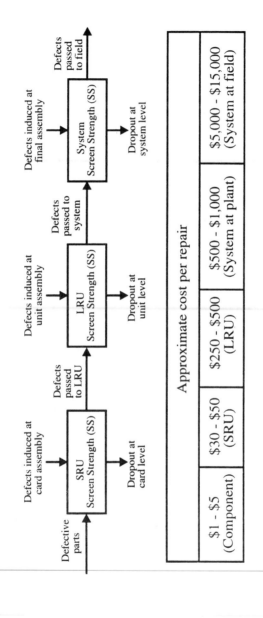

Fig. 16.15— *ESS* production flow model on which Subprogram 1 is based [12].

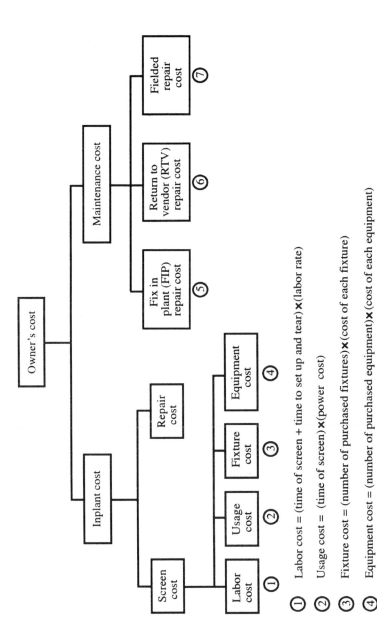

① Labor cost = (time of screen + time to set up and tear)×(labor rate)

② Usage cost = (time of screen)×(power cost)

③ Fixture cost = (number of purchased fixtures)×(cost of each fixture)

④ Equipment cost = (number of purchased equipment)×(cost of each equipment)

⑤ FIP = (defects remaining)×(FIP allocation factor)×(local repair cost)

⑥ RTV = (defects remaining)×(RTV allocation factor)×(local repair cost)

⑦ Field cost = (defects remaining)×(field allocation factor)×(repair cost)

Fig. 16.16– Breakdown of the owner's cost for Subprogram 1 [12].

Subprogram 2 estimates the expected number of life-cycle field failures and the failure number reduction using an increased $MTBF$ resulting from ESS implementation.

Subprogram 3 uses the same methodology as that for Subprogram 1.

16.16.4 PROGRAM INPUTS AND OUTPUTS

The required data inputs for these three subprograms are summarized in Tables 16.11 through 16.13, respectively.

A sample output for Subprogram 1 is provided in Tables 16.14 and 16.15 for the M1 Thermal Imaging System/Laser Range Finder (TIS/LRF) consisting of approximately 4,000 piece parts. The number of levels subjected to ESS requirements were two: circuit card assembly (CCA) and line replaceable unit (LRU).

Table 16.16 provides a sample output for Subprogram 2. Table 16.17 provides a sample output for Subprogram 3.

16.16.5 PROGRAM APPLICATION AND ACHIEVEMENTS

A number of systems have been studied by the U.S. Army Armament, Munitions and Chemical Command (AMCCOM) as listed in Table 16.18. These studies yielded over $1.1 billion in life-cycle cost savings, 80 to 1 for return on investment (ROI), and 3 to 1 for field reliability improvements of production systems.

16.17 *ESS* AS APPLIED TO A NEW COMPUTER

A new computer had several new technologies and features which manufacturing had never encountered in the past, namely, liquid cystal displays, micro-floppy disks, surface-mounted IC's, and total portability. Customers were complaining that the failures were approaching 50%. A decision was made to test 100%. This testing was done prior to customer shipment at ambient temperature for 24 hours. This produced random failures throughout the entire 24 hours.

A decision was made to use ESS with a two-fold purpose to (1) deliver a steady-state product to the Distribution Center and (2) eliminate defects. This required (1) an effective screening technique and (2) the identification and elimination of process problems.

Through screening failures were detected. When root caused, these failure mechanisms, physical, chemical, etc., showed the failure modes: shorts, opens, etc. This allowed the process to be modified to detect these failure modes.

A reasonable process of testing was established with the ability to measure each step's effectiveness within the process. A burn-in was required after the process to measure the effectiveness of the actual process was established.

TABLE 16.11– Data inputs for U.S. Army's *ESS* Subprogram 1 [12].

No.	Name of input	Input value				
1	Number of screen levels					
2	Thermal power cost ($/hr)					
3	Vibration power cost ($/hr)					
4	Labor rate ($/hr)					
5	Production rate (units/day)					
6	Production hours per day					
7	Units on contract					
8	Fix inplant allocation factor					
9	Fix inplant repair cost					
10	Return to vendor allocation factor					
11	Return to vendor repair cost					
12	Defects to field allocation factor					
13	Defects in field repair cost					
14	Existing thermal chambers					
15	Cost of thermal chamber					
16	Existing random vibrators					
17	Cost of random vibration table					

No.	Name of input	Input value at				
		Level 1	Level 2	Level 3	Level 4	Level 5
18	Repair cost					
19	Defects detectable by thermal cycling					
20	Defects detectable by vibration					
21	Assemblies per unit					
22	Test time of thermal screen					
23	Dwell time					
24	Temperature maximum rate of change					
25	Maximum range (°C)					
26	Maximum number of cycles					
27	Time (min) to set up power on					
28	Time (min) to set up power off					
29	Number of power-on fixtures					
30	Number of power-off fixtures					
31	Capacity of chambers					
32	Cost of basic equipment					
33	Cost of monitoring equipment vibration screen					
34	Maximum time of vibration (min)					
35	Force of vibration (*G*)					
36	Time (min) to set up power on					
37	Time (min) to set up power off					
38	Number of power-on fixtures					
39	Number of power-off fixtures					
40	Capacity of vibration table					
41	Cost of basic equipment					
42	Cost of monitoring equipment					

**TABLE 16.12– Data inputs for U.S. Army's *ESS* Subprogram
2 [12].**

No.	Name of input	Input value
	Non-*ESS* units	
1	Initial number fielded	
2	Mean time between failures	
3	a. Constant fielding rate	
	Time interval	
	Number to be fielded	
	Start time	
	Finish time	
	b. Variable fielding plan	
	Date	
	Number	
	c. No more will be fielded	
	Non-*ESS* spares	
4	Initial number fielded	
5	Mean time between failures	
6	a. Constant fielding rate	
	Time interval	
	Number to be fielded	
	Start time	
	Finish time	
	b. Variable fielding plan	
	Date	
	Number	
	c. No more will be fielded	
	***ESS* units**	
7	Initial number fielded	
8	Mean time between failures	
9	a. Constant fielding rate	
	Time interval	
	Number to be fielded	
	Start time	
	Finish time	
	b. Variable fielding plan	
	Date	
	Number	
	c. No more will be fielded	
	***ESS* spares**	
10	Mean time between failures	
	All units	
11	Operating time per year	
12	Number of years projected	

TABLE 16.13– Data inputs for U.S. Army's *ESS* Subprogram 3 [12].

No.	Name of input	Input value at				
		Level 1	Level 2	Level 3	Level 4	Level 5
1	Defects (thermal)					
2	Defects (vibration)					
3	Repair cost					
4	Thermal screen					
5	Power on					
6	Thermal test cost					
7	Number cycles					
8	Temperature range					
9	Rate of temp. change					
10	Vibration screen					
11	Type (0:NS, 1:RV, 2:SS)					
12	Power on					
13	Vibration test cost					
14	Length of vibration					
15	Force of vibration					
No.	Name of input	Allocation factor		Failure cost		
16	Fix in plant					
17	Return to vendor					
18	Received in field					

TABLE 16.14– A sample output of screen optimization from U.S. Army's *ESS* Subprogram 1, Part 1: screen parameters [12].

Level	1	2
Thermal test cost ($)	$27,279.20	$45,245.91
Number of thermal chambers	3.00	2.00
Number of thermal cycles	29.00	7.00
Temperature range (°C)	150.00	120.00
Rate of temperature change	15.00	8.00
Thermal cycle	3.00	3.00
1– None		
2– Power off		
3– Power on		
Vibration test cost ($)	0.00	$31,182.75
Number of vibration tables	0.00	1.00
Length of vibration (min)	0.00	20.00
Force of vibration (*grms*)	0.00	6.00
Random vibration	1.00	3.00
1– None		
2– Power off		
3– Power on		

TABLE 16.15– A sample output of screen optimization from U.S. Army's *ESS* Subprogram 1, Part 2: in-plant and field cost summary [12].

Level	Screen type	Defects detected	Test cost, $	Repair cost, $	Screen strength
1	Thermal	754.95	27,279.20	37,747.59	0.8338
2	Thermal	150.33	45,245.91	37,583.00	0.5693
2	Vibration	76.56	31,182.75	19,139.02	0.6028
FIP	N/A	77.16	N/A	115,347.03	N/A
RTV	N/A	27.91	N/A	75,684.31	N/A
Field	N/A	59.10	N/A	536,074.25	N/A

Total screen strength = 0.9568.
Pre-DD250 cost/100 units = $198,177.47.
Post-DD250 cost/100 units = $727,105.59.
Total cost of ownership/100 units = $925,283.06.

TABLE 16.16– A sample output of life-cycle failures from U.S. Army's *ESS* Subprogram 2 [12].

Year	Failures without *ESS*	Failures with *ESS*	Failure reduction
1	3,089	2,752	338
2	7,466	6,272	1,194
3	13,130	10,785	2,345
4	20,078	15,951	4,127
5	28,314	19,373	8,941
6	37,836	22,552	15.284
7	48,644	26,033	22,611
8	60,738	29,907	30,831
9	74,119	34,190	39,929
10	88,786	38,883	49,903
11	104,739	43,988	60,751
12	121,978	49,505	71,473
13	140,504	55,433	85,071
14	160,316	61,773	98,543
15	181,414	68,524	112,890
16	203,798	75,686	128,112
17	227,469	83,261	144,208
18	252,425	91,247	161,178
19	278,668	99,645	179,023
20	306,198	108,454	197,744

TABLE 16.17– A sample output of owner's cost estimation from U.S. Army's *ESS* Subprogram 3 [12].

Screen	Contractor — Temperature cycling				Contractor — Random vibration				Tank plant		Field	
	Screen strength	Screen cost, $	Rework cost, $	Defects remain	Screen strength	Screen cost, $	Rework cost, $	Defects remain	Flip cost, $	RTV cost, $	Defects to field	Rework cost, $
Present	0.38/0.41	2,650/8,600	10,450/34,612	341/201	0.03	550	1,762	195	136,434	93,605	69	625,899
					Alternatives							
(1). TC* (4 cycles)	0.38/0.51	2,650/8,600	10,450/43,478	341/167	0.03	550	1,250	162	113,345	77,764	58	526,118
(2). RV** (PWR 10 min)	0.38/0.51	2,650/8,600	10,450/43,478	341/167	0.35	2,500	14,500	109	76,263	52,323	39	353,769
(3). RV (PWR 20 min)	0.38/0.51	2,650/8,600	10,450/43,478	341/167	0.41	2,500	17,000	99	69,266	47,522	35	317,485
(4). TC (PWR)	0.76/0.51	8,300/8,600	20,900/16,750	132/65	0.41	2,500	6,750	38	26,587	18,241	14	126,994
(5). TC (8 cycles)	0.76/0.63	8,300/10,600	20,900/20,790	132/49	0.41	2,500	5,000	29	20,290	13,921	10	90,710

* TC – Temperature cycling screen.
** RV – Random vibration screen.

TABLE 16.18– Studies conducted by the U.S. Army Armament, Munitions, and Chemical Command
(AMCCOM) and the achieved life-cycle cost
savings [12].

Studies	Life-cycle savings, 10^6\$
M1 tank (TIS/LAF)	496.0
Bradley (ISU)	239.0
Bradley (Turret drive)	110.1
AH-64A Apache	179.0
Cam chemical alarm	46.2
M712 155 mm copperhead	15.6
Small arms transmitter (SAT)	1.2
Remote target system (RETS)	3.8
Multiple integrated laser equipment (MILES)	2.0
Conduct of fire trainer (COFT)	6.8
Integrated ι ɘlmet and display sighting system (IHADSS)	14.8
Vulcan air defense system (VADS)	20.0
Product improved VADS (PIVADS)	10.0
Total	1,144.5

No accurate field data existed on the effectiveness of this process because of its newness; therefore, a customer simulation test was established. The purpose of this test was to discover deficiencies in the newly implemented process. This involved randomly selecting 25 units which passed the process, and having these 25 units carried in and out of work by actual employees. They took them home every day to simulate a typical customer's application. All failures were root-caused, allowing the process to be fine-tuned to detect these failures. This test brought about the institution of a 3-inch drop test and the relocation of the vibration test to after the temperature/relative humidity *ESS* portion of the process. The following process resulted:

Test Engineering recommended changes to the process. Through fine-tuning of this process yields increased. In Table 16.19 the changes implemented in both the *PCB* and the system level process flows are given. Many failures were encountered, and the feedback of these data allowed resolution of the problems, as listed in Table 16.19. Table 16.20 lists the assembly steps and the potential defects observed in them.

As a result, reliability growth advanced at a rapid rate during the first several months. It took the same amount of time to get from 70% to 96% yield, as it took to get from 96% to 99% yield. This is a general behavior. The closer to 100% yield a process is, the more difficult it is to approach and reach 100%.

Once the yields were 99%, the product went directly to Distribution Center stock. Lots from the Distribution Center stock were sample-tested on a regular basis to ensure the stability of the process and of the product.

REFERENCES

1. Tustin, W., "Recipe for Reliability: Shake and Bake," *IEEE Spectrum*, pp. 37-42, December 1986.

2. Tustin, W., "IBM Stress Screens New Printer," *Test Engineering and Management*, pp. 12-13, December/January, 1988.

3. Chesney, K. E., "Step Stress Analysis of a Printer," *Proceedings of the Annual Reliability and Maintainability Symposium*, Las Vegas, Nevada, pp. 26-27, 1986.

4. Mandel, C. E., *Environmental Stress Screening*, A Tutorial, Screening Systems, Inc., 82 pp., 1991.

5. Mandel, C. E., "Environmental Stress Screening (ESS)," *Proceedings of the Institute of Environmental Sciences*, Orlando, Florida, pp. 294-302, 1984.

6. Schinner, C., "Board Electronic STRIFE Test (B.E.S.T.) Program," *Reliability Review*, Vol. 8, pp. 3-6, 1988.

TABLE 16.19– *ESS* **precipitated problems and their solutions.**

Liquid Crystal Display (LCD)	
Problem	Solution
Excessive pixel failures of vendor's LCD's recorded.	Vendor was removed from Authorized Vendors List (AVL).
Root cause revealed internal dendrite growth.	Vendor conformal coated all LCD screen integrated circuits (IC's).
Excessive puddles on another vendor's LCD's revealed.	Electro-coupled oscillator (ECO) by vendor removed excessive pressure on LCD glass sandwich, drastically reducing puddle failures.
LCD lenses produced a rainbow effect, termed Newton rings, caused by air gaps defracting light rays.	Lens thickness was reduced. Cleaning technique for assembly was implemented. Glue was applied at the corners of the lens to keep it stationary.
Disk	
Problem	Solution
3.5" micro floppy disk problems: Computer Readout Correction (CRC) errors during read. Requested sector not found. Address mark not found.	Major redesign of key areas: Spindle coated with teflon. Drive Read/Write (RW) timing changed. Faulty media design was resolved.

TABLE 16.19– (continued).

Interconnection	
Problem	Solution
Interconnection problem between LCD screen and main Printer Circuit Board (PCB); Root cause revealed fretting corrosion.	NYO gel 1759G was used which is a light-yellow grease made from synthetic hydrocarbon oil. This was applied to the connector to form a protective shield over the tin-lead fingers to prevent fretting corrosion.
FPC connectors (Berileum clips) inside the unit were falling off causing electrical short circuits.	A bad lot had been received. Stock purged.
Workmanship	
Problem	Solution
Solder shorts, flakes, and cold solder joints caused between 5% and 7% of all failures.	This problem was addressed with training and process changes.

TABLE 16.20– Assembly steps and potential defects observed on PCB's.

Step no.	Operation	Potential defects
1	Parts selection and kitting	Wrong part; Damaged part.
2	Parts insertion	Damaged lead; Damaged part; Damaged PCB; Contamination.
3	Lead trimming	Damaged lead; Lead too long; Damaged PCB.
4	PCB cleaning	Contamination; Damaged part; Damaged PCB.
5	Flux application	Contamination; Damaged part; Damaged PCB.
6	Soldering	Cold solder joint; Solder bridging; Thermal damage to part; Thermal damage to PCB.
7	Cleaning	Contamination; Damaged part; Damaged PCB.
8	Assembly test	Damaged part; Electrical overstress.
9	Final inspection and test	Damaged part; Damaged PCB; Electrical overstress.

7. Tersteeg, D. J., "Reliability Demonstration Testing and Strife Testing – Is There a Correlation?" *Proceedings of Power Electronics Conference*, Convention Center, Long Beach, California, 4 pp., 1990.

8. "STRIFE Helps Zytec Win Baldrige," *Test & and Measurement World*, pp. 46-52, April 1992.

9. Nagle, A. L., "Cost-Effectiveness of Environmental Stress Screening (ESS)," *Journal of the IES*, pp. 35-38, November/December, 1991.

10. Blake, R. K., "Random Vibration Screening of Six Militarized Avionic Computers," *Journal of the IES*, pp. 15-24, March/April, 1982.

11. Parker, T. P., "*ESS* Case Study of a High Density Surface Mount Circuit Card," *Proceedings of the Institute of Environmental Sciences*, pp. 393-402, 1991.

12. Huizinga, M. A., "Optimized Environmental Stress Screening of U.S. Army Electronic Hardware," *Proceedings of the Institute of Environmental Sciences*, pp. 209-216, 1987.

ADDITIONAL REFERENCES

1. Bohan, E. M. and McGrath, J. D., "Shake and Bake – Shape Your Future," *Proceedings of the Institute of Environmental Sciences*, Los Angeles, California, pp. 234-238, 1983.

2. Capitano, J. L., "Innovative Stimulus Testing at the Lowest Level of Assembly to Reduce Costs and Induce Reliability," *Proceedings of the Institute of Environmental Sciences*, Orlando, Florida, pp. 303-305, 1984.

3. Capitano, J. L. and Feinstein, J. H., "Environmental Stress Screening (*ESS*) demonstrates Its Value in the Field," *Proceedings of the Annual Reliability and Maintainability Symposium*, Las Vegas, Nevada, pp. 31-35, 1986.

4. Caruso, H., "Significant Subtleties of Stress Screening," *Proceedings of the Annual Reliability and Maintainability Symposium*, Orlando, Florida, pp. 154-158, 1983.

5. Caruso, H., "Generic Environmental Stress Screening Requirements for Statements of Work," *Proceedings of the Institute of Environmental Sciences*, San Diego, California, pp. 410-414, 1991.

6. Caruso, H., "Environmental Stress Screening of Spares and Repairs," *Proceedings of the Institute of Environmental Sciences*, Nashville, Tennessee, pp. 476-481, 1992.

7. Cerasuolo, D., "Development of Military Spares Screening Program Including Investigation of Subassembly Energization," *Proceedings of the Institute of Environmental Sciences*, Anaheim, California, pp. 207-212, 1989.

8. Chenoweth, H. B. and Bell, J. M., "Semiconductor Industry Screening Performance," *Proceedings of the Institute of Environmental Sciences*, Los Angeles, California, pp. 253-257, 1983.

9. Curtis, A. J., *Reliability Testing and Environmental Stress Screening*, Hughes Aircraft Company, 34 pp., 1989.

10. Dane, A. J., "Profitability of Planning for Stress Screening," *Proceedings of the Institute of Environmental Sciences*, Los Angeles, California, pp. 230-233, 1983.

11. DeCristoforo, R. J., "Environmental Stress Screening – Lessons Learned," *Proceedings of the Annual Reliability and Maintainability Symposium*, San Francisco, California, pp. 129-133, 1984.

12. Diekema, J., "Beyond ESSEH," *Evaluation Engineering*, pp. 84-89, March 1991.

13. Douglas, P., Johnson, A. and Mixer, M., "Continuous-Flow *ESS* Minimizes Defects," *Electronics Test*, pp. 42-44, August 1990.

14. Drees, D. and Winski, G., "Low Level Environmental Stress Screening," *Proceedings of the Institute of Environmental Sciences*, Los Angeles, California, pp. 242-246, 1983.

15. Emerson, D. A. and Buck, R. A., "Non-operating Temperature Cycling - an Effective Screen," *Proceedings of the Institute of Environmental Sciences*, Los Angeles, California, pp. 239-241, 1983.

16. "Environmental Stress Screening Guidelines for Assemblies," Panel Discussion, *Journal of the Institute of Environmental Sciences*, pp. 33-47, September/October, 1990.

17. Eustace, R. H., "Environmental Stress Screening (*ESS*) – Enthusiasm with Limited Understanding?," *Proceedings of the Institute of Environmental Sciences*, San Diego, California, pp. 384-392, 1991.

18. Fedraw, K. and Becker, J., "Impact of Thermal Cycling on Computer Reliability," *Proceedings of the Annual Reliability and Maintainability Symposium*, Orlando, Florida, pp. 149-153, 1983.

19. Garry, W. J., "Developing an *ESS* Automation Tool," *Proceedings of the Annual Reliability and Maintainability Symposium*, Atlanta, Georgia, pp. 495-501, 1989.

20. Geniaux, B., "*ESS* at Final Assembly Level – Actual Results – Optimization Method," *Proceedings of the Institute of Environmental Sciences*, San Jose, California, pp. 195-201, 1987.

21. "Global Competitive Streamlining of *ESS* for the 1990's," Executive Panel Discussion on *ESS*, *Journal of the Institute of Environmental Sciences*, pp. 35-44, May/June, 1991.

22. Golshan, S. and Oxford, D. B., "A Quality Correlation Program," *Proceedings of the Institute of Environmental Sciences*, King of Prussia, Pennsylvania, pp. 58-61, 1988.

23. Gould, D., "Tough Environments Call for Tough Tests," *Test & Measurement World*, pp. 49-56, February 1990.

24. Hnatek, E. R., *Effectiveness of Various Environmental Stress Screens*, Viking Labs/Honeywell, Mountain View, California, 56 pp., 1990.

25. Hobbs, G. K., "Evaluation of Stress Screens," *Proceedings of the Institute of Environmental Sciences*, King of Prussia, Pennsylvania, pp. 47-49, 1988.

26. Hobbs, G. K., "Modern Methods in Stress Screening," *Test & Measurement World*, pp. 47-48, June 1991.

27. Hobbs, G. K., "Highly Accelerated Stress Screens – HASS," *Proceedings of the Institute of Environmental Sciences*, Nashville, Tennessee, pp. 451-457, 1992.

28. Jacob, G., "Benefiting from *ESS* Experience," *Evaluation Engineering*, pp. 45-48, April 1992.

29. Jacob, G., "The Many Faces of *ESS*," *Evaluation Engineering*, pp. 58-59, April 1992.

30. Jawaid, S. and Crook, K., "Linear Ramp Chambers and Thermal *ESS*," *Evaluation Engineering*, pp. 70-79, June 1992.

31. Kallis, J. M., et al, "Techniques for Avionics Thermal/Power Cycling Reliability Testing," *Proceedings of the Institute of Environmental Sciences*, Nashville, Tennessee, pp. 427-436, 1992.

32. Killion, R. E., "An Overview and Critique of Environmental Stress Screening Theory and Practice," *Proceedings of the Institute of Environmental Sciences*, Orlando, Florida, pp. 289-293, 1984.

33. Lascaro, C. P. and DiGiovanni, F., "Environmental Stress Screening (*ESS*) for VRC-12 Overhaul Process, A Case Study," *Proceedings of the Institute of Environmental Sciences*, Anaheim, California, pp. 217-224, 1989.

34. McLean, H., "Highly Accelerated Stressing of Products with Very Low Failure Rates," *Proceedings of the Institute of Environmental Sciences*, Nashville, Tennessee, pp. 443-450, 1992.

35. Meyers, R. and Randazzo, E., "Analytical Spares Screening Evaluation Technique (ASSET)," *Proceedings of the Annual Reliability and Maintainability Symposium*, Philadelphia, Pennsylvania, pp. 120-124, 1987.

36. Moen, G., "Trials and Tribulations of Implementing *ESS*," *Proceedings of the Institute of Environmental Sciences*, pp. 437-442, 1992.

37. Nagle, A. L., "Cost-Effectiveness of Environmental Stress Screening (*ESS*)," *Journal of the Institute of Environmental Sciences*, pp. 35-37, November/December, 1991.

38. Neumann, B., "Automation and Integration of Environmental Stress Screening in Continuous Flow Production," *Proceedings of the Institute of Environmental Sciences*, San Diego, California, pp. 403-409, 1991.

39. Pellicione, V. H. and Popolo, J., *Improved Operational Readiness Through Environmental Stress Screening*, Final Technical Report RADC-TR-87-225, C-51 pp., November 1987.

40. Perlstein, H. J. and Bazovsky, I., "Identification of Early Failures in *ESS* Data," *Proceedings of the Institute of Environmental Sciences*, Anaheim, California, pp. 194-197, 1989.

41. Peterson, J., "Testing for Equipment Reliability with AC Power Cycling," *Evaluation Engineering*, pp. 70-73, April 1983.

42. Phaller, L. J., "How Much Environmental Stress Screening is Really Required?" *Proceedings of the Institute of Environmental Sciences*, Orlando, Florida, pp. 306-311, 1984.

43. Punches, K., "Stress Screening Failure Sources," *Proceedings of the Institute of Environmental Sciences*, San Jose, California, pp. 192-194, 1987.

44. Rawal, B. S. et al, "Reliability of Multilayer Ceramic Capacitors after Thermal Shock," *AVX Technical Information Series*, 4 pp., 1981.

45. Rawal, B. S. and Chan, N. H., "Conduction and Failure Mechanisms in Barium Titanate Based Ceramics under Highly Accelerated Conditions," *AVX Technical Information Series*, 6 pp., 1982.

46. Reeve, P., "Whole Lotta' Shakin' Goin' On," *Test & Measurement World*, pp. 45-50, February 1991.

47. Screening Systems, Inc., *Environmental Stress Screening (ESS) Seminar*, 7 Argonaut, Laguna Hills, California 92656, 1991.

48. Screening Systems, Inc., *Environmental Stress Screening, A Tutorial*, 7 Argonaut, Laguna Hills, CA 92656, 69 pp., Revised May 1987.

49. Screening Systems, Inc., *Environmental Stress Screening, A Tutorial*, 7 Argonaut, Laguna Hills, CA 92656, 77 pp., Revised April 1990.

50. Screening Systems, Inc., *Environmental Stress Screening, A Tutorial*, 7 Argonaut, Laguna Hills, CA 92656, 82 pp., Revised April 1991.

51. Shalvoy, C. E., "Finding the Most Effective Stress-Screening Strategy," *Electronics Test*, pp. 42-46, March 1989.

52. Sly, L. D., "Stress Screening Improves AN/AYK-14(V) Computer Productivity and Reliability," *Proceedings of the Institute of Environmental Sciences*, Los Angeles, California, pp. 247-252, 1983.

53. Smithson, S. A., "Shock Response Spectrum Analysis for *ESS* and STRIFE/HALT Measurement," *Test Engineering & Management*, pp. 10-14, December/January 1991-1992.

54. Turner, R. M., "Study of a Commercial Power Supply," *Proceedings of the Institute of Environmental Sciences*, Nashville, Tennessee, pp. 466-475, 1992.

55. Tustin, W., "Stress Screening: Its Role in Electronics Reliability," *Quality Progress*, pp. 18-22, June 1982.

56. Tustin, W., "Shake and Bake the Bugs Out," *Quality Progress*, pp. 61-64, September 1990.

57. Tustin, W., "Starting Up *ESS*," *Quality*, pp. 57-60, April 1992.

58. Williams, C. L., "Product Verification While Stress Screening," *Proceedings of the Institute of Environmental Sciences*, Anaheim, California, pp. 213-217, 1989.

59. Wong, K. L., "Unified Field (Failure) Theory – Demise of the Bathtub Curve," *Proceedings of the Annual Reliability and Maintainability Symposium*, Philadelphia, Pennsylvania, pp. 402-407, 1981.

60. Wong, K. L., "Off the Bathtub onto the Roller-Coaster Curve," *Proceedings of the Annual Reliability and Maintainability Symposium*, Los Angeles, California, pp. 356-363, 1988.

61. Wong, K. L., "A New Environmental Stress Screening Theory for Electronics," *Proceedings of the Institute of Environmental Sciences*, Anaheim, California, pp. 218-224, 1989.

62. Wong, K. L., "Demonstrating Reliability and Reliability Growth with Environmental Stress Screening Data," *Proceedings of the Annual Reliability and Maintainability Symposium*, Los Angeles, California, pp. 47-52, 1990.

63. Zimmerman, W., "Screening Tests to Monitor Early Life Failures," *Proceedings of the Annual Reliability and Maintainability Symposium,* Orlando, Florida, pp. 443-447, 1983.

Index

ABOUT THE AUTHORS

Dr. Dimitri Kececioglu, P.E., a Fullbright Scholar, received his B.S.M.E. from Robert College, Istanbul, Turkey in 1942, and his M.S. in Industrial Engineering in 1948 and his Ph.D. in Engineering Mechanics in 1953, both from Purdue University, Lafayette, Indiana. He is currently a Professor in the Department of Aerospace and Mechanical Engineering, The University of Arizona; Professor-in-Charge of a unique 10-course Reliability Engineering program leading to the Master of Science degree in Reliability Engineering; Director of the Annual Reliability Engineering and Management Institute; Director of the Annual Reliability Testing Institute; and a Reliability and Maintainability Engineering consultant. This book is based on the following extensive experience of the author in Reliability Engineering & Life Testing:

1. He initiated and was the Director of the Corporate Reliability Engineering Program at the Allis-Chalmers Manufacturing Co., Milwaukee, Wisconsin, from 1960 to 1963.

2. He started the Reliability Engineering Instructional Program at The University of Arizona in 1963, which now has more than 10 courses in it. A Master's Degree with a Reliability Engineering Option is currently being offered in the Aerospace and Mechanical Engineering Department at The University of Arizona. He started this option in 1969. A Master's Degree in Reliability Engineering is also being offered in the Systems and Industrial Engineering Department at The University of Arizona. This degree started in January 1987. Since 1991 this became an all encompassing Master's Degree in Reliability and Quality Engineering.

3. He conceived and directed the first two Summer Institutes for College Teachers in reliability engineering ever to be supported by the National Science Foundation. The first was in 1965 and the second in 1966, for 30 college and university faculty each summer. These faculty started teaching reliability engineering courses at their respective universities and/or incorporating reliability engineering concepts into their courses.

4. He helped initiate The Professional Certificate Award in Reliability and Quality Engineering program at The University of Arizona in 1991. This is a 15-unit program. The certificate's requirements are met via videotapes of the VIDEOCAMPUS organization. No participant need to be present on the campus of The University of Arizona to get this certificate.

5. In 1963 he conceived, initiated, and has directed since then the now internationally famous *Annual Reliability Engineering and Management Institute* at The University of Arizona.

6. In 1975 he conceived, initiated, and has directed since then the now internationally famous *Annual Reliability Testing Institute* at The University of Arizona.

7. He has lectured extensively and conducted over 300 training courses, short courses and seminars worldwide, and has exposed over 10,000 reliability, maintainability, test, design, and product assurance engineers to the concepts in this book.

8. He has been the principal investigator of mechanical reliability research for the NASA-Lewis Research Center, the Office of Naval Research, and the Naval Weapons Engineering Support Activity for 10 years.

9. He has been consulted extensively by over 82 industries and government agencies worldwide on reliability engineering, reliability and life testing, maintainability engineering, and mechanical reliability matters.

10. He has been active in the Reliability and Maintainability Symposia and Conferences dealing with reliability engineering since 1963.

11. He founded the Tucson Chapter of the Society of Reliability Engineers in 1974 and was its first President. He also founded the first and very active Student Chapter of the Society of Reliability Engineers at The University of Arizona.

12. He has authored or co-authored over 125 papers and articles, of which over 115 are in all areas of reliability engineering.

13. In addition to this book, he authored or contributed to the following books:

    (1) *Bibliography on Plasticity – Theory and Applications,* Dimitri Kececioglu, published by the American Society of Mechanical Engineers, New York, 191 pp., 1950.

    (2) *Manufacturing, Planning and Estimating Handbook,* Dimitri Kececioglu and Lawrence Karvonen contributed part of Chapter 19, pp. 19-1 to 19-12, published by McGraw-Hill Book Co., New York, 864 pp., 1963.

    (3) *Introduction to Probabilistic Design for Reliability,* Dimitri Kececioglu, published by the US Army Management Engineering Training Agency, Rock Island, Illinois, contributed Chapter 7 of 109 pp., and Chapter 8 of 137 pp., May 1974.

    (4) *Manual of Product Assurance Films on Reliability Engineering and Management, Reliability Testing, Maintainability, and Quality Control,* Dimitri Kececioglu, printed by Dr. Dimitri Kececioglu, 7340 N. La Oesta Avenue, Tucson, Arizona 85704-3119, 178 pp., 1976.

(5) *Manual of Product Assurance Films and Videotapes,* Dimitri Kececioglu, printed by Dimitri Kececioglu, 7340 N. La Oesta Avenue, Tucson, Arizona 85704-3119, 327 pp., 1980.

(6) *Reliability Engineering Handbook,* Dimitri Kececioglu, Prentice Hall, Englewood Cliffs, New Jersey 07632, Vol. 1, 720 pp., May 1991, and its fourth printing in July 1995.

(7) *Reliability Engineering Handbook,* Dimitri Kececioglu, Prentice Hall, Englewood Cliffs, New Jersey 07632, Vol. 2, 568 pp., June 1991, and its fourth printing in December 1994.

(8) The 1992-1994 *Reliability, Maintainability and Availability Software Handbook,* Dimitri Kececioglu and Pantelis Vassiliou, 7340 N. La Oesta Avenue, Tucson, Arizona 85704–3119, 118 pp., November 1992.

(9) *Reliability & Life Testing Handbook,* Dimitri Kececioglu, Prentice Hall, Englewood Cliffs, New Jersey 07632, Vol. 1, 960 pp., 1993.

(10) *Reliability & Life Testing Handbook,* Dimitri Kececioglu, Prentice Hall, Englewood Cliffs, New Jersey 07632, Vol. 2, 900 pp., 1994.

(11) *Advanced Methods in Reliability Testing,* Dimitri Kececioglu, Prentice Hall, Englewood Cliffs, New Jersey 07632, to be published in 1995.

(12) *Maintainability, Availability and Operational Readiness Engineering Handbook,* Dimitri Kececioglu, Prentice Hall, Englewood Cliffs, New Jersey 07632, Vol. 1, to be published in 1995.

(13) *Burn-in Testing – Its Quantification and Optimization,* Dimitri Kececioglu and Feng-Bin Sun, Prentice Hall, Englewood Cliffs, New Jersey 07632, 400 pp., to be published in 1995.

14. He has received over 80 prestigious awards and has been recognized for his valuable contributions to the field of reliability engineering and testing. Among these are the following: (1) Fulbright Scholar in 1971; (2) Ralph Teetor Award of the Society of Automotive Engineers as "Outstanding Engineering Educator" in 1977; (3) Certificate of Excellence by the Society of Reliability Engineers for his "personal contributions made toward the advancement of the philosophy and principles of reliability engineering" in 1978; (4) ASQC-Reliability Division, Reliability Education Advancement Award for his "outstanding contributions to the development and presentation of meritorious reliability educational programs" in 1980; (5) ASQC Allen Chop Award for his "outstanding contributions to Reliability Science and Technology" in 1981; (6) The University of Arizona College of Engineering Anderson Prize for "engineering the Master's Degree program in the Reliability Engineering Option" in 1983; (7) Designation of "Senior Extension Teacher" by

Dr. Leonard Freeman, Dean, Continuing Education and University Extension, University of California, Los Angeles in 1983; (8) Honorary Member, Golden Key National Honor Society in 1984; (9) Honorary Professor, Shanghai University of Technology in 1984; (10) Honorary Professor, Phi Kappa Phi Honor Society in 1988; (11) The American Hellenic Educational Progressive Association "Academy of Achievement Award in Education" in 1992; (12) On the occasion of "The 30th Annual Reliability Engineering and Management Institute," the President of The University of Arizona, Dr. Manuel T. Pacheco, presented him a plaque inscribed:

"Your reputation as an outstanding teacher and advocate of reliability and quality engineering is well established in the international engineering community. In your capacity as Director of this Institute, as well as the Reliability Testing Institute, you have provided the forum in which many hundreds of our nation's engineers and students of engineering have received training in reliability and quality engineering.

I particularly acknowledge your efforts in establishing and developing funding for the endowment which bears your name and which will support worthy students in the future. The 'Dr. Dimitri Basil Kececioglu Reliability Engineering Research Fellowships Endowment Fund' will help to ensure that The University of Arizona remains in the forefront of engineering education and continues to provide engineering graduates to support our nation's industries. In this highly competitive world the quality and the reliability of American products are essential to retaining our position of world economic leadership. The University of Arizona is proud to be an important part of that effort and can take justifiable pride in your own very significant contribution."

15. He conceived and established *The Dr. Dimitri Basil Kececioglu Reliability Engineering Research Fellowships Endowment Fund* in 1987. The cosponsors of his institutes, mentioned in Items 5 and 6, have contributed generously to this fund which has now crossed the $335,000 mark.

16. He has been granted five patents.

Mr. Feng-Bin Sun received his B.S.M.E. specializing in Electro-mechanical Structural Design from the Southeast University, the former Nanjing Institute of Technology, Nanjing, China, in 1984. He earned his M.S.M.E. specializing in Reliability Engineering from the Shanghai University, the former Shanghai University of Technology, Shanghai, China, in 1987. He is currently a Ph.D. candidate, a research associate and a guest lecturer under Dr. Dimitri Kececioglu in the Department of Aerospace and Mechanical Engineering at The University of Arizona.

His introduction to Reliability Engineering was at the Shanghai University of Technology during 1984-1987 where he earned his M.S.M.E. specializing in Reliability Engineering and enjoyed his first practical reliability and maintainability experience in industry: a big iron and steel plant. The following extensive experience has qualified him to contribute major segments of this book:

1. During 1985-1986, he worked at the Shanghai First Iron and Steel Plant, Shanghai, China, where he conducted extensive research on the reliability and maintainability of a steel rolling production line.

2. During 1986-1990, he offered over 20 lectures and training courses in Reliability and Maintainability Engineering for managers, chief and senior engineers, design and quality engineers from China's Ministry of Electromechanical Industry and its subordinate factories and research institutes. He also lectured at and was consulted by two big metallurgical and chemical companies in Shanghai, China, on reliability, preventive maintenance decision making and spares provisioning.

3. As a lecturer at the Shanghai University of Technology during 1987-1990, he taught several courses on Reliability and Maintainability including Reliability Engineering and Management, Reliability Life Testing, Maintainability Engineering and Probabilistic Mechanical Design. These courses were open to both undergraduate and graduate students.

4. During 1987-1990, he supervised 15 undergraduate and 4 graduate students in the Mechanical Engineering Department, Shanghai University of Technology, for their graduate projects and theses in reliability engineering.

5. During 1988-1990, he cooperated with the Beijing Machine Tool Research Institute, Beijing, China, and was the person-in-charge of a research grant on Reliability Assessment of Numerical Control Systems with emphasis on burn-in and *ESS* data analysis, which was funded by the Ministry of China's Electromechanical Industry.

6. During 1988-1990, he was in charge of another research grant on replacement policies for large-scale industrial systems which was funded by China's National Natural Science Foundation. He proposed a new

approach of system classification for the purpose of maintenance and replacement, and advanced the multimode replacement policy idea for large-scale industrial systems.

7. During 1989-1990, he cooperated with the Shanghai Machine Tool Research Institute and was the person-in-charge of a research grant on reliability evaluation of Numerically Controlled Machine Tools, which was funded by the Shanghai Municipal Bureau of Electromechanical Industry.

8. As a guest lecturer under Dr. Dimitri Kececioglu in the Department of Aerospace and Mechanical Engineering at The University of Arizona, he has been lecturing on Reliability Engineering, Reliability Engineering and Quality Analysis, Reliability & Life Testing and Maintainability Engineering since 1991.

9. In 1991 he conducted research with Dr. Dimitri Kececioglu on *MTBF* demonstration test plans of electronic units for ITT Power Systems Corporation.

10. In 1992 he conducted research with Dr. Dimitri Kececioglu on burn-in test planning and statistical analysis of multi-time-terminated burn-in data for Brooktree Corporation.

11. He has co-authored over 10 papers in various areas of reliability, maintainability and quality engineering, such as maintenance and replacement policies, system reliability, maintainability and availability evaluation, burn-in and *ESS* data analysis, mechanical reliability, quality loss functions, etc.

12. In addition to this book, he co-authored or contributed to the following books:

    (1) *Maintainability Design*, by Zhi-Zhong Mou, contributed two chapters on advanced replacement policies and a practical application in metalurgical industry, published by Shanghai Science and Technology Press, Shanghai, China, 300 pp., 1990.

    (2) *Reliability Design*, co-authored with Zhi-Zhong Mou, et al, contributed Chapter 10 on reliability design for electromechanical equipment, pp. 238-272, published by China Machinery Industry Press, Beijing, China, 366 pp., 1993.

    (3) *Reliability & Life Testing Handbook*, by Dimitri Kececioglu, contributed four chapters on modified matching moment parameter estimation, modified K-S, Anderson-Darling and Cramer-von Mises goodness-of-fit tests, proposed jointly with Dr. Dimitri Kececioglu a new outlier test method called *Rank limit method*, published by Prentice Hall, Englewood Cliffs, New Jersey 07632, Vol. 1, 960 pp., 1993.

(4) *Advanced Methods in Reliability Testing*, by Dimitri Kececioglu, contributed three chapters on burn-in, *ESS*, and reliability growth, respectively, to be published in 1995 by Prentice Hall, Englewood Cliffs, New Jersey 07632.

(5) *Maintainability, Availability and Operational Readiness Engineering Handbook*, by Dimitri Kececioglu, contributed two chapters on renewal theory and additional replacement policies, respectively, Vol. 1 to be published in 1995 by Prentice Hall, Englewood Cliffs, New Jersey 07632.

(6) *Burn-in: Its Quantification and Optimization*, co-authored with Dr. Dimitri Kececioglu, 400 pp., to be published in 1995 by Prentice Hall, Englewood Cliffs, New Jersey 07632.